GEOCHEMICAL MODELING OF GROUNDWATER, VADOSE AND GEOTHERMAL SYSTEMS

Multiphysics Modeling

Series Editors

Jochen Bundschuh
University of Southern Queensland (USQ), Toowoomba, Australia
Royal Institute of Technology (KTH), Stockholm, Sweden

Mario César Suárez Arriaga
Department of Applied Mathematics and Earth Sciences,
Faculty of Physics and Mathematical Sciences, Michoacán University UMSNH,
Morelia, Michoacán, Mexico

ISSN: 1877-0274

Volume 5

Geochemical Modeling of Groundwater, Vadose and Geothermal Systems

Editors

Jochen Bundschuh

University of Southern Queensland (USQ), Toowoomba, Australia
Royal Institute of Technology (KTH), Stockholm, Sweden

Michael Zilberbrand

Hydrological Service of Israel, Research Division, Jerusalem, Israel

CRC Press
Taylor & Francis Group
Boca Raton London New York Leiden

CRC Press is an imprint of the
Taylor & Francis Group, an **informa** business

A BALKEMA BOOK

Figures 7.4 and 7.5 are reprinted with permission by Elsevier (License Number 2754460210991)

Figures 10.1 through 10.3 are reprinted with permission by Springer (License Number 2763011191327)

Figures 11.1 through 11.4 are reprinted with permission by Elsevier (License Number 2762950559281)

First issued in paperback 2017

CRC Press/Balkema is an imprint of the Taylor & Francis Group, an informa business

Typeset by Vikatan Publishing Solutions (P) Ltd., Chennai, India

Published by: CRC Press/Balkema
P.O. Box 447, 2300 AK Leiden, The Netherlands
e-mail: Pub.NL@taylorandfrancis.com
www.crcpress.com – www.taylorandfrancis.co.uk – www.balkema.nl

Library of Congress Cataloging-in-Publication Data

Geochemical modeling of groundwater, vadose, and geothermal systems / editors, Jochen Bundschuh, Michael Zilberbrand.
p. cm. -- (Multiphysics modeling, ISSN 1877-0274 ; v. 5)
Includes bibliographical references and indexes.
ISBN 978-0-415-66810-1 (hbk. : alk. paper) -- ISBN 978-0-203-12298-3 (ebook) 1. Groundwater--Simulation methods. 2. Zone of aeration--Simulation methods. 3. Geothermal resources--Simulation methods. 4. Geochemical modeling. 5. Hydrologic models. 6. Water chemistry. I. Bundschuh, Jochen. II. Zilberbrand, Michael.

GB1005.G425 2012
551.4901--dc23

2011043040

ISBN 13: 978-1-138-07444-6 (pbk)
ISBN 13: 978-0-415-66810-1 (hbk)

About the book series

Numerical modeling is the process of obtaining approximate solutions to complex problems of scientific and/or engineering interest. The book series addresses novel mathematical and numerical techniques with an interdisciplinary emphasis that cuts across all fields of science, engineering and technology. It focuses on breakthrough research in a richly varied range of applications in physical, chemical, biological, geoscientific, medical and other fields in response to the explosively growing interest in numerical modeling in general and its expansion to ever more sophisticated physics. The goal of this series is to bridge the knowledge gap among engineers, scientists, and software developers trained in a variety of disciplines and to improve knowledge transfer among these groups involved in research, development and/or education.

This book series offers a unique collection of worked problems in different fields of engineering and applied mathematics and science, with a welcome emphasis on coupling techniques. The book series fills a need for up-to-date information on numerical modeling. Faster computers and newly developed or improved numerical methods such as boundary element and meshless methods or genetic codes have made numerical modeling the most efficient state-of-art tool for integrating scientific and technological knowledge in the description of phenomena and processes in engineered and natural systems. In general, these challenging problems are fundamentally coupled processes that involve dynamically evolving fluid flow, mass transport, heat transfer, deformation of solids, and chemical and biological reactions.

This series provides an understanding of complicated coupled phenomena and processes, its forecasting, and approaches in problem solving for a diverse group of applications, including natural resources exploration and exploitation (e.g., water resources and geothermal and petroleum reservoirs), natural disaster risk reduction (earthquakes, volcanic eruptions, tsunamis), evaluation and mitigation of human-induced phenomena as climate change), and optimization of engineering systems (e.g., construction design, manufacturing processes).

Jochen Bundschuh
Mario César Suárez-Arriaga
(Series Editors)

Editorial board of the book series

Table of contents

Section 2: *Cases studies*

Contributors

Jay J. Ague (Chapter 9)	Department of Geology and Geophysics, Yale University, New Haven, Connecticut, USA and Yale Peabody Museum of Natural History, Yale University, New Haven, CT 06511, USA
Carlos Ayora (Chapters 3, 4 and 5)	Department of Geosciences, Institute of Environmental Assessment and Water Research (IDÆA-CSIC), c/Jordi Girona, 18–26, 08034 Barcelona, Spain
Peter Birkle (Chapter 7)	Instituto de Investigaciones Eléctricas (IIE), Gerencia de Geotermia, Reforma 113, Col. Palmira, Cuernavaca, Morelos, 62490 Mexico
Jochen Bundschuh (Chapter 1)	University of Southern Queensland, Toowoomba, Queensland 4350, Australia & Department of Land and Water Resources Engineering, Royal Institute of Technology, SE-100–44 Stockholm, Sweden
James G. Brown (Chapter 8)	Formerly with the US Geological Survey in Tucson, Arizona
Jesus Carrera Ramirez (Chapters 3, 4 and 5)	Institute of Environmental Assessment and Water Research (IDAEA), Spanish Council for Scientific Research, Jordi Girona 18, 08034 Barcelona, Spain
Simon Emmanuel (Chapter 9)	Institute of Earth Sciences, The Hebrew University of Jerusalem, Givat Ram, Jerusalem, 91904, Israel
Pierre Glynn (Chapter 8)	National Research Program/Eastern Branch, U.S. Geological Survey, 432 National Center, Reston, VA 20192, USA
Ricardo Hirata (Chapter 10)	Department of Sedimentary and Environmental Geology, Institute of Geosciences, University of São Paulo, Rua do Lago 562, Cidade Universitária, CEP 05508-900, São Paulo, SP, Brazil
Robinah Nakawunde Kulabako (Chapter 12)	Department of Civil and Environmental Engineering, Makerere University, P.O.Box 7062 Kampala, Uganda
Lionel Mercury (Chapter 2)	Institut des Sciences de la Terre d'Orléans, UMR 6113 CNRS/Université d'Orléans, 1 A rue 8 de la Férolerie, 45071 Orléans Cedex, France

Maimuna Nalubega (Chapter 12)	Department of Civil and Environmental Engineering, Makerere University, P.O. Box 7062 Kampala, Uganda
Maarten W. Saaltink (Chapters 3, 4 and 5)	GHS, Department of Geotechnical Engineering and Geosciences, Universitat Politecnica de Catalunya, UPC-BarcelonaTech, Jordi Girona 1–3, 08034 Barcelona, Spain
Ondra Sracek (Chapters 1, 10 and 11)	Department of Geology, Faculty of Science, Palacký University, 17. listopadu 12, 771 46 Olomouc, Czech Republic & OPV s.r.o. (Protection of Groundwater Ltd), Bělohorská 31, 169 00 Praha 6, Czech Republic
Leigh A. Soutter (Chapter 12)	Physics Logic LLC, 6333E Mockingbird Lane, Suite 147PMB 722, Dallas, TX 75214, USA
Roger Thunvik (Chapter 12)	Department of Land and Water Resources Engineering, Royal Institute of Technology, SE-100–44 Stockholm, Sweden
Zbynek Vencelides (Chapter 11)	OPV, Bělohorská 31, 169 00 Praha 6, Czech Republic
Olav Walderhaug (Chapter 9)	Statoil ASA, N-4035 Stavanger, Norway
Alexander Yakirevich (Chapter 4)	Zuckerberg Institute for Water Research (ZIWR), The Jacob Blaustein Institute for Desert Research, Ben-Gurion University of the Negev, Sede-Boqer Campus, 84990, Israel
Michael Zilberbrand (Chapters 2 and 6)	Hydrological Service of Israel, Yaffo Str. 234, Jerusalem 91063, Israel

Foreword

The world beneath our feet often remains unseen, and yet it provides us with water that feeds our streams, our ecosystems, and our water wells. Its shallowest parts, the soil and vadose zones, breathe gases in and out, help regulate moisture, fix nutrients, breakdown minerals, decompose and recycle organic debris and other waste, and sustain a rich microbial ecosystem that is vital to the sun-fed plants living on the surface. A bit deeper, in the saturated zone, permeable environments (aquifers) provide us with naturally filtered drinking water, and water for many other uses (e.g., irrigation). Deeper yet, permeability and storativity decrease, fluid residence times increase, dissolved salts increase, and selected groundwater environments provide an opportunity to store wastes (e.g., brines, hazardous chemicals, nuclear waste). Geochemical processes affect water quality as well as the mineral fabric and geologic framework of the subsurface, including its permeability and capacity to store water.

The subsurface also provides mineral and energy resources that are essential to human society. Water is often essential in the formation and development of those resources (e.g., hydrothermal mineral deposits, geothermal energy). Groundwater is also often affected by the presence of mineral or energy resources (e.g., oil, coal) or by their extraction (e.g., acid mine drainage). Elevated levels of toxic chemicals can occur either naturally (e.g., arsenic in shallow Bengal basin groundwaters) or as the result of human action (e.g., gasoline and oil spills). Geochemical and biogeochemical reactions drive the transformation, mobility, and often the relative toxicity of constituents of concern in groundwater environments.

Because of their large relevant volumes, relative stability, and distribution of residence times, groundwater environments also serve as an archive of climate and land-use change, including human-driven change. Dissolved gases, i.e., noble and nitrogen gas concentrations, in groundwaters around the world have recorded a shift of generally at least 4 to 5°C in recharge temperatures as the world became warmer after the last glacial maximum about 20,000 years ago. Secondary calcite precipitated in aquifers, such as found in Devil's Hole (NV, USA), has recorded the ^{18}O (i.e., temperature) and ^{13}C (i.e., vegetation distribution) history of recharging water over the last 500,000 years. On shorter timescales, shallow groundwaters have also recorded the history of land-use change (e.g., increased fertilizers and pesticides) as well as the diversity of anthropogenic constituents (e.g., chlorofluorocarbon compounds, methyl tert-butyl ether) introduced in the environment over the last 50 years.

Groundwater provides a ubiquitous control on water availability and water quality in ecosystems, often in more ways than the lay public realizes. The relatively stable temperatures of groundwaters and their often dominant contribution to streamflow helps moderate temperature and flow variability and helps stabilize aquatic habitats. The lagged response of groundwater systems means that human actions that affect surface conditions have a delayed effect on the water quality and availability of groundwater, and therefore on the surface waters and environment that it feeds. Conversely, the stability, lagged response and heterogeneity of groundwater systems make them difficult and time consuming to remediate, once contaminated or otherwise altered.

Pervasive heterogeneity, from the scale of pores and mineral grains to the regional scale of geologic formations and facies, is a key characteristic of groundwater environments. In turn, characterizing and modeling that physical, geochemical, and microbiological heterogeneity, and its effects on the transport and transformation of groundwater solutes, are major challenges for hydrogeologists and geochemists. The challenges are great enough to warrant the application of a wide variety of geophysical, geochemical, and numerical modeling tools and tracers (e.g., heat and solutes).

So, why do we need geochemical and biogeochemical modeling of groundwater systems? A first answer is that geochemical and microbial processes are active processes that help trace the origin, flowpaths, residence times, and evolution of waters in the subsurface; and these processes also often affect the permeability and physical properties of geologic materials controlling flow and transport. Given the complexity of groundwater systems, their heterogeneity and inaccessibility, the path to enhanced understanding of subsurface systems lies in making use of all information and simulation tools available—not just physical flow and conservative transport models. Modeling, in all its forms, helps organize and provide a framework for the information that is available, helps recognize information that is missing and needed to answer specific questions, helps explore (and sometimes forecast) a diversity of future scenarios, and generally helps test hypotheses and gain understanding.

A second answer is that geochemical and biogeochemical processes transform not only geologic materials and their interfaces with subsurface fluids, but also affect the nature, transport, and fate of solutes, colloids, and microbes in the subsurface. Modeling and understanding these processes and their temperature dependence can potentially result in better management and regulatory decisions regarding (i) human activities on the land surface, (ii) appropriate extraction and use of subsurface resources (water, energy, minerals), (iii) suitable practices for waste disposal, and (iv) cost-effective contaminant remediation. Additionally, modeling and understanding these processes can help explore and document the archive of climate, environmental, and land-surface changes recorded in groundwater systems—from the distant past when human activities were not a major influence, to the present when they often are.

The twelve chapters in this handbook provide an excellent introduction and a highly useful reference on the methods and challenges of groundwater modeling, with an emphasis on the modeling of geochemical and biogeochemical processes. The principles of aqueous chemistry and the basic theories describing chemical reactions and the physics of fluid, solute, and heat transport are discussed, together with the algorithms used in numerical simulations of groundwater geochemical modeling. Practical issues, such as useful sampling methods and analytical characterization techniques, are covered. A broad array of applications and case studies of geochemical modeling are also discussed, including geochemical evolution in the Guarani regional aquifer (in Brazil), porosity development, geothermal applications, and a diversity of modeling approaches and studies relating to reactive transport at sites contaminated by petroleum hydrocarbons, excess phosphorus, and acidic heavy-metal-contaminated waters.

The diversity of relevant topics, and the basic and advanced methods and approaches to geochemical and biogeochemical modeling covered in this handbook, make it a useful reference for practicing hydrogeologists and geochemists around the world. The importance of discovering, studying, utilizing, preserving, and/or remediating the world beneath our feet has never been greater for society's welfare.

Pierre Glynn
Chief, National Research Program/Eastern Branch
U.S. Geological Survey, Reston, VA

Editors' preface

"The grand aim of all science is to cover the greatest number of empirical facts by logical deduction from the smallest number of hypotheses or axioms."

Albert Einstein

"The purpose of science is not to analyze or describe but to make useful models of the world. A model is useful if it allows us to get use out of it."

Edward de Bono

Geochemical modeling is an important tool in environmental studies, and in the areas of subsurface and surface hydrology, pedology, water resources management, mining geology, geothermics, hydrocarbon geology, and related areas dealing with the exploration and extraction of natural resources. Geochemical modeling simulates the chemical and physical processes affecting the distribution of chemical species in liquid, gas, and solid phases. The reactions and processes, and their coupled interactions, are dependent on a number of environmental variables (e.g., temperature, pressure, ionic strength), and are also affected by the dynamics of matter and energy flows, including fluid, solute and heat flow.

This book seeks to make easily available to a broad readership some basic knowledge and fundamental concepts regarding the mathematical modeling of geochemical and groundwater processes, from the shallow vadose zone to deep geothermal systems. Since it is impossible to cover this subject given the space limitations, a selection of essential topics and case studies has been made. However, almost all the developments described herein are discussed in detail. Fundamental concepts, and the physical laws and equations needed to model different processes are described and presented in a simple and logical manner.

The book explains in a didactic manner the different applications of geochemical modeling, the existing conceptual methods, and the mathematical and numerical tools that can provide useful solutions. The reader will also receive a thorough understanding of (i) the physical laws describing the flow of mass and energy and the transport of solutes, (ii) the partial differential or algebraic equations representing these laws, and (iii) the principal numerical methods that allow approximate solutions of these equations and their corresponding mathematical models.

In the last 20 years, significant progress has been made in the use of different computational methods for geochemical and groundwater modeling. New concepts, methods and important findings have been presented in numerous publications. However, these often cover either specific issues or are restricted to specific systems or applications. Other publications related to the topics presented in this book may be restricted to describing some modeling software, or alternatively, the knowledge contained may not be applicable or transferable to groundwater and geochemical modeling without major modifications. Many of the publications do not necessarily describe the basic physical, chemical, mathematical and numerical theories and principles used. We saw a need for a synoptic compendium on the fundamentals of groundwater and geochemical modeling that would also have broad applicability to a diversity of environments. Our compendium clearly reveals the need for further research and development, that to be most useful, should be informed by additional field studies and practical applications of numerical modeling. Hopefully, this handbook will stimulate its readership to address some of these needs.

This book is meant to help overcome some of the barriers that hinder the use (or the correct use) of geochemical modeling. It presents applications of geochemical modeling in real case studies. Additionally, we emphasize the need for critical consideration and review of geochemical modeling and simulation results. A well-designed application will (i) start with a judicious examination of all available primary information, including the methods and techniques used for extraction and analysis of water, gas and solid samples, (ii) will carefully consider the applicable processes and domain boundaries of the simulations, and (iii) will test the simulation results using sensitivity analyses and a variety of validation/verification techniques. The reader will learn that the quality of the simulation results depends on the preparation of the model, the quality of the input data, the knowledge of the geological or pedological situation, the knowledge of the hydrogeochemical parameters and data, and the knowledge of the initial values and boundary conditions for the given subsurface system.

Our handbook provides a review of physical, chemical, mathematical, and numerical theory, and describes the correct use of computational methods in simulating chemical reaction processes in low- and high-temperature aqueous systems such as groundwater, petroleum and geothermal systems. The book also presents new and stimulating ideas for other possible simulations and applications.

Chapter 1 gives an introduction to the best practices in sampling and sample preservation, which together with accurate analyses play a vital role and are a precondition for geochemical modeling. The chapter also provides a summary of the basic principles of aqueous geochemistry and thermodynamics; this section is kept purposely short since more detailed information is given in several excellent text books to which we refer. In Chapter 2, the thermodynamics of gas and mineral solubility in the unsaturated-zone are treated in more detail since these aspects are not fully dealt with in the previously mentioned textbooks. Chapter 3 describes the governing equations and solution algorithms for geochemical modeling; it contains the basic concepts for the mathematical formulation of homogeneous and heterogeneous chemical reactions, including their kinetic simulation, and for the calculation of species concentrations. Chapter 4 provides the physical laws and mathematical equations describing fluid heat, and reactive solute transport, and presents the relevant computational solution algorithms for geochemical modeling; the numerical solution methods for the reactive transport equations are separately discussed in Chapter 5. Chapter 6 discusses how a conceptual model is implemented into a mathematical/numerical model, and indicates the processes, model parameters and data needs relevant for individual field scenarios and modeling tasks. Limitations and problems of existing thermodynamic databases are discussed. The principle types of geochemical models (speciation, reaction-path or forward, inverse- and reactive-transport models) are described together with examples of the most common codes. Chapter 7 deals with the advances in geochemical modeling for geothermal applications. The second part of the book (Chapters 8–12) discusses the application of geochemical models in different scientific areas and environmental settings. The focus is on the practical aspects of modeling, by the use of case studies of real-world environmental problems, including (i) inverse and forward modeling of heavy metal transport in an aquifer under acidic conditions, (ii) modeling and measurements of porosity and permeability evolution in a sandstone aquifer, (iii) geochemical modeling of water chemistry evolution along groundwater flow paths, (iv) modeling of reactive solute transport at a site contaminated by petroleum hydrocarbons, and (v) modeling for preliminary assessment of natural remediation of phosphorus in variably saturated soil.

This book is addressed to students, teachers, other professionals, and to the institutions involved in water resources and environmental management. We hope that it will give them an introduction on the practical use of geochemical modeling in their fields, and that it will beneficially contribute to policy-making. We also hope that this book will provide a reference used by educational and scientific institutions. The book should prove useful to senior undergraduate and graduate students, postgraduates, professional geologists and geophysicists, engineers, environmental scientists, soil scientists, hydrochemists, and others interested in groundwater and geochemistry.

The book fills a gap in the literature through its presentation of insights in geochemical modeling as applied to a diversity of subsurface systems from close to the Earth's surface, down to deep-seated geothermal reservoirs. It also encourages a broad community of environmental scientists, geologists, chemists, pedologists, hydrologists, engineers and applied mathematicians to join together to help better protect and manage our freshwaters and our environment.

Finally, through its 2012 publication date, this book reminds us of the 50th anniversary of a key milestone in geochemical modeling. In 1962, R.M. Garrels and M.E. Thompson of Harvard University published the first paper that used thermodynamic modeling to predict the speciation of seawater.

Jochen Bundschuh
Michael Zilberbrand
January 2012

About the editors

Jochen Bundschuh (1960, Germany) received his PhD from the University of Tübingen in 1990 after completing a thesis on the numerical modeling of heat transport in aquifers in 1990. He works in the areas of geothermics, subsurface and surface hydrology, integrated water resources management, and associated disciplines. From 1993 to 1999 he served as an expert for the German Agency of Technical Cooperation (GTZ), and as a long-term professor for the DAAD (German Academic Exchange Service) in Argentine. From 2001 to 2008, he served as an adviser to the Costa Rican government at the Instituto Costarricense de Electricidad (ICE) within the framework of the German governmental cooperation program (Integrated Expert Program of CIM; GTZ/BA). He assisted Costa Rica in the evaluation and development of its huge low-enthalpy geothermal resources for power generation. Starting in 2005, he was appointed Affiliate Professor at the Royal Institute of Technology, Stockholm, Sweden. In 2006, he was elected Vice-President of the International Society of Groundwater for Sustainable Development ISGSD. From 2009 through 2011, he was Visiting Professor at the Department of Earth Sciences at the National Cheng Kung University, Tainan, Taiwan. At the end of 2011, he was appointed as Professor of Hydrogeology at the University of Southern Queensland, Toowoomba, Australia.

Dr. Bundschuh has authored the books "Low-Enthalpy Geothermal Resources for Power Generation" (2008) and "Introduction to the Numerical Modeling of Groundwater and Geothermal Systems: Fundamentals of Mass, Energy and Solute Transport in Poroelastic Rocks" (2010) (both CRC Press/Balkema). He also edited the books "Geothermal Energy Resources for Developing Countries" (2002), "Natural Arsenic in Groundwater" (2005), and the two-volume monograph "Central America: Geology, Resources and Hazards" (2007), "Groundwater for Sustainable Development" (2008), "Natural Arsenic in Groundwater of Latin America (2008). Dr. Bundschuh also serves as an editor for several book series: "Multiphysics Modeling", "Arsenic in the Environment", and "Sustainable Energy Developments" (all CRC Press/Balkema).

Michael Zilberbrand was born in 1949 in the Ukraine. He received his PhD in 1987 from the Institute of Geological Sciences of the Academy of Sciences of the Ukraine. His thesis focused on physical and numerical modeling of water and salt transport in the unsaturated zone. He has taught a course on unsaturated zone processes at the Hebrew University of Jerusalem. Since 1996, he has worked as a research scientist in the Hydrological Service of Israel, applying his knowledge of hydrogeological and hydrogeochemical modeling. He is a member of the Israeli Geological Society.

Dr. Zilberbrand is a co-author of an approved patent on a device for estimating the rate and parameters of phase transition and vapor transfer in the air. He has conducted numerous field and laboratory studies and modeling of geochemical processes in the vadose zone and in groundwater systems, and has also conducted fundamental research of aqueous speciation in the vadose zone.

Dr. Zilberbrand has published 17 papers and 6 book chapters, including a publication in the Encyclopaedia of Surface and Colloid Science. He has reviewed numerous manuscripts for the United State–Israel Binational Science Foundation (BSF), Geochimica et Cosmochimica Acta, Journal of Hydrology, Chemical Geology and other international scientific journals.

Acknowledgements

This book would be incomplete without an expression of our sincere and deep sense of gratitude to Pierre Glynn at US Geological Survey (Reston, VA), for his careful reading; his valuable comments and suggestions greatly improved chapters 1, 3, 4 and 5.

We thank David Parkhurst of the USGS (U.S. Geological Survey, Denver Federal Center, Denver, CO, USA) for his valuable remarks on Chapter 2. We express our gratitude to Mario César Arriaga Suarez (Dept. of Applied Mathematics and Earth Sciences, Michoacán University UMSNH, Morelia, Mexico) who provided important inputs to improve the mathematical descriptions in Chapters 3 to 5.

We thank the following scientists who reviewed the other chapters and made a number of very useful suggestions for its improvement: Nicolas F. Spycher (Earth Sciences Division, Lawrence Berkeley National Laboratory, Berkeley, CA; Chapter 7); David Parkhurst (U.S. Geological Survey, Denver, CO; Chapter 8); Mark Fuhrman (Office of Nuclear Regulatory Research, U.S. Nuclear Regulatory Commission, Rockville, MD; Chapters 8 and 9); Alexander Yakirevich (Jacob Blaustein Institute for Desert Research, Ben-Gurion University of the Negev, Sede-Boqer Campus, Israel; Chapters 10 and 11); Simon Emmanuel (Institute of Earth Sciences, The Hebrew University of Jerusalem, Jerusalem, Israel; Chapters 10 and 11); Maarten W. Saaltink (Dept. Geotechnical Engineering and Geosciences, Universitat Politecnica de Catalunya, Barcelona, Spain; Chapter 12); and Micòl Mastrocicco (Earth Sciences Department, University of Ferrara, Ferrara, Italy). We wish to express our sincere thanks to them, whose efforts contributed to the high quality of the book.

The editors thank also the technical people of Taylor & Francis Group, for the excellent typesetting of the manuscript. Finally, we thank the board members of the book series Multiphysics Modeling for their editorial suggestions.

Jochen Bundschuh
Michael Zilberbrand
January, 2012

Section 1: Introduction to groundwater geochemistry and fundamentals of hydrogeochemical modeling

CHAPTER 1

Hydrogeochemistry principles for geochemical modeling

Jochen Bundschuh & Ondra Sracek

> *"Learn from yesterday, live for today, hope for tomorrow. The important thing is not to stop questioning."*
>
> Albert Einstein

A wide diversity of physical and chemical processes control the distribution of species in waters in the vadose zone above the water table, and in the saturated zone below. The mineralogical composition of rocks or sediments, chemical reactions between solid, aqueous and gas phases, and oxidation/reduction (redox) processes are principal factors that influence the chemical composition of vadose-zone, ground- and surface waters, and the concentrations and mobilities of individual species.

This chapter discusses the sampling of water, solids and gases, key steps in the investigation of groundwater systems, and then provides a brief summary of the thermodynamic fundamentals relevant for understanding geochemical processes in the vadose zone and in the aquifers and aquitards below it. This is followed by a short overview of important hydrogeochemical processes such as dissolution, precipitation, exsolution and redox processes, as well as sorption and ion exchange processes. These last two sections will help the reader understand the subsequent chapters on geochemical modeling. For more information on the fundamentals of aqueous and groundwater geochemistry, the reader is referred to standard textbooks such as Stumm and Morgan (1996) and Appelo and Postma (2005).

1.1 SAMPLING AND ANALYSIS OF WATER, SOLIDS AND GASES

Proper sampling procedures are critical to the success of geochemical investigations. When the integrity of water, solid and gas phases samples is compromised, and/or when the samples taken are not representative of the system studied, geochemical assessments and modeling may be misleading or incorrect. The investigator's objective is to obtain and analyze a water sample with the same chemical and isotopic composition as that of the water in its original environment (e.g., aquifer, surface water body, vadose zone) prior to disturbances caused by processes like sampling. When groundwater is sampled in wells with long open screens, water from several different redox zones may mix and irreversible chemical processes may occur. For example, when waters close to the water table are oxic (i.e., dissolved oxygen is present), and waters in deeper zones are reducing and contain dissolved iron, the mixing of these waters in the sample taken will result in the loss of both oxygen and iron according to the reaction:

$$Fe^{2+} + 0.25O_2(g) + 2.5H_2O \rightleftarrows Fe(OH)_3(s) + 2H^+ \tag{1.1}$$

Not only are iron and dissolved oxygen lost, but other metals like Zn and Pb may also be lost because of their adsorption on the precipitated ferric hydroxide.

Sampling of a mixture of water compositions is a problem even when no heterogeneous reactions occur (i.e., reactions involving the transfer of chemical elements between different phases, such as occurs in mineral precipitation or gas exsolution). Indeed, conservative mixing may result in the possible dilution (or concentration) of various species of interest, and the measurement of certain properties such as *pH* and *Eh* (or *pe*) or thermodynamic activities may be even more of a problem because of their non-linear response to the mixing. For example, when a chloride plume (i.e., a non-reactive "conservative" tracer) has a vertical extent of about 4 m and the screen zone is 10 m thick, the chloride concentration in the sampled water may be significantly lower than that present in the plume, and the chemistry of the water sample may differ significantly from the actual concentrations found in the aquifer, either in the background water or in the chloride plume. Similarly, it is well known that even though two waters may, individually, have measured *pH* values and calculated calcium and carbonate activities that reflect their equilibrium with calcite, the *pH* value and calcium and carbonate activities of a sampled mixture of the two waters may show a significant departure from such equilibrium, specifically because *pH* and thermodynamic activities have non-linear responses to "conservative" mixing.

The solution to obtaining water samples that are characteristic of their original subsurface environment is to use sampling techniques that offer vertically discrete resolution. For example, one method is to use a nest of piezometers with short screen zones open at different depths (Appelo and Postma, 2005), or to use packers to isolate and sample different discrete levels in a long-screen well or unscreened borehole (e.g., in a fractured rock environment). The cost of these sampling devices or procedures is their principal limitation. Other methods that can sometimes be used, and that may be cheaper or easier, include (i) a multi-pump strategy, for example where one low-flow pump is placed near the bottom of a well screen and a second pump with a higher flow rate is placed above it to ensure upwards flow in the well, past the first sampling pump, or (ii) the slow lowering of a sampling tube in a long narrow well. This second technique is difficult to perform, but has been used successfully in fractured basalt and crystalline rock environments. Plummer *et al.* (2011, in press), Glynn and Plummer (2005), Glynn and Voss (1999), and many references therein, discuss strategies and sampling methods that can be used to characterize groundwaters in different geologic environments.

Degassing (or ingassing) of sampled waters, for example by exposure and partial or full equilibration with the atmosphere or vadose zone gases can be another problem that needs to be overcome in groundwater characterization. For example, degassing of CO_2 often occurs because the partial pressure of CO_2 (p_{CO_2}) dissolved in groundwater is usually several orders of magnitude higher than the atmospheric value of $10^{-3.5}$ atm. Thus, when a groundwater sample is equilibrated with the atmosphere, the reaction:

$$Ca^{2+} + 2HCO_3^- \rightleftarrows CaCO_3(s) + CO_2(g) + H_2O \qquad (1.2)$$

shifts to the right and calcite precipitates. Some calcium and dissolved inorganic carbon (*DIC*) are lost from the water sample. Exposure of the sample to a higher temperature, caused perhaps by delayed measurements, can also accelerate this degassing process. Degassing may proceed not just because a sample is exposed to the atmosphere and partially equilibrated with it. Dissolved gas partial pressures in groundwater environments often reflect the product of individual dissolved gas concentrations and the confining hydrostatic pressure, which increases with depth below the water table at the rate of about one atmosphere of pressure for every 10 m. Samples brought up to atmospheric pressure from depth may naturally either degas to the atmosphere or exsolve and form a separate gas phase in their sampling container. Additionally, the degassing of CO_2 may entrain the loss of other dissolved gases present in the water sample (e.g., Ar, N_2, O_2, H_2S, CH_4).

There is a controversy related to a question of how many volumes of a well or piezometer should be evacuated prior to collection of groundwater sample(s). It is generally assumed that 3 volumes should be enough, but sometimes this requirement is not practical and cannot be fulfilled (e.g., in a case of large diameter domestic wells). In addition, the fast pumping of large amounts of water from an aquifer can also significantly perturb the groundwater flow field in the vicinity of the pumping well. Thus, it is often recommended to pump water until parameters like *pH* and electrical conductivity (*EC*) are stabilized and then start the sampling (Appelo and Postma, 2005), although it should be realized that the time-stability of some commonly measured parameters such as EC and *pH* may not mean that *all* relevant parameters or concentrations are necessarily stable. For example, the concentrations of trace substances, redox-sensitive ions, or any depth-sensitive concentrations or parameters are notoriously unstable (Glynn and Plummer, 2005, Glynn and Voss, 1999), and even when stable, these parameters may reflect the achievement of some hydrologic steady state that does not reflect the original groundwater environment prior to the sampling and hydrogeologic characterization process. When possible, it is recommended to use low pumping rates (<1 L min^{-1}) so as to minimally disturb the flow field around the pumping well, and also to avoid the entrainment of colloids or fine particulates from well packing materials near the screen (Puls and Barcelona, 1996). This low pumping requirement may be particularly helpful in the investigation of contaminant plumes, because the uncontaminated background water could otherwise dilute the samples. Generally, investigations of the effect of different sampling and pumping strategies, and different pumping rates and times, on observed groundwater chemistry are recommended to determine optimal sampling procedures for different subsurface environments.

1.1.1 *Measurement of field parameters*

Several parameters have to be measured either immediately after sampling in the field or ideally *in-situ* in the well because of their instability. Delayed measurements of these parameters in the field or back at the laboratory may otherwise make the measurements almost meaningless.

Temperature: this parameter changes quickly and has an impact on many other parameters such as *pH* and *Eh*. Furthermore, temperature influences geochemical speciation calculation. Ideally, temperature should be measured directly in a well by a down-hole probe. If this is impossible, then the measurement should be performed in a flow-through cell that minimizes contact with the atmosphere and exposure/equilibration to surface temperatures (i.e., overly long, small diameter tubing made of relatively gas-porous plastics can sometimes be a problem). A flow-through cell is typically a plexiglass cylinder which is connected to the pumping device and has holes that allow the insertion of measurement electrodes (e.g., temperature, *pH*, redox potential *Eh*, electrical conductivity *EC*, dissolved oxygen *DO*, ion-selective electrodes).

Hydrogen ion activity (pH): *pH* is an essential parameter because most geochemical processes are *pH*-sensitive. The *pH* value is often defined by the distribution of carbonate species in water and, thus, it is strongly affected by the contact and equilibration of samples with the atmosphere. Because of CO_2 degassing (eq. 1.2), the measured *pH* (especially if measured later in the laboratory) may be higher than its natural value. In contrast, oxidation of dissolved H_2S and oxidation of Fe(II) and precipitation of Fe-oxyhydroxides, or precipitation of other metal oxyhydroxides (e.g., Al, Mn) will generally decrease the natural *pH*. Thus, measurements without contact with the atmosphere are required.

Redox potential (Eh): *Eh* values are generally lower in groundwater than in surface waters. Indeed, the contact of a sample with the atmosphere ($\log p_{O_2} = -0.68$ atm) causes changes in speciation following oxidation, for example, of Fe(II) to Fe(III) (eq. 1.1). This means that any contact of the sample with atmosphere has to be avoided. Measurement of *Eh* should also be performed in a flow-through cell. The *Eh* value measured in the field is performed with respect to Ag-AgCl or Hg-HgCl (calomel) electrodes and the measured value has to be converted to the calculated potential, E_{H2}, relative to the standard hydrogen electrode:

$$E_{H2} = Eh_{field} + Eh_{correction} \tag{1.3}$$

where $Eh_{correction}$ depends on the type of electrode and the temperature and is indicated by the manufacturer. $Eh_{correction}$ can also be obtained through the measurement of *Eh* in both the field sample and in a "Zobell's" solution with known E_{H_2} value (Kehew, 2000). The correction for a calomel electrode at a temperature of 25°C is about +0.245 V.

Two different variables can be used to represent the redox potential of a solution. The *Eh* redox potential discussed above, measured in volts and expressed with respect to the standard hydrogen electrode, is commonly used. The relation between the *Eh* redox potential and the aqueous activity ratio of a redox couple is expressed through the Nernst equation. For example, the equation for the Fe^{3+}/Fe^{2+} redox couple is:

$$Eh = Eh^0 + \frac{RT}{nF}\ln\frac{[Fe^{3+}]}{[Fe^{2+}]} \quad E^0 = +0.77\text{V} \tag{1.4}$$

where E^0 is the standard potential (in volts, tabulated for different half-reactions), n is a number of transferred electrons, *F* is the Faraday's constant and terms in squared brackets are activities. E^0 values for other redox reactions can be found, for example, in Appelo and Postma (2005).

Alternatively, the negative log of the activity of electrons, *pe*, is sometimes used by analogy with *pH*: $pe = -\log[e^-]$. This notation is convenient for thermodynamic analyses and representations but offers only a formalized representation because electrons do not exist as separate species in solution. A *pe* value reflects a hypothetical electron activity at equilibrium, but it can be related to the activity ratio of two different redox species. For example, the equation relating *pe* to the Fe^{3+}/Fe^{2+} redox couple is:

$$pe = pe^0 + \log\frac{[Fe^{3+}]}{[Fe^{2+}]} \tag{1.5}$$

where under standard conditions at 25°C, $pe^0 = 13.0$.

Thus, the *pe* notation connects the redox potential of a solution with the equilibrium distribution of species in different redox states, and through its analogy with *pH*, facilitates the construction of *pe*/*pH* stability diagrams to represent that distribution.

pe is converted to *Eh* redox potential notation by:

$$Eh\,[\text{V}] = [RT\,(\ln 10)/F]\,pe \qquad \text{or at 25°C, } Eh = 0.059\,pe \tag{1.6}$$

Determination of oxidation state partitioning of redox sensitive elements: different forms of the same element in different oxidation states can have completely different behavior and toxicity. For example, Fe(II) species are much more more soluble than Fe(III) species and Cr(VI) species are much more toxic than Cr(III) species. Therefore, information about element oxidation states is essential in investigating water contamination. This information can be obtained either:

a. Directly, preferably through *in-situ* measurements or through analytical determinations at the field site of relevant redox state concentrations for all redoxactive elements in solution. Alternatively, and only if absolutely not possible in the field, the analytical determinations may also be done soon after return or shipping to a laboratory on carefully preserved (filtered and acidified in the field) samples. This approach is the best way to determine redox distributions of redoxactive elements, but it incurs significant costs.
b. By analytical determinations of total aqueous concentrations (e.g., Fe_{total}) of the redoxactive elements in solution, and then by the calculation of "equilibrium" aqueous distribution of the relevant redox states (e.g., Fe(II) and Fe(III)) on the basis of the measured field *Eh* value and the Nernst equation. This determination is problematic because the measured *Eh* value, at best, relates to the most active redox couple(s) (i.e., the highest and most kinetically dynamic activities) that the *Eh* electrode responds to. Often (but not always) the field *Eh* measurement dominantly reflects the activities of the Fe(III)/Fe(II) couple and

it may not represent the concentrations/activities of other redox couples and species. For example, there is often a poor match between measured *Eh* values and the O_2/H_2O couple (which can be determined through a dissolved oxygen electrode measurement or even better, through an analytical determination).

c. By calculating the *Eh* or *pe* value from one measured redox couple distribution (e.g., Fe(II)/Fe(III)) and then by using that calculated value to determine the activity ratios of other redox couples (e.g., Mn(II)/Mn(IV)) for which total element concentrations have been determined. This approach suffers from the same problem as the application of field *Eh*, i.e., different redox couples frequently are not in redox equilibrium. Geochemical programs like PHREEQC (Parkhurst and Appelo, 1999) allow the use of all the different input and calculation options mentioned above.

Alkalinity: this is a measure of the acid-neutralizing capacity of a solution. It is generally based on carbonate species, but other species can also contribute:

$$\text{Alkalinity} = HCO_3^- + 2CO_3^{2-} + H_3SiO_4^- + H_3BO_2^- + \dots \tag{1.7}$$

For example, dissolved organic matter with deprotonated carboxylic acid groups, R-COO$^-$, as often found near landfills, may also contribute to the alkalinity. If that is the case, the measured alkalinity cannot be used to calculate the distribution of carbonate species, at least without applying corrections (Deutsch, 1997). Ideally, alkalinity is determined in the field by titration with an acid (e.g., HCl). It is reported in units of mg $CaCO_3$ per L or mg HCO_3 per L. When determination in the field is impossible, then a special sample should be collected for later determination in the laboratory. When a sub-sample is taken from a larger sample later, there can be loss of alkalinity due to precipitation of calcite (eq. 1.2). Fritz (1994) concluded that most positive charge-balance errors (i.e., missing anions) were caused by errors in the measurement of alkalinity.

Dissolved oxygen (DO): this is a very useful parameter, although the O_2/H_2O redox pair can generally not be used for quantitative calculations. Detectable DO values indicate oxic environments where, for example, dissolved nitrate should be stable and ferrous iron should be absent. Measurements are often performed with an oxygen electrode, but more precise methods such as a "Winkler" titration are also available. The presence of *DO* in waters that should not have it can sometimes indicate that a mixture of different waters is being sampled, or alternatively, that the sampling process is allowing gas exchange with the collected water.

Electrical conductivity (EC): this parameter roughly correlates with total dissolved solids (*TDS*) concentration. It can often be used to delineate contaminant plumes, salt water intrusion, etc. It is measured through the use of a conductivity meter, typically in units of $\mu S\ cm^{-1}$. According to Appelo and Postma (2005), the *EC* divided by 100 gives a good estimate of the sum of either anions or cations (in meq L^{-1}):

$$\sum \text{anions [meq } L^{-1}] = \sum \text{cations [meq } L^{-1}] = EC/100\ [\mu S\ cm^{-1}] \tag{1.8}$$

This relation holds up to *EC* values of about 2000 $\mu S\ cm^{-1}$. A new method that allows the calculation of *EC* based on electrolytes found in ground waters was introduced by McCleskey *et al.* (2011). This method can be used to check the accuracy of water analyses.

1.1.2 *Filtration and preservation of water samples*

Proper collection and preservation of samples is very important. While it is generally recommended to conduct analyses as soon as possible or directly at the field site, this is often not possible. Additionally, sample archives may often prove useful for later analysis of constituents or characteristics initially considered unimportant. Proper filtration of samples should be carefully considered because of its importance both for field measurements and/or for later measurements at a laboratory. There is some controversy regarding filtration but it is generally recommended when taking samples for cation and metal analyses. Typically, 0.45 μm filters are used, but 0.1 μm filters might sometimes be more suitable because some colloids (e.g., Al-or Fe-hydroxide colloids) can pass through 0.45 μm filters. Filtering using the same filter size is also recommended for samples collected for anion analyses because it can help to preserve the samples for analysis: organic matter and some bacteria that could cause reactions (e.g., sulfate reduction) after the collection of the sample may be removed by the filtration. Filters are generally located on the output pipe of a flow-through cell (i.e., as an in-line filter). Samples for heavy-metal analysis should be acidified with acid such as HCl to *pH* less than 2.0, for example to avoid precipitation of iron (eq. 1.1). Samples for anion analysis are left unacidified, but filtration is still recommended.

If the presence of colloids is of interest, two samples are taken: one sample filtered through 0.1 μm filter and acidified, and a second one unfiltered and acidified. If colloid concentrations are important, concentrations in the second sample should be higher.

After collection, samples should be kept in completely filled closed vessels at low temperature (about 4°C) and should analyzed as soon as possible. Conservation of samples is indicated in Table 1.1, based on Appelo and Postma (2005).

1.1.3 *Sampling of solid materials*

Sampling and analysis of solid materials obtained from drilling of boreholes and from natural outcrops and stream sediments should be a common part of projects in hydrogeology and environmental geochemistry. The presence of reactive solids has a significant impact on concentrations of some contaminants in water. Also, compositions and abundances of reactive minerals are often needed for geochemical modeling.

As a general rule, contact with the atmosphere should be avoided after collection of solid samples. Samples should be preserved in gas-tight barrels, and possibly in a nitrogen atmosphere (to help prevent contact with oxygen). Oxidation of pyrite can occur in samples exposed to atmospheric oxygen and can result in the precipitation of ferric hydroxide. Thus, the reducing capacity of solids could be underestimated and the adsorption capacity could be overestimated. Also, the drying of samples (or loss of moisture) should be prevented because it could result in the formation of secondary minerals not initially present in the samples. Additionally, the exsolution of gases from the pore waters of a solid sample should also, ideally, be prevented. For example, degassing of CO_2 from pore water could cause precipitation of calcite, thus, artificially increasing the concentration of calcite in the sample.

The following mineral phases are often of interest.

Calcite: often determines the neutralization capacity of a system. It plays a significant role in the mitigation of acid mine drainage. Dissolution is generally fast (unless it gets armored

Table 1.1. Conservation of samples (adapted from Appelo and Postma, 2005).

Parameter	Conservation of sample
Ca, Mg, K, Na, Si, PO_4	Acidified to $pH < 2.0$ in polyethylene
Heavy metals	Acidified to $pH < 2.0$
SO_4, Cl	No conservation
NH_4, NO_3, NO_2	Store at 4°C and add bactericide like thymol
H_2S	Zn-acetate conservation or spectrophotometric field measurement
TIC	Dilute sample to $TIC < 0.4$ mmol L^{-1} to prevent CO_2 escape
Alkalinity	Field titration using Gran method
Fe^{2+}	Spectrophotometric measurement on acidified sample
pH, T, Eh, O_2	Field measurement

by metal oxyhydroxides or other secondary minerals). Determination is by X-ray diffraction analysis (for content > 5.0 wt%) and/or by acid digestion in the "carbonate step" in a sequential extraction process.

Dolomite: like calcite, also contributes to neutralization capacity, but its dissolution is slower. Determination methods are similar to those used for calcite.

Iron oxyhydroxides: these are significant adsorbents of metals, and may play a role in the neutralization of acid mine drainage at low *pH* values (above 3.0). They also are significant electron acceptors in hydrocarbon and landfill plumes. Their adsorption capacity is strongly *pH*-dependent. They can sometimes be determined by X-ray diffraction (although they are often amorphous or poorly crystalline), or more generally, by the "oxalate step" in a sequential extraction. Leaching in 0.5 M HCl can also be used to determine the readily available fraction of ferric oxides and hydroxides (Christensen *et al.*, 2000). In general, ferric minerals and Mn(IV) minerals play a principal role in the oxidation capacity (*OXC*) of solid materials. Methods for *OXC* determination are described in Heron *et al.* (1994) and Christensen *et al.* (2000).

Pyrite: contributes to the reducing capacity of aquifer solids (for example, may be an electron donor for denitrification) and can be determined by X-ray diffraction and by scanning electron microscopy (SEM) or electron microprobe (EMP). Pyrite and organic matter content often control the reduction capacity (*RDC*) of sediments (see discussion in Christensen *et al.*, 2000).

Organic matter: contributes to the reducing capacity and to the *pH*-buffering capacity of aquifer solids. The fraction of organic carbon (f_{oc}) is especially important in determining the adsorption potential for organic contaminants (Fetter, 1999). Determination is based on burning of solid samples and detection of CO_2 released during combustion. However, removal of carbonates by acid digestion must be conducted prior to this analysis.

Clays (kaolinite, smectite): contribute to the adsorption capacity of solid materials and generally represent the *pH*-independent component of adsorption (especially smectite). Clays can also buffer acid mine drainage (Langmuir, 1997). They are determined by X-ray diffraction or by differential thermal analysis (DTA).

1.1.4 *Sampling of gases*

Sampling and analysis of gases is necessary for modeling geochemical processes in the vadose zone (especially, at contaminated sites and landfills). They can also be essential for dating groundwaters (*cf.* Glynn and Plummer, 2005; or Plummer *et al.*, in press), or for conducting research on oil-gas geochemistry. Dissolved gas analyses can provide very useful information on the origin, temperature, elevation and timing of groundwater recharge, on the type of recharge process (e.g., slow *vs.* resulting from flash flooding), and on the reaction processes (e.g., denitrification, organic fermentation, biodegradation and other redox processes, carbonate dissolution) that have affected the groundwaters during their transport through the vadose zone and through the saturated zone.

Reactive gases of interest in the vadose zone are usually CO_2 and O_2. In groundwater, other dissolved gases of interest also include N_2, Ar (and other noble gases), atmospheric tracers of groundwater age such chlorofluorocarbons or sulfur hexafluoride, and also reactive gases such as CH_4, H_2, NH_3, H_2S as well as volatile organic contaminants.

Gas sampling in deep vadose zones is usually performed by continuous pumping from uncased wells. An interval of sampling is isolated from the adjacent well portions by special packers (Affek *et al.*, 1998). When sampling shallower soil gas, the gas well is usually filled by soil previously taken out from it and left for several days for equilibration the system. Pumping rates for soil gas should be kept low to prevent atmospheric gas inflow (Schumacher, 2007).

Sampling of dissolved gases in groundwater requires special devices and sampling techniques. Field samples need to be collected without headspace and with a seal to prevent gas loss. Detailed sampling procedures are published by the USGS Reston Chlorofluorcarbon

Laboratory (USGS, 2006; http://water.usgs.gov/lab/dissolved-gas/sampling/). Gas samples or water samples for dissolved gas analysis are typically collected in hermetically closed vessels and sent to a laboratory for analysis. Sometimes the analyses are performed directly in the field. However, most real-time soil, vadose zone or groundwater monitoring usually measures concentrations of a single gas and is usually not as sensitive or as accurate as a complete laboratory analysis.

1.2 INTRODUCTION TO THERMODYNAMICS

In this section we introduce some basic thermodynamic concepts that are relevant to geochemical modeling and to calculations of the occurrence and distribution of chemical species in the aqueous, gas or solid phases. Indeed, in the subsurface, chemical species can be present either, (i) in aqueous form, (ii) as gases (either free or dissolved), (iii) as solid species (as suspended or colloidal particles in the aqueous phase or as part of the solid matrix), or (iv) as species sorbed onto solid surfaces. The chemical composition of a water depends on the interactions between these phases and their interfaces.

First we will discuss master variables which dominantly control water chemistry, and then we will introduce some essential thermodynamic terms and concepts.

Master variables: these are considered the most important controls and descriptors of water chemistry. They typically include *pH*, redox potential *Eh*, ionic strength *I*, and major ion concentrations (which define ionic strength), and occasionally the concentrations of some minor, but important and strongly reactive, constituents. These master variables control speciation (i.e., the distribution of free ion and complex species concentrations) and consequently, the behavior of dissolved species, including their mobility and toxicity. Major components occur in relatively high concentrations and influence the other master parameters of water, such as *pH* and *Eh*. The concentration of sulfate and Fe^{2+} in acid mine drainage, for example, has an impact on ionic strength (see later) and, thus, on activity coefficients; Fe^{2+} concentration also has an impact on redox potential, *Eh*. Major components influence behavior of minor components and trace elements (e.g., Pb^{2+}, Zn^{2+}) through their impact on ionic strength, activity coefficients and the formation of complexes and ion pairs. When Pb^{2+} enters groundwater, it typically has no influence on master variables or on other minor components. The formation of complexes like $PbSO_4^0$ has a strong impact on lead solubility and behavior, but it does not significantly affect sulfate behavior because lead concentrations are typically negligible compared to sulfate concentrations. When contamination is caused only by the presence of minor components and/or trace elements, changes in master variables are generally not affected by the contamination.

Concentrations in aqueous chemistry are commonly expressed as molarity *M*, defined as mass in moles per liter of solution; and molality *m*, defined as mass in moles per kilogram of solvent, i.e., pure water. In dilute solutions, molarity is approximately equal to molality. Concentrations are also sometimes expressed in miliequivalents per liter, that is concentrations in milimoles per liter multiplied by the charge of an aqueous ion.

Activity a_i is the "thermodynamic concentration" or the fraction of total concentration that participates in geochemical reactions. It is calculated as a product of an activity coefficient γ_i and a molal concentration m_i:

$$a_i = \gamma_i m_i \tag{1.9}$$

In very dilute solutions, activity coefficients are ≅ 1.0 and activities equals concentrations.

Activity coefficients are a function of ionic strength *I*, which is calculated as:

$$I = 1/2\sum m_i z_i^2 \tag{1.10}$$

where m_i are the molal concentrations and z_i is the charge of ionic species *i*.

Ionic strength essentially corresponds to the degree of mineralization of a solution. At moderate concentrations, when ionic strength increases, activity coefficients decrease (Fig. 1.1). The decreasing trend is related to the "cage effect", an effect caused by the isolation of individual ions through the concentration of oppositely charged ions around them. There is a reversal of the trend in concentrated solutions (e.g., brines), usually starting around ionic strengths of about 0.7 (i.e., seawater) to 1.0 mol kg^{-1} H_2O, depending on the ionic charge. The reversal is related to the decreasing amount of free water available: most of the water molecules are already bound around the other dissolved molecular species. In the case of uncharged species like $H_2CO_3^0$, $H_4SiO_4^0$ or $H_3AsO_3^0$ only the second effect takes place and their activity coefficients are always ≥1.0.

Activity coefficients are typically calculated using the Debye-Hückel equation for ionic $I < 0.1$, the Davies equation for $I < 0.5$, and the Pitzer ion interaction model for higher ionic strengths (Langmuir, 1997; Drever, 1997). However, Pitzer's parameters are generally available for only 25°C and 1 atm pressure, and not for all elements and ion associations.

The Debye-Hückel equation in extended form is:

$$\log \gamma_i = -Az_i^2 \frac{\sqrt{I}}{1 + Ba_i\sqrt{I}} \tag{1.11}$$

where γ_i is the activity coefficient of ion i, z_i is its charge, A, B are temperature-dependent constants, and a is an ion-size parameter.

For higher ionic strength ($0.5 > I > 0.1$), the Davies equation is often used,

$$\log \gamma_i = \frac{-Az_i^2\sqrt{I}}{1+\sqrt{I}} + 0.3I \tag{1.12}$$

where A is temperature-dependent constant, and the final term is purely empirical. Constants and ion sizes used in both equations can be found, for example, in Stumm and Morgan (1996), and in Langmuir (1997). There are also widely used alternative forms of the extended Debye-Hückel equation (e.g., the B-dot and Truesdell-Jones equations) that add a variable positive term (similar to the $0.3I$ term in the Davies equation) and that are often used up to ionic strengths of 0.7 and sometimes even beyond that. For higher ionic strength solutions, such as brines, an entirely different approach to activity calculations based on the Pitzer ion interaction model is used, but this is beyond the scope of the present text.

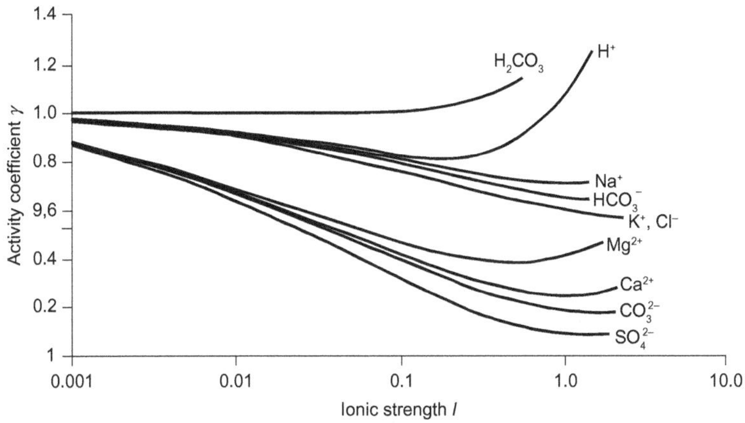

Figure 1.1. Relation between ionic strength and activity coefficients (modified from Freeze and Cherry, 1979).

Hand calculation of activity coefficients and species activities becomes tedious when multiple species are involved. Thus, aqueous speciation programs like PHREEQC (Parkhurst and Appelo, 1999) are typically used to calculate the distribution of total concentrations (and activities) among free ions and complexes.

The law of mass action is a fundamental principle of thermodynamics that relates the activities of the *products* of a reaction to the activities of its *reactants*. Consider a reaction

$$aA + bB \rightleftarrows cC + dD \tag{1.13}$$

where the small letters indicate stoichiometric coefficients and capital letters are the concentrations. Following the law of mass action, the Gibbs free energy, the driving force of the reaction, is expressed as:

$$\Delta G_R = \Delta G_R^0 + RT \ln \frac{[C]^c [D]^d}{[A]^a [B]^b} \tag{1.14}$$

where ΔG_R^0 is the standard Gibbs free energy, R is the universal gas constant, and T is the temperature. At thermodynamic equilibrium, $\Delta G_R = 0$ and therefore:

$$\Delta G_R^0 = -R\,T \ln K \tag{1.15}$$

where K is the equilibrium constant for the reaction. The standard Gibbs free energy of a reaction is calculated as:

$$\Delta G_R^0 = \Delta G_{f\ products}^0 - \Delta G_{f\ reactants}^0 \tag{1.16}$$

where G_f^0 is the Gibbs free energy of formation (determined from thermodynamic tables).

At equilibrium, reactant and product concentrations are invariant and $\Delta G_R = 0$. If $\Delta G_R < 0$, then the reaction proceeds from the left to the right, if $\Delta G_R > 0$, then the reaction proceeds from right to left.

The Van't Hoff equation is used to correct equilibrium constants for temperature changes: equilibrium constants are generally reported for a standard temperature of 25°C. When a constant is K_{T_1} for temperature T_1, its value at temperature T_2 is:

$$\log K_{T_2} = \log K_{T_1} + \frac{\Delta H_R^0}{2.303R}\left(\frac{1}{T_1} - \frac{1}{T_2}\right) \tag{1.17}$$

where ΔH_R^0 is the enthalpy of reaction (from thermodynamic tables).

The solubility product K_{sp} is an equilibrium constant for the dissolution/precipitation of a mineral. For example, consider the dissolution of gypsum in water:

$$CaSO_4 \cdot 2H_2O(s) \rightleftarrows Ca^{2+} + SO_4^{2-} + 2H_2O \tag{1.18}$$

The law of mass action states that:

$$[Ca^{2+}]\,[SO_4^{2-}]\,[H_2O]^2 / [CaSO_4 \cdot 2H_2O] = K_{equil} \tag{1.19}$$

where terms in brackets are activities K_{equil}. By definition, activities of pure solid phases and of water (except for very mineralized solutions and unsaturated-zone water) are equal to 1, and therefore:

$$[Ca^{2+}]\,[SO_4^{2-}] = K_{equil} = K_{sp} = 10^{-4.60} \quad \text{for } 25°C \tag{1.20}$$

where K_{sp} is the equilibrium constant, also known as the solubility product. This constant applies for the reaction as it is written, i.e., usually as a dissolution reaction.

The saturation index SI indicates the degree of saturation of an aqueous solution with respect to a given mineral. It is defined as:

$$SI = \log\frac{IAP}{K_{sp}} \tag{1.21}$$

where IAP is the ion activity product. For example, the gypsum IAP would be written as:

$$IAP = [Ca^{2+}]\,[SO_4^{2-}]\,[H_2O]^2 \tag{1.22}$$

where brackets denote activities. Sometimes, the degree of saturation $\Omega = IAP/K_{sp}$ is used instead of SI.

When $IAP = K_{sp}$, the SI value is 0 and the aqueous solution is at equilibrium with respect to the given mineral. If $SI > 0$, the solution is supersaturated with respect to the mineral and it should precipitate; if $SI < 0$, the solution is undersaturated and the mineral should dissolve (only of course, if the mineral is in contact with the water). It has to be emphasized that the SI value indicates direction, but does not provide any information about the rate of a reaction. Furthermore, there are reactions which never attain equilibrium, where reactants are gradually but irreversibly transformed into products. Oxidation of organic matter provides an example:

$$CH_2O + O_2 \rightarrow CO_2 + H_2O \tag{1.23}$$

In this reaction, organic matter is consumed and even when the concentration of products such as CO_2 increases, organic matter cannot be re-created. This reaction is irreversible (i.e., it proceeds only in the forward direction) and it requires a kinetic description.

Only a few minerals dissolve or precipitate relatively fast: they are called *reactive minerals*. In their case, calculated equilibrium solubilities may actually reflect observed concentrations, as long as they don't have highly variable crystallinities that strongly affect their thermodynamic stability (such as is the case for many metal oxyhydroxides). Some fast reacting minerals are (adopted from Deutsch, 1997):

- Carbonates: calcite, $CaCO_3$; dolomite, $CaMg(CO_3)_2$; siderite, $FeCO_3$; rhodochrosite, $MnCO_3$;
- Sulfates and other evaporate minerals: gypsum, $CaSO_4 \cdot 2H_2O$; jarosite, $KFe_3(SO_4)_2(OH)_6$; melanterite, $FeSO_4 \cdot 7H_2O$; epsomite, $MgSO_4 \cdot 7H_2O$; barite, $BaSO_4$; alunite $KAl_3(SO_4)_2(OH)_6$, halite, NaCl, sylvite, KCl, etc.;
- Oxides and hydroxides: ferrihydrite, $Fe(OH)_3$; goethite, FeOOH; gibbsite, $Al(OH)_3$; manganite, MnOOH; amorphous silica, SiO_2(am) or chalcedony, SiO_2; brucite, $Mg(OH)_2$, etc.

The minerals listed above are assumed to be pure because this facilitates the thermodynamic description of their solubility and reactions, and often provides fairly good approximation of their behavior in subsurface environments. The reality though is that minerals almost always have substitutional and other impurities that have some effect on their stability. A discussion of these effects is beyond the scope of this chapter; readers are referred to Glynn (2000), Prieto (2009) and a Nuclear Energy Agency volume (OECD-NEA, 2007) for data compilations and reviews of these thermodynamic effects.

Two types of mineral dissolution can be distinguished:

Congruent dissolution, where the ratios between the ions dissolving into solution are identical to those of the ions present in the dissolving mineral. For example, gypsum dissolves congruently, because the ratio of Ca^{2+} to SO_4^{2-} going into solution is the same as that in the bulk mineral.

Congruent dissolution typically occurs in the dissolution of carbonate, sulfate, phosphate and halite minerals; some examples are:

gibbsite	$Al_2O_3 \cdot 3H_2O$	$\rightleftarrows$	$2Al^{3+} + 6OH^-$
quartz	$SiO_2 + 2H_2O$	$\rightleftarrows$	$Si(OH)_4$
hydroxyapatite	$Ca_5OH(PO_4)_3$	$\rightleftarrows$	$5Ca^{2+} + 3PO_4^{3-} + OH^-$
dolomite	$CaMg(CO_3)_2$	$\rightleftarrows$	$Ca^{2+} + Mg^{2+} + CO_3^{2-}$
calcite	$CaCO_3$	$\rightleftarrows$	$Ca^{2+} + CO_3^{2-}$
gypsum	$CaSO_4 \cdot 2H_2O$	$\rightleftarrows$	$Ca^{2+} + SO_4^{2-} + 2H_2O$
halite	$NaCl$	$\rightleftarrows$	$Na^+ + Cl^-$

Incongruent dissolution, where the ratios between the ions dissolving into solution differs from those of the ions present in the primary mineral. This typically occurs for the dissolution of silicates when secondary minerals are formed. For example, the dissolution of orthoclase proceeds incongruently with the formation of secondary kaolinite:

$$CaAl_2Si_2O_8 + 2H^+ + H_2O \rightleftarrows Al_2Si_2O_5(OH)_4 + Ca^{2+} \tag{1.24}$$

The ratio Ca^{2+}:Al^{3+}:Si^{4+} is 1:2:2 in orthoclase, but 1:0:0 in water because Al^{3+} and Si^{4+} are retained in the secondary mineral kaolinite.

The formation of aqueous ion complexes influences mineral dissolution. For example calcium can be in solution as the free ion Ca^{2+}, and also in the form of complexes like $CaSO_4^0$, $CaHCO_3^+$ etc. The formation of these complexes increases the solubility of calcium-bearing minerals, which would be lower if only Ca^{2+} were present in the water.

Let us consider the dissolution of anglesite,

$$PbSO_4(s) \rightleftarrows Pb^{2+} + SO_4^{2-} \quad K_{sp} = 10^{-7.79} \tag{1.25}$$

If Mg^{2+} and Cl^- are present in the water, complexes such as $MgSO_4^0$, $PbCl^+$ (and also $PbSO_4^0$) will form:

$$Mg^{2+} + SO_4^{2-} \rightleftarrows MgSO_4^0 \quad \text{and} \quad Pb^{2+} + Cl^- \rightleftarrows PbCl^+ \tag{1.26}$$

and the total concentration of dissolved lead will be:

$$(Pb)_{total} = (Pb^{2+}) + (PbSO_4^0) + (PbCl^+) \tag{1.27}$$

Ionic strength also affects the dissolution of anglesite. In water with Mg^{2+} and Cl^-, activity coefficients for Pb^{2+} and SO_4^{2-} will be lower than in distilled water and more anglesite will dissolve. In that case:

$$\gamma_{Pb^{2+}}\, m_{Pb^{2+}}\, \gamma_{SO_4^{2-}}\, m_{SO_4^{2-}} = K_{anglesite} \tag{1.28}$$

The value of $K_{anglesite}$ is constant. Therefore, a decrease in the activity coefficient values γ_i as a consequence of an increased ionic strength means that the concentrations m_i have to increase to compensate.

Common ion effect: if a solution is at equilibrium with a relatively less soluble mineral and is then equilibrated with a more soluble mineral that contains a "common" ion from the first mineral, then the first mineral will precipitate (Drever, 1997). For example, the dissolution of gypsum added to a solution at equilibrium with calcite would likely cause calcite precipitation: the common ion, Ca^{2+}, controls equilibration between the solution and both minerals:

$$Ca^{2+} + HCO_3^- + CaSO_4 \cdot 2H_2O(s) \rightleftarrows Ca^{2+} + SO_4^{2-} + CaCO_3(s) + H^+ + 2H_2O \tag{1.29}$$

When a water of Ca-HCO_3 type at equilibrium with calcite encounters gypsum, the resulting water becomes of Ca-SO_4 type. Calcite is less soluble than gypsum and dissolution of gypsum with input of large quantity of Ca^{2+} forces the precipitation of calcite and a reduction in the bicarbonate concentration.

The order of encounter refers to the order in which strata with different minerals are encountered by a flowing groundwater. Even when the same strata with the same mineral assemblage are encountered, the resulting groundwater chemistry at the outflow will be different if the order of strata encountered differs (Palmer and Cherry, 1984). For example, consider a first sequence that contains first a shale layer with Na-montmorillonite and organic matter, followed by a limestone layer, and then a gypsum layer. Then consider a second sequence that contains first a limestone layer, followed by a gypsum layer, and then a shale layer with Na-montmorillonite and organic matter. Water having gone through the first sequence will have high Ca^{2+} and SO_4^{2-} concentrations and low Na^+ concentration. In contrast, water having gone through the second sequence will have high Na^+, and low SO_4^{2-} and Ca^{2+} concentrations. These differences result from the common ion effect, cation exchange, sulfate reduction and the irreversible oxidation of organic matter acting in different order in the two sequences.

1.3 CHEMICAL COMPOSITION OF PRECIPITATION

Before reaching the Earth surface as rain or snow, water comes in contact and equilibrates with gases, principally with nitrogen, oxygen, carbon dioxide, and other minor natural and anthropogenic gases. Anthropogenic inputs and natural processes can lead to local, or global, increases in the concentrations of individual gases.

Important chemical reactions between water and gases in the atmosphere include:

$$\begin{aligned} &CO_{2\,gas} + H_2O \rightleftarrows H_2CO_3 \\ &\qquad H_2CO_3 + H_2O \rightleftarrows H_3O^+ + HCO_3^- \end{aligned} \tag{1.30}$$

The above reaction is typical for rain water equilibrating with an uncontaminated atmosphere: it results in *pH* values of rainwater between 5 and 6 under natural conditions.

$$\begin{aligned} &SO_2 + 0.5O_2 \rightleftarrows SO_3 \\ &\qquad SO_3 + H_2O \rightleftarrows H_2SO_4 \\ &\qquad\qquad H_2SO_4 + 2H_2O \rightleftarrows 2H_3O^+ + SO_4^{2-} \end{aligned} \tag{1.31}$$

This is a typical reaction between rainwater and the atmosphere in areas with contaminant SO_x gases: it results in acidic rain with *pH* values down to 3.

Other reactions include:

$$\begin{aligned} &3NO_2^- + H_2O \rightleftarrows NO + 2HNO_3 \\ &\qquad HNO_3 + H_2O \rightleftarrows H_3O^+ + NO_3^- \end{aligned} \tag{1.32}$$

$$NH_3 + H_2O \rightleftarrows NH_4^+ + OH^- \tag{1.33}$$

$$HCl + H_2O \rightleftarrows H_3O^+ + Cl^- \tag{1.34}$$

Atmospheric chemical species can also form "dry deposition" on vegetation during dry periods, especially in regions where the atmosphere has some anthropogenic contamination. This "dry deposition" can then later be mobilized during rainfall. This results in an increased mineralization of the rainwater and the soil water.

1.4 HYDROCHEMICAL PROCESSES

1.4.1 *Introduction*

A portion of rainwater infiltrates into the ground and percolates through the unsaturated (vadose) zone before reaching the aquifer. In both the vadose zone and the aquifer, the water comes into contact with rocks, their weathering products, organic materials, gases and microbes. Chemical and biochemical reactions and other processes occur. Interactions between percolating water or flowing groundwater and solid and gas phases result in a change of the chemical composition and physicochemical properties (such as *pH* and *Eh*) of the infiltrating or flowing water. In the unsaturated zone, the most important processes to consider are the disintegration of organic matter by aerobic respiration, mineral dissolution and precipitation, nitrification and denitrification, equilibration with soil gases, changes in moisture content, and temperature and pressure changes. These processes generally affect the concentrations of O_2 and CO_2 in the soil gas phase, and this affects other reactions and the composition of the water and solid phases. For example, the presence of O_2 in the air of the vadose zone may cause oxidation of iron and the formation of iron oxides. O_2 is also typically consumed by the oxidation of organic matter (aerobic respiration) and by nitrification reactions.

The air in the vadose zone usually contains between 0.05 and 1.5% CO_2 (i.e., a couple of orders of magnitude greater concentration than that of the atmosphere: 0.03%). CO_2 production in the soil occurs mostly through decomposition of organic matter by aerobic respiration; a smaller amount is also formed by respiration of plant roots. The presence of CO_2 induces the dissolution of minerals, such as carbonates. Many of these hydrogeochemical processes also occur in the saturated zone below the water table (i.e., in aquifers and aquitards), but may differ in the vadose zone, due to its high gas content and the specific properties of capillary water. In the saturated zone, the primary geochemical processes include: dissolution and precipitation reactions, sorption and ion exchange, redox reactions, and biogeochemical reactions such as those caused or catalyzed by microbial activity. Although microbes and microbially-driven reactions occur in a wide range of subsurface environments, they are especially active at shallower depths. The following sections will address some of these geochemical processes.

1.4.2 *Oxidation-reduction reactions*

Oxidation-reduction reactions, also known as redox reactions, involve a change of elemental oxidation states. Oxidation is the loss of electrons by an element, and reduction is the gain of electrons. Electrons cannot freely exist in solution, thus there is no oxidation without reduction and *vice versa*. An example is the oxidation of Fe^{2+} and the precipitation of ferric hydroxide (eq. 1.1). In this reaction iron is oxidized from the +2 state to the +3 state and oxygen is reduced from the 0 state to the −2 state. On the other hand, the dissolution of calcite,

$$CaCO_3(s) + CO_2 + H_2O \leftrightharpoons Ca^{2+} + 2HCO_3^- \qquad (1.35)$$

is not a redox reaction because the oxidation numbers of calcium and carbon are +2 and +4, respectively, and they do not change during the reaction.

A redox reaction can be written as a combination of 2 half-reactions, an oxidation half-reaction and a reduction half-reaction. For example, the oxidation of Fe^0 by oxygen can be written as:

$$Fe^0(s) \rightarrow Fe^{2+} + 2e^- \quad \text{oxidation of } Fe^0 \text{ to } Fe^{2+} \qquad (1.36)$$

$$0.5O_2 + 2H^+ + 2e^- \rightarrow H_2O \quad \text{reduction of } O^0 \text{ to } O^{2-} \qquad (1.37)$$

$$Fe^0(s) + 0.5O_2 + 2H^+ \rightarrow Fe^{2+} + H_2O \quad \text{complete reaction} \qquad (1.38)$$

In this reaction, Fe^0 is oxidized and is a donor of electrons (can be also called a reducing agent), and oxygen is an acceptor of electrons and is called an oxidizing agent.

Redox sequence (called also redox ladder): there is a natural sequence of redox reactions during the oxidation of organic matter. Dissolved oxygen is consumed first and methane is generated last (Drever, 1997). The sequence is: $O_2 \rightarrow NO_3^- \rightarrow Mn(IV) \rightarrow Fe(III) \rightarrow SO_4^{2-} \rightarrow CH_4$. The sequence often occurs according to the free energy released by each redox reaction. A similar sequence often occurs along flow paths in uncontaminated groundwater systems, i.e., dissolved oxygen is generally found close to the recharge area, and methane, if present, downgradient in the system. In the case of sanitary landfills and hydrocarbon contaminated plumes (e.g., from oil spills), the sequence is reversed, e.g., methane is found close to (or directly in) the most contaminated source areas (close to or in the landfill; or directly below the free phase oil zone) and dissolved oxygen is found at the periphery of the plume.

1.4.3 *Organic matter decomposition, photosynthesis and aerobic respiration*

A simple formula for organic matter decomposition by aerobic respiration can be written as:

$$(CH_2O)_n + nO_2 \rightarrow nCO_2 + nH_2O \tag{1.39}$$

Unlike most reactions previously discussed in this section, organic matter biodegradation by aerobic respiration (eq. 1.39) is an irreversible reaction which does not reach thermodynamic equilibrium. Similarly, the sunlight-mediated transformation of CO_2 and water into organic matter through the process of plant or algal photosynthesis is also a reaction that cannot be considered to reach equilibrium, even though it is the functional opposite of aerobic respiration. Plants and algae incorporate both processes (i.e., aerobic respiration and photosynthesis) in their biochemical reaction and growth/decay systems, assimilating or releasing not only C and water, but also other nutrients such as NO_3^-, HPO_4^{2-} and NH_4^+.

The assimilation of NH_4^+ during the photosynthetic formation of organic matter results in the production of H^+, which causes a decrease of alkalinity. The corresponding reaction can be written as:

$$106CO_2 + 16NH_4^+ + HPO_4^{2-} + 106H_2O \rightarrow C_{106}H_{263}O_{110}N_{16}P_1 + 106O_2 + 14H^+ \tag{1.40}$$

Conversely, the assimilation of NO_3^- results in the consumption of H^+ and correspondingly in an increase of alkalinity according to:

$$106CO_2 + 16NO_3^- + HPO_4^{2-} + 122H_2O + 18H^+ \rightarrow C_{106}H_{263}O_{110}N_{16}P_1 + 138O_2 \tag{1.41}$$

1.4.4 *Nitrification and denitrification*

Nitrification and denitrification are key processes in the biochemical-hydrochemical N-cycle and together with aerobic respiration are responsible for organic matter decomposition.

In the nitrification process NH_4^+ is converted to NO_3^-, and nitrogen is oxidized from N(–III) to N(+V):

$$NH_4^+ + 2O_2 \rightarrow NO_3^- + H_2O + 2H^+ \tag{1.42}$$

This process causes the production of protons, and increases the acidity of the vadose-zone water.

During denitrification, the NO_3^- ion is reduced from N(+V) to N(0) and N_2 gas is created following the reaction (Appelo and Postma, 2005):

$$5CH_2O + 4NO_3^- \rightarrow 2N_2 + 4HCO_3^- + H_2CO_3 + 2H_2O \tag{1.43}$$

which results in an increase of alkalinity.

1.4.5 *Sorption*

Adsorption refers to the bonding of solutes on adsorbent surfaces (clay minerals, hydroxides etc.), whereas absorption refers to penetration and bonding of solutes into the structure of a mineral. Exchange involves the replacement of a charged ionic species by another at the solid surface. All these processes are included in the general term sorption, and in many cases it is difficult to distinguish between them.

Clays generally have a sorption capacity that is relatively independent of *pH*, at least in the *pH* range of most natural waters. On the other hand, metal oxyhydroxides and soil organic matter have sorption capacities that are highly dependent on *pH*. Clays have large specific surface areas, and their surfaces are generally negatively charged at close to neutral *pH* values (due in part to the substitution of Si^{4+} and Al^{3+} in their crystal lattice by ions with lower valence).

Cation exchange capacity (CEC) characterizes the sorption capacity that a mineral (usually a clay) may have for cations. It is determined by the saturation of exchange sites by a cation and then by the determination of its concentration after expulsion from exchange sites by another cation. Values of *CEC* are reported in meq/100 g of soil for a specified *pH* (generally 7.0). Some example properties of clay minerals are given in Table 1.2.

Montmorillonite has the largest specific surface and the highest CEC values, and kaolinite exhibits the lowest values. Some sorbing minerals, such as metal oxyhydroxides, have a variable sorption capacity with *pH* because their surface protonate (i.e., become more positively charged and/or less negatively charged) at lower *pH* values. This often allows them to sorb anionic species from low *pH* solutions. In contrast, the bulk sorption capacity of most clays is *pH*-independent, with only a small fraction of the sorption sites (usually on clay edges) being *pH*-dependent (e.g., for montmorillonite, this fraction is only about 10% of the total sorption capacity). Sand-sized mineral grains have relatively low *CEC* values, generally less than 1 meq/100 g. However, even this value is by no means negligible.

Solid-phase organic matter (SOM) plays an important role in the sorption of non-polar organic contaminants, but it can also be an important sorbent for inorganic contaminants. The sorptive properties of solid organic matter are *pH*-dependent and at neutral *pH* values generally have an overall negative surface charge caused by the deprotonation of carboxylic (–COOH) and phenolic groups (–OH). Soil humus commonly has a *CEC* of 150–400 meq/100 g (Appelo and Postma, 2005) and can adsorb a significant amount of pollutants. Breeuwsma *et al.* (1986) present an equation for calculation of CEC on the basis of both clay content and organic matter content:

$$CEC\,[\text{meq/100 g}] = 0.7 \times (\%\ \text{clay}) + 3.5 \times (\%\ \text{C}) \tag{1.44}$$

Ion exchange involves the replacement of one species sorbed on a surface exchange site with another species present in the aqueous solution. Cation exchange is generally much more important than anion exchange because the surface charge of minerals in aquifers is generally negative under usual *pH* conditions. Ions with higher charge are preferentially adsorbed, but ion exchange also depends on concentrations of ions in solution. For example, Ca^{2+} has a higher affinity for adsorption than Na^+ when dissolved concentrations are comparable, but Na^+ is preferentially adsorbed relative to Ca^{2+} in seawater solutions. They have much higher

Table 1.2. Properties of clay minerals (adapted from Deutsch, 1997).

Property/clay mineral	Montmorillonite	Illite	Kaolinite
Particle size [μm]	0.01–1.0	0.1–2.0	0.1–5.0
Ion substitution	high	intermediate	low
Surface [$m^2\ g^{-1}$]	600–800	100–120	5–20
CEC [meq/100 g]	80–100	15–40	3–15

Na^+ to Ca^{2+} concentration ratios. Adsorption affinity decreases in the sequence $Al^{3+} > Ca^{2+} > Mg^{2+} > K^+ > Na^+$. When both the charge and the concentration of two competing sorbing ions are similar, then the ion with the lower hydrated radius will generally be preferred for adsorption. Hydrated radii are smaller in ions with higher ionic radii because their charge density (the ratio of the charge to the ionic volume) is lower. For example, Ca^{2+} has a higher ionic radius and a lower hydrated radius than Mg^{2+} and therefore generally shows preferential sorption relative to Mg^{2+}.

There are often colloids (particles of size <1 μm) present in water. The ion distribution near a charged particle can be characterized by the Gouy-Chapmen theory (Appelo and Postma, 2005). This theory assumes that there is a "*Stern layer*" with tightly held ions (generally cations) close to the surface of colloids and a "*Gouy layer*" with mobile ions (where counterions predominate) further from surface, between the Stern layer and the bulk solution. The stability of colloids in solution depends on the thickness of the Gouy layer which in turn depends on the ionic strength of the solution. When ionic strength increases, there is compression of this layer and flocculation of the colloids occurs (for example, during mixing of river water and sea water in estuaries).

Surface complexation reactions include surface hydrolysis and the formation of coordination bonds at the surface with metals and ligands.

In the case of Fe- and Mn-oxyhydroxides the sorption is strongly *pH*-dependent. When *pH* increases, deprotonation of the surface occurs,

$$\equiv SOH_2^+ \rightleftarrows \equiv SOH^0 + H^+ \tag{1.45}$$

$$\equiv SOH^0 \rightleftarrows \equiv SO^- + H^+ \tag{1.46}$$

where $\equiv SOH_2^+$ and $\equiv SOH^0$ are surface functional groups.

Adsorbed ions also change surface charge of adsorbent, for example, in the case of Zn^{2+} adsorption we can write:

$$\equiv SO^- + Zn^{2+} \rightleftarrows \equiv SOZn^+ \tag{1.47}$$

In equations above, the sign ≡ indicates adsorption sites on solid phase and S indicates a surface site (in the case of ferric hydroxide it is represented by Fe). When the average overall charge on the adsorbent surface is equal to zero, the positively charged and negatively charged surface species balance each other out. Neutrally charged species can also exist but are not part of the charge balance equation. For example, if a surface has only protonated and deprotonated sites/species, we can write:

$$\equiv SOH_2^+ \rightleftarrows \equiv SO^- + 2H^+ \tag{1.48}$$

The value of *pH* at the point of zero charge is called the pH_{ZPC}. Typical pH_{ZPC} values are in the following ranges: clays 2.0–4.0, feldspars 2.0–2.5, $Fe(OH)_3$ 8.3, and calcite 8.4–9.5 (Drever 1997, Appelo and Postma 2005).

When the *pH* of a solution changes, there also is a change of electrostatic component of adsorption related to adsorption/desorption reactions. This component of adsorption is expressed by the coulombic (Boltzmann) term K_{coul},

$$K_{coul} = \exp\frac{-\Delta ZF\Psi}{RT} \tag{1.49}$$

where ΔZ is change of surface charge, F is the Faraday constant, Ψ is surface potential, R is the universal gas constant, and T is temperature. The bulk adsorption constant is a product of the intrinsic (chemical) constant K_{int} and the Coulombic constant:

$$K_{bulk} = K_{int}\,K_{coul} \tag{1.50}$$

At pH_{ZPC} the numbers of positively and negatively charged sites are identical, and the bulk surface charge is equal to zero. When the pH is increased above the pH_{ZPC}, negatively charged sites start to predominate, more cations are adsorbed and anions are desorbed. This sorption/desorption behavior with changing pH is called an adsorption edge.

However, it has to be emphasized that even at $pH < pH_{ZPC}$ some negatively charged sites exist, but they are less abundant than positively charged sites. Cations like Pb^{2+} are often adsorbed at pH values much lower than the pH_{ZPC}, usually around $pH < 6.5$, because their concentrations in water are generally small and they have strong affinity for adsorption.

Modeling of adsorption: the most common approach is based on the application of adsorption isotherms. They are empirical (not true thermodynamic) constants determined from batch experiments. In batch experiments a flask with a known quantity of water and solid phase is spiked with a known concentration of a contaminant. After equilibration (usually after 24 hours) the partitioning between solid phase and water is determined. The adsorbed amount of contaminant (generally in mg g^{-1}) is plotted as a function of the aqueous concentration of the contaminant (generally in mg L^{-1}). The adsorbed amount is calculated as:

$$S = \frac{(C_i - C_r)V}{M} \tag{1.51}$$

where C_{equil} is equilibrium concentration, C_{init} is initial concentration, V is volume of water and M is mass of solid phase adsorbents.

The linear distribution coefficient (isotherm) K_d is determined as the slope in the graph $S = f(C)$, (Fig. 1.2). One problem with linear adsorption isotherms is the lack of an upper limit for adsorption with an increasing contaminant concentration in water. In fact, the number of adsorption sites is limited, and at some point, there is a complete saturation of the adsorption sites when the aqueous concentration is high enough.

An alternative model, the Freundlich adsorption isotherm, assumes the following relation:

$$S = K_F\, C^n \tag{1.52}$$

It allows for the decreasing adsorption with increasing aqueous concentration of a given constituent. The exponent n, which is generally < 1.0, changes the slope of straight line at higher concentrations of solutes. However, there is still no maximum adsorption limit in the Freundlich model. (A linear distribution coefficient K_d is a special case of the Freundlich isotherm for $n = 1.0$).

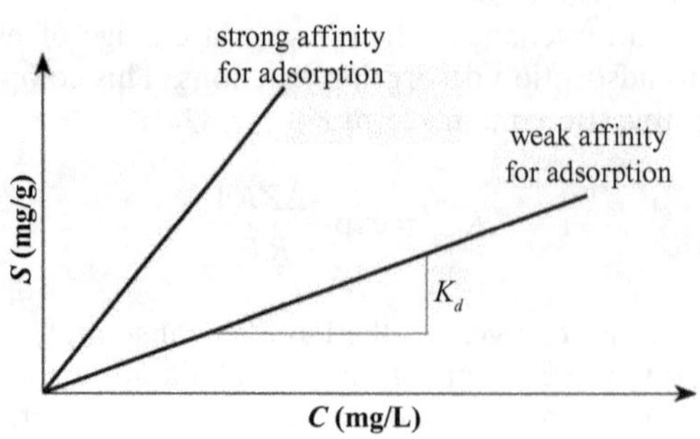

Figure 1.2. Linear adsorption isotherm K_d.

A third model, the Langmuir adsorption isotherm, is expressed as:

$$S = \frac{S_{\max} K_L C}{1 + K_L C} \tag{1.53}$$

where S_{max} is the maximum amount of a solute which can be adsorbed and K_L is an empirical constant. When the aqueous concentration of the given sorbing constituent of interest is high, the product $K_L\, C >> 1$, and $S = S_{max}$, which is the asymptote of the graph $S = f(C)$ or the adsorption limit.

Adsorption isotherm models are easily implemented into the advection-dispersion equation (ADE) for solute transport commonly used in numerical simulation models. A retardation factor, R, is typically defined in these models, where:

$$R = \frac{v_{water}}{v_{contam}} = 1 + \left(\frac{\rho_b}{n}\right)\frac{dS}{dC} \tag{1.54}$$

v_{water} and v_{contam} are the velocities of groundwater flow (or of a conservative, non-adsorbed tracer migration) and of the reactive (i.e., sorbing) contaminant, respectively; ρ_b is the dry bulk mass density of porous media and n is the porosity. The retardation factor for a linear isotherm can be written as (Fetter, 1999):

$$R = 1 + \left(\frac{\rho_b}{n}\right)K_d \tag{1.55}$$

For a Freundlich isotherm, the retardation factor is:

$$R = 1 + \left(\frac{\rho_b}{n}\right)K_F n C^{(n-1)} \tag{1.56}$$

The retardation factor is constant in the case of a linear isotherm, but depends on concentration when either the Freundlich or Langmuir isotherms are used.

Surface complexation modeling takes into account surface charge changes caused by changes of pH and water chemistry, and even more importantly, it provides a theory for calculating the speciation of a wide variety of sorbed species on a solid surface, possibly involving different types of sorption sites and characteristics. Surface complexation calculations are in many respect analogous to calculating the speciation of an aqueous solution. The diffuse double layer (DDL) theory is the most common surface complexation theory used. Commonly, due to lack of better information, applications of the theory often assume that sorption occurs on a hydrous ferric hydroxide (HFO) that has the composition of goethite, a molecular weight of 89 g mol^{-1}, a surface area of 600 $m^2\ g^{-1}$, and a density of adsorption sites of 0.2 mol/mol HFO. It is possible to change these parameters. For example, when the adsorption of trace metals is important), a second "strong" type of adsorption site can also be added (typically with a density of 0.005 mol/mol HFO; Dzombak and Morel, 1990; Stumm, 1992).

We have discussed so far the modeling of single ion sorption on a solid phase based on the application of isotherms or on surface complexation theory. Selectivity coefficients describe the exchange of two ions (say, A^+ and B^{2+}) for each other in reactions such as:

$$2\text{A-clay} + B^{2+} \rightleftarrows \text{B-clay} + 2A^+ \tag{1.57}$$

with the selectivity coefficient (equilibrium constant) $K_{A\text{-}B} = [A^+]^2[\text{B-clay}]/([\text{A-clay}]^2[B^{2+}])$. The selectivity coefficient $K_{A\text{-}B}$ is not a real thermodynamic constant,

Table 1.3. Selectivity coefficients (Appelo and Postma, 2005).

Ion I^+	$K_{Na/I}$	Ion I^{2+}	$K_{Na/I}$
Li^+	1.2 (0.95–1.2)	Mg^{2+}	0.50 (0.4–0.6)
K^+	0.2 (0.15–0.25)	Ca^{2+}	0.40 (0.3–0.6)
NH_4^+	0.25 (0.2–0.3)	Sr^{2+}	0.35 (0.3–0.6)
Rb^+	0.10	Ba^{2+}	0.35 (0.3–0.6)
Cs^+	0.08	Fe^{2+}	0.6
–	–	Cu^{2+}	0.5
–	–	Zn^{2+}	0.4 (0.3–0.6)
Ion I^{3+}	–	Cd^{2+}	0.4 (0.3–0.6)
Al^{3+}	0.6 (0.5–0.9)	Pb^{2+}	0.3

because there is no thermodynamic description of adsorbed/exchanged ion behavior. It can be defined differently following different conventions for the calculation of activities of adsorbed ions (Appelo and Postma, 2005).

Values of selectivity coefficients are valid only for a specified *pH* and water chemistry. In the case of the same concentrations in water, divalent Ca^{2+} is preferentially adsorbed over monovalent Na^+. However, as already mentioned, in seawater with high Na^+ concentrations there is much more Na^+ adsorbed on the solid phase than Ca^{2+}. Applications of selectivity coefficients are discussed in Appelo and Postma, (2005). Table 1.3 gives selectivity coefficients for cation I exchange with Na^+,

$$Na^+ + 1/i\,I{-}X_i \rightleftarrows Na{-}X + 1/i\,I^+ \tag{1.58}$$

where X represents X fraction of ion I on adsorption sites, and i is charge of the ion I. When exchange for a cation another than Na^+ is considered, selectivity coefficients are determined by dividing appropriate values for Na^+ exchange, for example $K_{K/Ca} = K_{K/Na}/K_{Ca/Na} = 0.2/0.4 = 0.5$. The convention used in equation (1.58) is the Gaines-Thomas notation. There are other types of conventions (Gapon, Vanselow, etc.) that are discussed in detail in Appelo and Postma, (2005).

1.5 KINETICS

Many geochemical reactions rarely reach thermodynamic equilibrium, usually because those reactions are slow especially when compared to dynamic processes that may be advecting solutes or otherwise altering the mix of reactants in contact with each other. Kinetically slow reactions include many heterogeneous reactions and redox reactions, where there are a large number of elements or species involved in the overall reaction, where there are many mechanistically different "steps" involved in the overall reaction, or alternatively in the case of redox reactions, where a large electron transfer is required to change an element from one oxidation state to another. Figure 1.3 gives an overview of approximate reaction times (half times assuming first-order reactions) needed for different types of reactions to reach equilibrium, and compares them with typical residence times for different hydrologic reservoirs.

Kinetic theory is used to describe the progress of reactions with time. Consider the dissolution of halite, NaCl, in a beaker. When the concentration of Na^+ is plotted *vs.* time (Fig. 1.4), the concentration increases initially, but at time $t = t_2$ it reaches a maximum, and then remains constant. After time t_2, the concentration of Na^+ does not depend on time and is determined by the solubility product of halite.

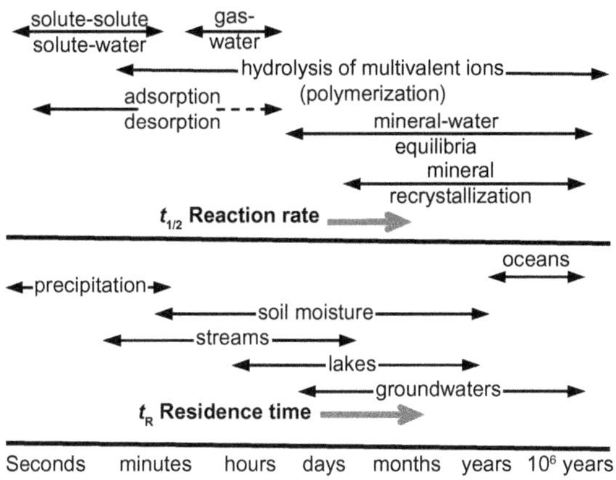

Figure 1.3. Schematic comparison of reaction times (represented as reaction half times to achieve equilibrium) and residence time (contact time) in different environments of the hydrosphere. Modified from Langmuir and Mahoney (1985).

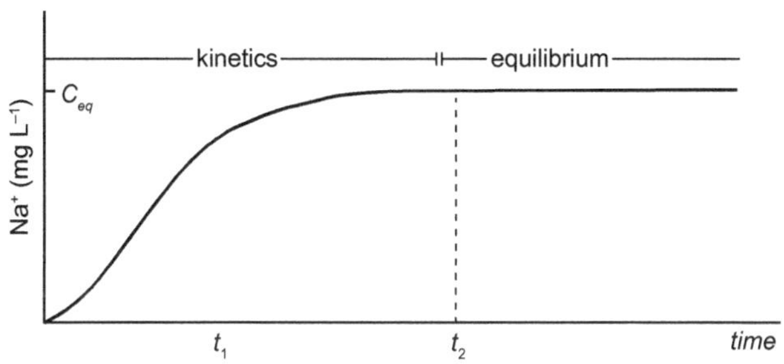

Figure 1.4. Dissolution of halite (adapted from Appelo and Postma, 2005).

The Na^+ concentration in the region where $t > t_2$ is described by equilibrium thermodynamics, whereas the concentration change in the region where $t < t_2$ is described by reaction kinetics. The time t_2 necessary to attain equilibrium differs for different minerals. It is relatively short for reactive minerals, like calcite and gypsum, and long for non-reactive minerals like aluminosilicates and quartz. However, whether a reaction is "fast" or "slow", and whether thermodynamic equilibrium is reached or not, depends also on the velocity of groundwater flow (and the removal of reaction products). The faster the flow is, the less likely it is that thermodynamic equilibrium will be reached. Indeed, imagine that fresh dilute water is being added at a constant rate to an imaginary beaker with dissolving halite, and that the well-mixed solution is being removed from the beaker at the same flow rate. The solution in this imaginary beaker might never reach thermodynamic equilibrium with halite if the inflow of fresh water was fast enough (although some other steady state concentration might be reached); additionally the constant inflow of fresh water and the limited amount of halite in the beaker might eventually result in the complete dissolution of all the halite in the beaker.

In most cases, reaction rates also depend on temperature, pressure and reactant concentrations. The manner in which the reaction rate varies with the concentration of reacting substances is characterized by the reaction order. Usually the reaction rate is expressed as proportional to n-th power of concentration.

Order of reaction refers to the exponent of a concentration in a rate equation. For example, the equation:

$$-\frac{dC}{dt} = kC \tag{1.59}$$

refers to a first order rate equation, because the exponent of the concentration C is one. In the equation, t is time and k is the reaction constant. The time needed to halve the reactant concentration, i.e., the half-life of the reaction, $t_{1/2}$, can be calculated from $t_{1/2} = \ln 2/k$. The solution of equation (1.59) is obtained by integration, assuming that $C = C_0$ at some initial time t_0:

$$C_t = C_0 e^{-kt} \tag{1.60}$$

and that C_t is the concentration at time t.

Equation (1.60) describes radioactive decay of isotopes (like ^{14}C and tritium), and is also used for description of biodegradation of petroleum products (like benzene and toluene) and chlorinated solvents (like PCE and TCE). The rate of decay/biodegradation in these reactions decreases as the concentrations decrease.

By comparison, a zero order rate reaction would be expressed as:

$$-\frac{dC}{dt} = k \tag{1.61}$$

and

$$C_t = C_0 - kt \tag{1.62}$$

Zero-order reactions are often observed for thermal chemical decomposition. The rate of a zero order reaction is always the same and does not depend on concentration.

In the case of second order kinetic reactions, the concentration is squared,

$$-\frac{dC}{dt} = kC^2 \tag{1.63}$$

The oxidation of Fe^{2+} by oxygen provides an example of a second order reaction:

$$-\frac{dm_{Fe^{2+}}}{dt} = km_{Fe^{2+}} \left(\mathrm{OH^-}\right)^2 P_{O_2} \tag{1.64}$$

The reaction is 2nd order with respect to OH^- (and therefore pH), but 1st order with respect to Fe^{2+} concentration and to the partial pressure of O_2.

There are other types of kinetic models, for example Michaelis-Menton kinetics and Monod kinetics, which describe the rates of enzymatic reactions in biodegradation of organic contaminants. These are beyond the scope of this chapter, but are discussed for example in Appelo and Postma (2005).

Temperature has a strong effect on reaction rates, i.e., the rates increase with increasing temperature. As a rule of thumb, the rates speed up by a factor of two for every 10°C. More accurately, the relation between reaction kinetic constants and temperature is given by the Arrhenius equation,

$$k = A\exp\frac{-E_a}{RT} \tag{1.65}$$

where A is the pre-exponential (Arrhenius) factor, E_a is activation energy, R is the universal gas constant, and T is temperature. If a reaction kinetics constant k is known as a function of temperature, a graph of log $k = f(1/T)$ can be used to determine the activation energy: the slope equals $-E_a/2.3\,R$.

REFERENCES

Affek, H.P., Ronen, D. & Yakir, D.: Production of CO_2 in the capillary fringe of a deep phreatic aquifer. *Water Resour. Res.* 34:5 (1998), pp. 989–996.

Appelo, C.A.J. & Postma, D.: *Geochemistry, groundwater and pollution.* 3rd edition, A.A. Balkema, Rotterdam/Brookfield, The Netherlands, 2005.

Breeuwsma, A., Wösten, J.H.M., Vleeshouwer, J.J., Van Slobbe, A.M. & Bouma, J.: Derivation of land qualities to assess environmental problems from soil surveys. *Soil Sci. Soc. Amer. J.* 50 (1986), pp. 186–190.

Christensen, T.H., Bjerg, P.L., Banwart, S.A., Jakobsen, R., Heron, G. & Albrechtsen, H.-J.: Characterization of redox conditions in groundwater contaminant plumes. *J. Contam. Hydrol.* 45 (2000), pp. 165–241.

Deutsch, W.J.: *Groundwater geochemistry, fundamentals and applications to contamination.* Lewis Publishers, Boca Raton, Florida, 1997.

Drever, J.I.: *The geochemistry of natural waters, surface and groundwater environments.* 3rd edition, Prentice Hall, Upper Saddle River, NJ, 1997.

Dzombak, D.A. & Morel, F.F.M.: *Surface complexation modeling: hydrous ferric oxide.* John Wiley & Sons, NewYork, NY, 1990.

Fetter, C.W.: *Contaminant hydrogeology.* Prentice Hall, Upper Saddle River, NJ, 1999.

Freeze, A.R. & Cherry, J.A.: *Groundwater.* Prentice Hall, Englewood Cliffs, NJ, 1979.

Fritz, S.J.: A survey of charge-balance errors on published analyses of potable ground and surface waters. *Ground Water* 32 (1994), pp. 539–546.

Glynn, P.D. & Plummer, L.N.: Geochemistry and the understanding of ground-water systems. *Hydrogeology J.* 13 (2005), pp. 263–287.

Glynn, P.D. & Voss, C.I.: Geochemical characterization of Simpevarp ground waters near the Äspö Hard Rock Laboratory. Swedish Nuclear Power Inspectorate (SKI), SKI report 96:29, 1999.

Glynn, P.D.: Solid-solution solubilities and thermodynamics: sulfates, carbonate and halides. In: C.N. Alpers, J.L. Jambor & D.K. Nordstrom (eds): Sulfate minerals-crystallography, geochemistry and environmental significance. *Reviews in Mineralogy and Geochemistry* 40, 2000, pp. 481–511.

Heron, G., Christensen, T.H. & Tjell, J.C.: Oxidation capacity of aquifer sediments. *Envir. Sci. Technol.* 28 (1994), pp. 153–158.

Kehew, A.E.: *Applied chemical hydrogeology.* Prentice Hall, Upper Saddle River, NJ, 2000.

Langmuir, D.: *Aqueous environmental geochemistry.* Prentice Hall, Upper Saddle River, NJ, 1997.

Langmuir, D. & Mahoney, J.: Chemical equilibrium in and kinetics of geochemical processes in ground water studies. In: B. Hitchon & E.I. Wallick (eds): *Practical applications of ground water geochemistry.* Proceedings 1st Canadian/American conference on hydrogeology, Worthington, OH, Natl. Water Well Assoc., 1985, pp. 69–95.

McCleskey, R.B., Nordstrom, D.K. & Ryan, J.N.: Electrical conductivity method for natural waters. *Appl. Geochem.* 26, Suppl. 1, S227–S229, 2011.

OECD-NEA: Chemical thermodynamics: Chemical thermodynamics of solid solutions of interest in radioactive waste management. Edited by: Edited by: F.J. Mompean, M. Illemass & J. Perrone; publisbed by: Organization for Economic Cooperation & Development OECD Nuclear Energy Agency NEA, 2007.

Palmer, C.D. & Cherry, J.A.: Geochemical evolution of groundwater in sequences of sedimentary rocks. *J. Hydrol.* 75 (1984), pp. 27–65.

Parkhurst, D.L. & Appelo, C.A.J.: PHREEQC-2, a Hydrogeochemical computer program. U.S. Geological Survey, Water Res. Inv., 1999.

Plummer, L.N., Sanford, W.E. & Glynn, P.D.: Characterization and conceptualization of groundwater-flow systems. *Guidebook on dating of old groundwaters*, Chapter 2. International Atomic Energy Agency, Vienna, Austria, in press.

Prieto, M.: Thermodynamics of solid solution-aqueous solution systems. *Reviews in Mineralogy & Geochemistry* 70, 2009, pp. 47–85.

Puls, R.W. & Barcelona, M.J.: Low flow (minimal drawdown) ground-water sampling procedures, EPA/540/S-95/504, 1996.

Schumacher, B.A.: Final project report for the development of an active soil gas sampling method. EPA/600/R-07/076. San Diego, CA, 2007.

Stumm, W.: *Chemistry of the solid-water interface*. John Wiley & Sons, Hoboken, NJ, 1992.

Stumm, W. & Morgan, J.J.: *Aquatic chemistry*. 3rd edition, John Wiley & Sons Inc., Hoboken, NJ, 1996.

US Geological Survey, 2006. 25 Feb 2009 http://water.usgs.gov/lab/dissolved-gas/sampling/ (accessed September 2011).

Wagner, W. & Kruse, A.: Properties of water and steam, the industrial standard IAPWS-IF97 for the thermodynamic properties and supplementary equations for other properties. Springer-Verlag, Berlin, Germany, 1998.

CHAPTER 2

Thermodynamics of gas and mineral solubility in the unsaturated-zone water

Lionel Mercury & Michael Zilberbrand

> *"How do we assure ourselves that our models are appropriate? There is of course no way to be sure. The construction of models useful in understanding nature is the essence of science and relies as much upon creativity and imagination as any painting or musical composition."**
>
> Greg M. Anderson (2005)

2.1 INTRODUCTION

One of the particularities of the unsaturated zone (UZ) is that it places in physical contact (i) the atmospheric air; (ii) a solution of variable composition (that depends on both depth and interaction time with the solid matrix), either percolating as free water or held in pores by capillarity; and (iii) minerals and organic matter. Geochemically, the coexistence of two types of water (percolating *vs.* capillary), each having its own spatial and time dynamics of interactions with solids, organics, and air is a basic characteristic of the soil systems.

The transport models have very early integrated the physical parameters associated with the unsaturated state of the medium. Meanwhile, most attempts to develop a model coupling chemistry and transport through the UZ have assumed that properties of water and minerals are the same as those in the Saturated Zone (SZ). Recently, different tentative efforts have been done to calculate explicitly the chemical potential of the capillary water and its influence on chemical properties of the capillary water—mineral—soil atmosphere system (e.g., Zilberbrand, 1997, 1999; Mercury and Tardy, 1997a,b, 2001, 2004; Mercury *et al.*, 2003, 2004). The parameter proposed to control the capillary state within the UZ is the relative humidity of the close-contacting air. Mercury *et al.* (2003, 2004) determined the effect of capillarity on the aqueous speciation and the solubility of both minerals and gases in capillary solutions in general terms. Lassin *et al.* (2005) incorporated the osmotic aspects related to saline capillary solutions and tested this frame against some laboratory studies.

2.2 BACKGROUND

2.2.1 *Capillary water*

Plane water-air interface (like free water table of water in a well, in lakes, seas etc.) expresses a physical equilibrium: water and air are at the same pressure *P*, in general, the atmospheric pressure. Curved interfaces mean that pressure changes across them; in the case of the concave capillary menisci, the pressure decreases from the atmosphere to the liquid. This is the Young-Laplace law (Fig. 2.1):

$$\Delta P = P_{liquid} - P_{vapor} = 2\frac{\gamma}{r} \tag{2.1}$$

*Source: Greg M. Anderson: *The thermodynamics of natural systems.* Cambridge University Press, 2005.

where γ is the liquid-air surface tension [J m^{-2}] and r is the radius of the curved spherical interface [m].

Liquid-air interfaces obey a further thermodynamic condition: the liquid–to-vapor equilibrium. In a case of plane interfaces, it is easily calculated and usually plotted in the p-T diagram of water by the saturation curve (Fig. 2.2), where vapor and liquid are at the same chemical potential. As for the curved interfaces, the chemical potential of liquid water is modified by the pressure effect and so the saturation equilibrium is: this is the Kelvin's law (Fig. 2.1, see right vertical axis, and Fig. 2.3; e.g., Mercury and Tardy (2001) for more details):

$$\mu_{liquid} - \mu_{saturated\ liquid} = RT\ln\left(\frac{p}{p^o}\right) \tag{2.2a}$$

$$= \int_1^P (V_{H_2O})dP \tag{2.2b}$$

where μ is the chemical potential of the subscribed substance [J mol^{-1}], R: the ideal gas constant [J mol^{-1} K^{-1}], T: temperature [K], p: vapor pressure at equilibrium with a curved interface [MPa], $p°$: the saturated vapor pressure (plane interface) [MPa]; V_{H_2O}: molar volume of water as a function of pressure [cm^3 mol^{-1}], and P: internal water pressure [MPa].

Equations (2.2) mean that the concave curved liquid does not evaporate at vapor pressure lower than the saturated vapor pressure (in a dry atmosphere), down to a new "dry atmosphere saturation curve" (Fig. 2.3). We could say otherwise that water vapor (like gases in general) has an increasing solubility in capillary water and so a decreased pressure at equilibrium (e.g., Mercury *et al.*, 2004).

Equations (2.1) and (2.2) make clear that a pressure of the liquid decreases owing to the capillary phenomena, but this is restricted by certain limits. On the contrary of the usual feeling, the limiting pressure for condensed matter (liquids and solids) is not the zero pressure (which is the limiting pressure for gases), but the (P, T) spinodal pairs (Fig. 2.2) defined by the thermodynamic stability conditions (e.g., Debenedetti, 1996 or Mercury *et al.*, 2003). The P-T

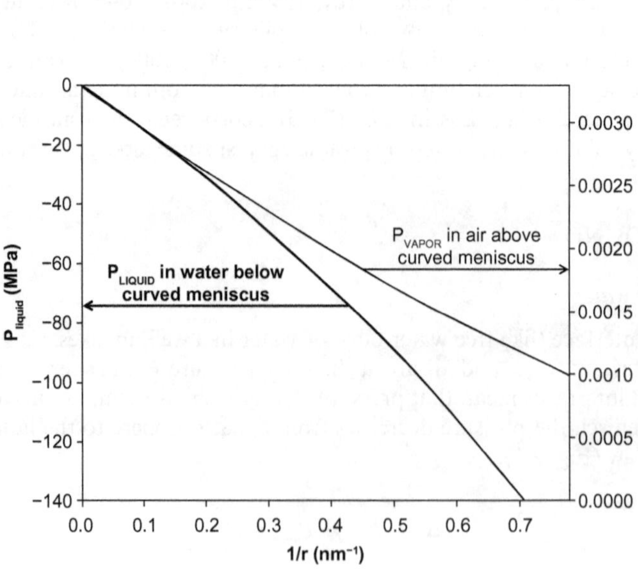

Figure 2.1. Vapor and liquid pressures at Young-Laplace equilibrium as a function of liquid-air interface radius r (from plane to highly concave interface), at 25°C.

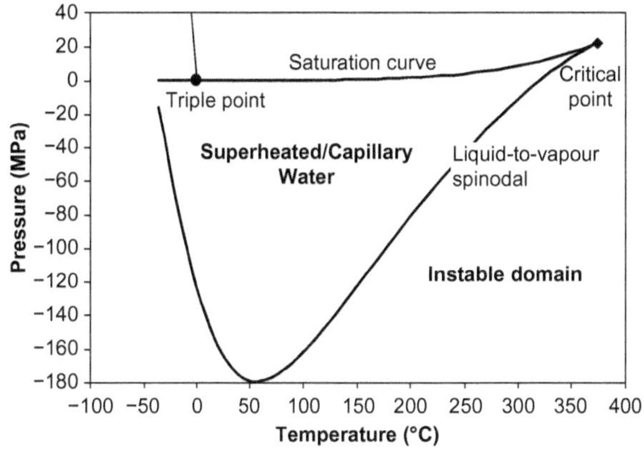

Figure 2.2. *p-T* phase diagram of pure water including its extension to the metastable domain.

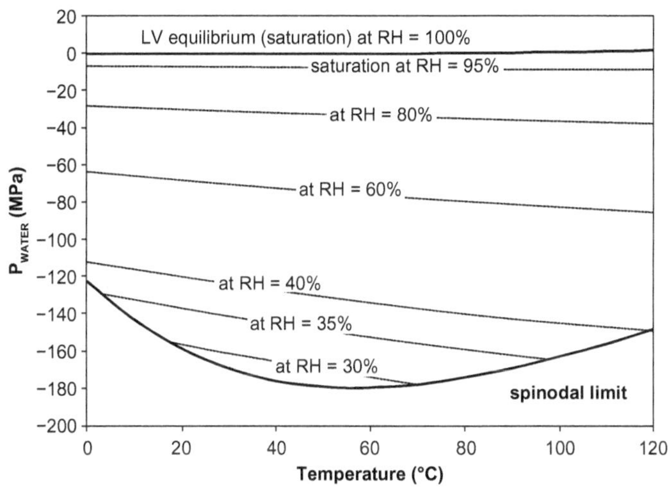

Figure 2.3. Internal pressure of capillary water as a function of relative humidity *RH* along the metastable (superheated) liquid domain.

space limited by the saturation and the spinodal curves is the domain of the superheated (metastable) water, either heated in excess of the saturation temperature or decompressed below the saturation pressure. Indeed, the capillary water has a decreasing internal pressure and so plots inside the superheating domain. This fact offers a very useful perspective: capillary water has the same properties as the superheated water which can be easily calculated extrapolating the Equation of State (EoS) (e.g., Mercury and Tardy, 2001; Mercury *et al.*, 2003). We use the EoS currently recommended by the International Association for the Properties of Water and Steam, usually called the IAPWS-95 equation (Wagner and Pruss, 2002). When doing so, we assume explicitly that the IAPWS-95 equation retains its validity even being extrapolated beyond its normal domain. Indeed, Wagner and Pruss (2002) specifically tested the capabilities of their equation in metastable domains (superheating/supercooling) and found it to fit very well the scarce existing data.

What makes capillarity different from superheated water is the second characteristic involved in equation (2.1): the decreasing liquid pressure relates to a decreasing size of the liquid-air interface (controlled in real world by the thinness of the porous spaces). And the interface radius is rapidly nanometrically-sized when liquid pressure decreases (Figs. 2.1 and 2.3). Why the pore size makes capillarity different from superheating? Actually, the transformation from liquid to vapor requires a certain quantity of triggering energy (the so-called activation energy). A gas-liquid interface must be created (bubble nucleation) increasing the system energy by $4\pi r^2\gamma$ (for a spherical bubble with radius r, and the liquid-vapor surface tension γ), while the formation of the most stable phase brings back bulk energy ($4/3\pi r^3\ \Delta P$, where $\Delta P = P_{liquid} - P_{vapor}$). After the Classical Nucleation Theory (CNT), the competition between these two opposite effects results in a maximal energy barrier E_b (for more details, e.g., Debenedetti, 1996 or Caupin and Herbert, 2006) expressed as:

$$E_b = \left(\frac{16\pi}{3}\right)\left(\frac{\gamma^3}{\Delta P^2}\right) \tag{2.3}$$

E_b is calculated according to the $\Delta\Psi(P)$ reached inside a spherical bubble of "critical" radius r_C obeying the Young-Laplace law.

In other words, a bubble required for the capillary (superheated) liquid to vaporize is of the same size than the capillary container itself: the vaporization of capillary liquid obeying the Young-Laplace condition cannot occur due to geometrical restriction.

The key conclusions of this section and the previous equations are the following:

1. it is possible to "capillarize" liquid water (that is to decrease its internal pressure, even down to the negative pressure domain) by increasing the dryness of the atmosphere without evaporation;
2. if the capillary radius obeys the Young-Laplace law, capillary water is a stable liquid, because the internal tension of the liquid cannot be relaxed by nucleation due to the thinness of the pores.

2.2.2 *"Capillarizing" the water by the dryness of the soil atmosphere*

In UZ systems, the chemical potential of vapor in air drives the chemical potential of liquid water. The aqueous solution trapped in the UZ pores containing rather dry air, can avoid evaporation by decreasing its internal pressure and so becoming a capillary liquid. The soil air humidity is related to the dryness of the external atmosphere and its seasonal behavior, to the consequent dryness gradient between the top surface and the soil depth (which generates a high suction force), to the depth of decay of the evaporation, to distribution of the local water uptake by plants with depth, and also, in general, to the soil heterogeneity. Obviously the true question is how dry can a soil atmosphere be? Usually, basing on field measurements and because soil wetting is much more rapid than the soil drying, the soil humidity is considered close to saturation below the first centimeters of the soil even under arid conditions. We present now two types of counter-arguments.

The first type comes from some field measurements, pointing to a more contrasted situation than usually argued. Andraski (1997) measured the matric potential with psychrometers (valid range: $\Psi_m > -8$ MPa) and the water content by neutron probes in the Mojave Desert (USA) during a 5-years period. Water potentials decreased seasonally below the detection limit of the psychrometers, with corresponding volumetric water content ranging between 0.02 and 0.19 m^3 m^{-3} at depths between 0.15 and 1.25 m. Scanlon *et al.* (2003) measured the matric potential in the Chihuahuan desert (Texas) during May through September 1989: psychrometers went out of range ($\Psi_m < -8$ MPa) down to 0.8 m deep throughout most of the monitoring period. Finally, Flint *et al.* (2002) measured matric potentials in the Mojave Desert (Western USA) with a heat dissipation sensor, and recorded Ψ_m around -100 MPa at depths from 40 to 80 cm.

The second type of arguments comes from some indirect observations. First of all, numerous plant species adapted to arid climate (as is the Californian chaparral species) have their xylem pressures at low values, down to –10 MPa (Scholander *et al.*, 1965; Tyree and Zimmermann, 2002). That allow them to extract water kept by strong surface forces, what points to arid conditions in soil in the immediate closure to the roots along their depth (down to 10–15 m). Second, we consider the interpretation of pedological profiles made by Tardy and collaborators (see especially Tardy and Nahon, 1985; Trolard and Tardy, 1987, 1989; Tardy and Novikoff, 1988) in tropical soils old enough to expect a chemical evolution obeying the thermodynamic equilibrium. They observed that many mineralogical assemblages (what they called concretions), at the aggregate and at the soil profile scales, direct to local arid conditions, being coexisting at a short distance with assemblages (called excretions) characteristic of more humid conditions. Eventually, the experiments on waste rock pile (e.g., Stockwell *et al.*, 2006) showed intense weathering in the unsaturated pile together with strong variations of pore-water chemistry, evidencing the short-scale heterogeneity characterizing the geochemical behavior inside the pile.

These arguments point to a possible occurrence of strong capillarity in certain soils compartments possibly disconnected from the other ones. This disconnection of capillary water helps establishing capillary conditions which control then the route to attain the chemical equilibrium: a local equilibrium included in strongly disequilibrated systems.

2.2.3 *Capillarity and size of pores*

A direct consequence of the above capillary situation is that the liquid is now under a lower pressure than the atmospheric compartment, so that the capillary menisci are concave towards the atmosphere. This concavity depends quantitatively on the pressure difference between the two phases, so that it is possible to calculate the corresponding radius of the capillary meniscus. Additionally, it is possible to relate the meniscus radius to that of the trapping pore, taking care of the so-called contact angle and the thickness of adsorbed films that are always present at the solid surface (see Mercury and Tardy, 2004, for more details).

The fundamental interest of this relationship can be highlighted by stressing that the Young-Laplace radius in the equation (2.1) equals to the critical radius of the growing nucleus of the equation (2.2). Actually, the capillary liquid is "back-stabilized" by the geometrical restriction. The usual consequence is to assume that capillary water takes place only in such pores obeying the Young-Laplace law. For significant capillary suctions that corresponds to nanosized capillaries.

On the other hand, it must be stated that it is difficult to know whether the capillary systems obey the Young-Laplace equation, that is, to know how stable they are. For instance, it has been demonstrated that just the occurrence of metastable water in tensiometers enlarges the measurement range of these devices down to –0.14 MPa (Tamari *et al.*, 1993). Also, it is noticeable that aqueous solutions located in relatively large pores (30 μm and more) are able to reach high levels of superheating before nucleating (e.g., Green *et al.*, 1990; Zheng *et al.*, 1991; Alvarenga *et al.*, 1993; Shmulovich *et al.*, 2009). Therefore, even in such pores water could be metastable.

Pettenati *et al.* (2008) calculated the volume of capillary water held at high suctions using measured water retention curves treated with the model of Rossi and Nimmo (1994). Such calculation shows that several liters to several tens of liters of water can be held per cubic meter of soil by strong suction. For instance a "sand L-soil" held only 4.6 L m^{-3} whereas 34 L m^{-3} is held in the "Salkum silt loam". McCrea *et al.* (1990) measured in the laboratory the retention capacity of granitic material weathered from the Yilgarn craton in Southwest Australia and found 50 to 60 L m^{-3} of soil water at –125 MPa (*pF* 6.1). Interestingly, that a potential of –125 MPa corresponds to arid conditions (air $RH = 40\%$, at $T = 35°C$), close to the mean value in this region from November to March (it varies between 30 to 40% *RH* during this period; source: www.bom.gov.au). Kew and Gilkes (2006) measured the water content in the Darling Range bauxite at different high suctions: the measurements range from 40 up to 100 L m^{-3}. These relatively large volumes appear difficult to reconciliate with the very fine capillary sizes (1.6 nm) predicted by Young-Laplace law to sustain suction potentials of −125 MPa.

Therefore, we are led to take care that capillary water may disobey the Young-Laplace law, namely, may be located in larger pores than those predicted by equation (2.1). In this case, capillary (metastable) water can occupy the microporous spaces, with large associated volume. This conclusion is probably very important for soil physics: capillarity can occur also inside the micropores what extends considerably its potential domain of influence. However, such capillarized water disobeying Young-Laplace law is a metastable compound and could have a limited lifetime. The question is whether this lifetime is long enough for the capillary water to be of practical importance?

2.2.4 *Capillary water: stable or metastable?*

Whenever capillary liquid is located in pores larger than the nanopores (obeys Young-Laplace law), it acquires a metastable character which makes it able to nucleate vapor bubbles at any moment. Assessment of the lifetime of the metastable liquid could be performed in the following way.

According to the Classical Nucleation Theory (CNT), the nucleation rate of vapor bubble follows the Arrhenius law (e.g., Debenedetti, 1996):

$$J = J^{O} e^{\frac{-E_b}{kT}} \tag{2.4}$$

where J is a number of nuclei per m^3 and per second, J° the kinetic pre-factor, and E_b the nucleation barrier of equation (2.3). The nucleation probability p in a volume V during a time τ is:

$$p = 1 - e^{-JV\tau} \tag{2.5}$$

Nucleation is assumed to occur when $p = 1/2$. Then, the median value of duration τ_m required to get at least one vapor nucleus is expressed as:

$$\tau_m = \frac{\ln 2}{JV} \tag{2.6}$$

Then, a time necessary to create a bubble nucleus depends on the space available for the fluid. In other words, if one second is sufficient to create one nucleus in a volume of 1 L, more than 11 days are required for a volume of 1 mm^3, and statistically, almost 32,000 years are necessary for a volume of 1000 μm^3, which is a typical size for interstitial pores in most natural porous media. This calculation is made at constant J, namely at a given value of E_b, which depends both on the metastable "intensity" (that is, a distance from the stable state) and occurrence of heterogeneous nucleation (due either to chemicals other than water in the liquid, or to solid surface singularities). Eventually, at constant physico-chemical conditions, nucleation is an event that becomes rare in a very small container.

El Mekki *et al.* (2009) carried out experimental measurements on synthetic fluid inclusions to record the lifetime of differently superheated (that is capillarized) pure water. They found that liquid water trapped inside a porous volume of about 104 μm^3, can sustain at about −60 MPa during one century, and at about −45 MPa during one million years.

The key conclusion here is that the lifetime of metastable liquid can be long enough for the capillary liquid to influence on natural processes.

2.3 CAPILLARY THERMODYNAMICS

Now, the modeling of the geochemical "capillary reactions" requires to know the thermodynamic properties of the capillary compounds, and first, of the capillary water. We noted above that capillary water is in the same state as the superheated liquid, that is, a "normal" liquid put at pressures below the saturated ones (including the negative domain). Extrapolating the EoS to negative pressure range, the thermodynamic properties of capillary water can be readily calculated down to the extreme tensile stress that liquid water can intrinsically sustain. It is defined by the spinodal curve (e.g., Mercury and Tardy, 2001): around 22% *RH* at 0°C and 35% *RH* at 100°C (Figs. 2.2 and 2.3). At *RH* lower than the spinodal location, only the liquid water whose thermodynamic properties are not described by the EoS can continue to exist, for instance, the adsorbed water films.

2.3.1 *Capillary solutions and the gas-solutions equilibria*

The capillary state of liquid water is assumed to have a direct feed-back on the dissolved solutes (Mercury *et al.*, 2003), and so their thermodynamic properties are calculated extrapolating a convenient EoS (e.g., Tanger and Helgeson, 1988) down to negative pressures. This allows us to evaluate how the capillary state modifies the behavior of any reaction taking place in aqueous solutions.

The Henry's law gives the equilibrium distribution of a volatile species between liquid and gaseous phases according to the following reaction:

$$\text{gas A} \leftrightarrows \text{aqueous species A} \tag{2.7a}$$

whose chemical equilibrium requires:

$$k_H = \frac{p_{gas}}{X_{aqueous\ species}} \tag{2.7b}$$

where k_H is the Henry's law equilibrium constant, p_{gas} is a pressure of the studied gas in air, and $X_{aqueous\ species}$ is a mole fraction of the dissolved neutral species. Fugacity and activity are not considered here because we assume dilute conditions.

Practically speaking, soil is open towards the atmosphere in a more or less direct pathway. Therefore, the neutral gases (as N_2) in soil air are primarily controlled by external atmosphere. *In-situ* production may modify soil air composition, especially for biogenic gases, like CO_2, O_2, N_2O, or CH_4. Some mineral sources subjected to radioactive decay can supply He or Rn addition. Thus, gaseous phase equilibrating with soil solution may result from these three main sources that could be both global and local. In the case of deep marine (non-saturated) sediments (Clenell *et al.*, 1999; Henry *et al.*, 1999), gaseous atmosphere is controlled by gas-hydrates dissociation. Then, either methane, or propane (C_3H_8) together with water vapor predominate in equilibrium with the capillary solution.

Considering a given atmosphere, influence of the capillarity can be estimated by calculating the pressure dependence of Henry's law constant when pressure decreases:

$$\frac{k_H(T,P)}{k_H(T,0.1)} = -\frac{1}{RT}\int_{0.1}^{P}\left(\Delta_R V^o_{T,P}\right)dP \tag{2.7c}$$

where $k_H(T, P)$ and $k_H(T, 0.1)$ are equilibrium constants at a given T and pressures of P and 0.1 MPa (standard pressure), correspondingly; ΔR°_V is a volume change of the studied reaction and dP is a pressure change related to capillarity.

Actually, the volume change ΔR_V° corresponds only to a change in a volume of the dissolved aqueous species, because the solution is the only capillarized phase (the atmosphere remains at nearly constant pressure). It is what we call an anisobaric reaction: one component experiences a change of pressure which does not act on the other chemicals involved in the reaction.

2.3.2 *Solids in capillary situations*

The case of minerals in contact with capillary water deserves a special mention. Theoretically, the Young-Laplace law applies also to mineral-solution interface so that any combination of pressures may exist at each side of the interface. Three main cases are presently considered to calculate the stability of minerals (signed 2 and 3) in one capillary solution (signed 1).

The first mineral-solution configuration is called isobaric: all the system is under capillary pressure ($P_1 = P_2 = P_3$). A typical example of such situation arises when cooling a superheated fluid inclusion containing liquid water: the freezing temperature is increasing what means that ice and water are put under a negative pressure (Roedder, 1967). The isobaric case arises when the stretched water pulls sufficiently upon the surrounding solid to which it is bound, so as to distort its crystalline network (at least close to the surface) in a way similar to a mechanical traction (by clamps for instance). By analogy with the liquid-air meniscus, the flatness of the solution-solid interface points to a similar pressure from each side of the interface. In other words, a planar crystal face would point to isobaric conditions of the solution-solid interaction.

Two anisobaric configurations have been proposed: (i) solids pressure differs from water pressure ($P_1 \neq P_2 = P_3$); (ii) pressure of the weathering solid differs from that of the water and of the secondary solid ($P_1 = P_2 \neq P_3$) (Mercury *et al.*, 2003; Lassin *et al.*, 2005). A simple example of such anisobaric configuration takes place in porous media when measuring a decreasing freezing point of liquid water as a function of pore radius (Gibbs-Thompson equation). In this case, the solid and the solution pressures are different because their mutual interface can change according to the Young-Laplace equation. That amounts to apply the equation (2.2) to the solid-solution equilibrium, with solution-solid surface tension τ. Then, the curvature of the solution-solid interface is a function of the pressure disequilibrium at each side of the interface.

Therefore, the isobaric/anisobaric scenario relates to a shape of the solid, but this relationship leads to two difficulties. First, very often, solids are not isotropic as the liquids are. The faceting of a polyhedral crystal hints that the surface tension depends on the crystal faces.

Second, the solution-solid surface tension (or those per crystal faces) is not constant through the change of physico-chemical conditions. Therefore, a quantitative approach requires to know the aqueous solution-solid surface tensions (without speaking of those per crystal faces), as well as their dependency either on the solution pressure, or on the solution salinity, or still on the shape/size of the solids (e.g., the review by Wu and Nancollas, 1999). Unfortunately, all this information is not available in general, so that the "solid pressure" parameter must be considered rather as a simulation sketch than a field-based modeling choice. For the time being, we use the simplest approach of planar *vs.* curved surface of minerals which gives the model the isobaric (planar) or anisobaric (curved) parametrization.

2.3.3 *Thermodynamic modeling of reactions in capillary systems*

For the following equation:

$$X_{crystal} + Z_{gas} + H_2O \rightleftarrows Y_{crystal} + Cat^+ + An^- \tag{2.8a}$$

the variation of the equilibrium constant of reaction with pressure could be written as:

$$-RTd\left(\Delta_R G_P\right)=\int_{P_r}^{P_1}\left(V_{cat}+V_{an}-V_{H_2O}\right)dP+\int_{P_r}^{P_2}\left(V_{Y_{cristal}}\right)dP-\int_{P_r}^{P_1}\left(V_{X_{cristal}}\right)dP \tag{2.8b}$$

Equation (2.8b) enables parameterizing each constituent of the reaction to its own pressure conditions. *X* and *Y* crystals can be at pressures equal or different from each other, and also from that of the solution, which is taken into account through *VdP* function for H_2O, Cat^+ and An^-. When one constituent remains at constant pressure inside a globally changing system pressure (as is the *Z* gas), its own *dP* is put to 0, so it does not appear in (2.8b).

2.3.4 *Simplified modeling of salt solubility in capillary systems*

In the isobaric case, the influence of pressure on chemical equilibrium is given by the equation (Stumm and Morgan, 1996):

$$\frac{\partial(\ln K)}{\Delta P}=\frac{-\Delta V^o}{RT} \tag{2.9}$$

where *P* is a pressure in water solution [M L^{-1} T^{-2}] and ΔV^o is a change in the partial molar volume [L^3 mol^{-1}].

The partial molar volume for a solid is equal to the ratio of its molecular weight and density, whereas for gases it is expressed by the universal gas law, $V^o = RT/P$ (Morel, 1983). The calculated partial molar volumes of ions in solution are often negative (Shock and Helgeson, 1982; Krumgaltz *et al.*, 1996). Total change in the partial molar volume of the system as influenced by pressure equals the sum of the partial molar volume changes in each phase.

The simplest solution of (2.9), when ΔV^o does not depend on the pressure change (at low pressures), may be obtained as:

$$\ln\frac{K_p}{K_o}=\frac{-\Delta V^o \Delta P}{RT} \tag{2.10}$$

where K_p and K_o are equilibrium constants in pore solution at a given pore-water pressure and in free solution (capillary pressure is nil, and water pressure equals the atmospheric pressure), respectively; ΔP is the same as in equation (2.1).

At high pressures, constancy of the partial molar volume can not be assumed. In such cases, V^o must be expressed as a function of the pressure (Owen and Brinkley, 1941). It has been shown that the linear relation between V^o and pressure may be assumed at positive pressures below 100 MPa (Millero, 1982). If extrapolation to negative pressures is applicable, these relations should work reasonably well also in unsaturated soils or sediments up to their air-dry state (water pressure of minus 100 MPa). In such a situation, the solution of equation (2.9) can be given as:

$$\ln\frac{K_p}{K_o}=\frac{1}{RT}\left(-\Delta V^o \Delta P_c+0.5\Delta k^o \Delta P_c^2\right) \tag{2.11}$$

where $k^o = -\Delta V^o/\Delta P$ is the partial molar compressibility [M^{-1} L^4 T^2].

The total change of partial molar compressibility is equal to the sum of partial molar compressibilities of all constituents of the system (Owen and Brinkley 1941).

Results of the calculations performed using equations (2.10) and (2.11) (Zilberbrand, 1997, 1999; Mercury *et al.*, 1997a) show that below pore-water pressures of minus 1.5–10 MPa the

K_p/K_o ratio for dissolution reaction decreases drastically. This must cause a decrease in ion concentrations in bound water. At negative pressures typical for the completed first mono-molecular layer of water (–220 MPa) equilibrium constants for dissolution of common sulfates and calcium bicarbonate are equal to 0.5–4.5% of their values for free water. Therefore, owing to precipitation, concentration of ions of these salts in solution should be also small. A conclusion about Mg^{2+}, Ca^{2+}, SO_4^{2-} and HCO_3^- deficiency in hygroscopic water may be drawn. For chlorides, K_p decrease in the first mono-molecular layer of water is smaller (the equilibrium constant equals about 24% of the free-water value). The drastic change in the K_p/K_o ratio for dissolution-precipitation of common soluble salts at pore-water pressures below minus 1.5–10 MPa, implies that saturation of soil water with respect to different salts (and their precipitation) in dry soil may occur earlier (Zilberbrand, 1997; Mercury *et al.*, 1997b).

The calculation also shows that some geochemical processes in dry soil should also be influenced by specific water pressures. The dedolomitization process in unsaturated sediments is characterized by a non-trivial change of equilibrium constant with changing pore-water pressure (Zilberbrand, 1997). The drier sediment becomes (water pressure decreases), the larger is the equilibrium constant growth. Thus, dedolomitization must be easier (at larger Mg^{2+}/Ca^{2+} ratio) as sediment becomes drier. This may explain the relatively extensive dedolomitization in unsaturated sediments (Friedman and Sanders, 1967; Sha-Qingan *et al.*, 1979).

A trend of increasing equilibrium constants with a drop in pore-water pressure was found also for some types of cation exchange. For example, Na^+–K^+ exchange in air-dry soils and sediments may be enhanced by 55% (Zilberbrand, 1997). An excess of K^+ formed in soil water during dry periods may be further partially consumed by plants. Since the cation exchange process is reversible, remaining K^+ may be adsorbed during the following wet period.

Calculations indicate that adsorption of heavy metals, such as Zn^{2+}, must also be enhanced in dry soils and sediments (Zilberbrand, 1997). In rainy periods a portion of the adsorbed Zn^{2+} will be released (the equilibrium constant decreases) and may be flushed towards groundwater. The majority of heavy metals have a partial molar volume less than that of Ca^{2+} (Krumgaltz *et al.*, 1996), which usually dominates the exchangeable cation composition. Therefore, the described consequence of adsorption and release should be typical for heavy metals in unsaturated soil.

2.4 ILLUSTRATIONS IN NATURAL SETTINGS

2.4.1 *Capillarity and mineralogy of desert roses*

Shearman (1981) reported that grains of sand become embedded in gypsum crystals (desert roses) growing in Abu Dhabi sebkhas, in clearly defined zones. Gypsum minerals with "trapped" grains are found in temporary salt marshes, but not in nearby carbonate-water-saturated silt. This zonation is classically related to both gypsum growth rate (rapid evaporation leads to high supersaturation), and lack of cohesion (which facilitates the trapping in the growing crystals) of the grains of sand in the sebkhas compared to those in the silt.

The gypsum formation reaction can be written as:

$$Ca^{2+} + SO_4^{2-} + 2H_2O \rightleftarrows CaSO_4 \cdot 2H_2O \tag{2.12}$$

where saturation ratio is expressed as $\Omega = Q/K_{gypsum}$ and ion-activity product as:

$$Q = [CaSO_4 \cdot 2H_2O]/([Ca^{2+}][SO_4^{2-}][H_2O]^2) \tag{2.13}$$

When the salinity increases due to evaporation, water activity decreases (by inverse square law) and the Ω (the saturation ratio) increases causing supersaturation ($\Omega > 1$: supersaturation, $\Omega = 1$: equilibrium, $\Omega < 1$: undersaturation). This finally leads to gypsum precipitatation.

The situation could be considered as a solution-solid counterpart of the bubble nucleation with the similar metastable (supersaturation) state, before equilibrium is restored by nucleation.

The standard interpretation of desert roses's formation refers to the role of water evaporation for causing the supersaturation, a logical effect in presence of dry air. However, the grain engulfment of sand needs more delicate explanation, that is in general overlooked. Actually, when one crystal is growing, it can either push away a close particle or engulf it, depending on the ability of the system to suck (or not) liquid fast enough in the crystal-particle gap to prevent (or not) the two from joining (e.g., Corte, 1962; Kastner, 1970; Scherer, 1999). The key parameter is the growth rate which must exceed what is usually called the critical growth velocity for enabling the engulfment. It is well-known that increasing the supersaturation increases the crystal-growth rate. Because of this, a threshold effect exists: the supersaturation must be high enough to allow the growth rate to be greater than the critical velocity. In general, the nucleation (supersaturation) threshold of gypsum is rather low due to both its high solubility and the associated low solution-solid surface tension (Putnis *et al.*, 1995). Thus, evaporation appears to be able to concentrate the solution and make it supersaturated, but this effect may be not strong enough for gypsum to account for the grain engulfment associated with the desert roses.

Alternative interpretation of the process can invoke capillarity features. The low atmospheric RH values (from 23% in May to 44% in December, after Abu Dhabi NOAA meteorological station, http://www.climate-charts.com) can lead to very negative water pressures in microporous environments, even if we are in the supratidal zone of a coastal area. The working sketch is to consider localized low-RH micro-media where the capillary liquid water bridges the grains of sand. These bridges make the grains to "stick" together inside the globally non-cohesive sandy matrix, while maintaining a certain geometrical separation between these grains. If the solution does not evaporate, but becomes capillary, the equilibrium constant decreases due to low *RH* values which modifies the chemical potential of the solution, and then the supersaturation ($Q/K > 1$) rises automatically. This effect can be described by:

$$\ln \frac{K_{gypsum}(T,P)}{K_{gypsum}(T,0.1)} = \frac{1}{RT} \int_{0.1}^{P} \Delta_R V^o_{T,P}\, dP \tag{2.14}$$

where $K_{gypsum}(T,P)$ is the thermodynamic equilibrium constant of reaction (2.12) at 25°C and negative pressure P and $K_{gypsum}(T,0.1)$ is the thermodynamic equilibrium constant of the same reaction at the same temperature and standard pressure (0.1 MPa).

The consequences of the *RH* drop are here evaluated at constant salinity (namely, at constant Q in equation (2.13)). We can express the solid pressure/texture effect developing the $\Delta_R VdP$ term:

$$-RTd(\Delta_R G_P) = \int_{P_r}^{P_1} \left(V_{Ca^{2+}} + V_{SO_4^{2-}} - 2V_{H_2O}\right) dP_1 + \int_{P_{0.1}}^{P_2} \left(V_{gypsum}\right) dP_2 \tag{2.15}$$

The only solid involved in the reaction can be either at the same pressure as that of the solution ($dP_1 = dP_2$), or not (for instance: $dP_2 = 0$, solid pressure does not vary with the capillarization of the solution). The fact that the outer crystal faces of the desert roses extend as planar faces directs to the isobaric scenario, according to which the growing gypsum crystals and capillary solution both undergo capillary pressure (that is, $dP_1 = dP_2$). $K_{gypsum}(T,P)$ calculated at $P_w = P_{gypsum}$ results in a decreasing gypsum equilibrium constant with water pressure (Fig. 2.4): the capillary state generates supersaturation with respect to gypsum at constant Q by decreasing K_{gypsum}. Thus, as the bridging solution becomes supersaturated with respect to gypsum, its precipitation can envelop the grains the capillary solution was anchored to, a mechanical link which makes easy to engulf the sand particle: at the extreme, the engulfment becomes independent on the growth rate.

2.4.2 *Capillarity and the dissolution of gases*

Equation (2.7b) enables to calculate the variation of gases solubility in differently capillarized dilute solutions. Figure 2.5 shows that the Henry's law constant for the studied gases decreases when capillary pressure increases (thermodynamic data from Mercury *et al.*, 2003). As the molar fraction appears in the denominator in (2.7b), the concentration of gases dissolved in water increases with decreasing water pressure. It must be noted that the pressure effect on gas solubility depends on a mass of the studied gas, namely there is a peculiarity of behavior according to the nature of each gas with respect to what is known at the standard pressure (0.1 MPa). In the case of noble gases, this model can explain the observed mass fractionation from *Ne* to *Xe* at the high concentrations measured in some aquifers (Mercury *et al.*, 2004). For the sea-bottom sediments, Henry *et al.* (1999) advocated that the total methane content measured *in-situ* is 3 to 10 times larger than the expected methane concentration if the capillary effects are not considered. With our present model (see especially Fig. 2.5), we can evaluate that such enrichment with respect to results of usual calculations is attained in capillary water having very low pressure, lower than −70 MPa.

The capillary effect does not influence only on the gases storage capacity within the UZ. In a case of reactive gases, additional consequences will be found. For CO_2, the "gas-in" effect implies a change in the equilibrium *pH* in contact with the atmosphere (thermodynamic data from Mercury *et al.*, 2003): at 25°C and −100 MPa, the *pH* of pure water equilibrated with atmospheric CO_2 ($p_{CO_2} = 10^{-4.5}$ MPa) is 5.7, instead of 5.5 in free water. Nevertheless, the potential aggressiveness of this soil water does not decrease because, at the same time, the neutral *pH* of pure water increases from 7 to 7.2. In soil, where p_{CO_2} may reach 10^{-2} MPa (at presence of living or inert organic matter), *pH* increases from 4.3 in free water to 4.5 at 100 MPa, but the acidity magnitude is not modified because of the correlated increase of the neutral *pH* under tension. On the other hand, CO_2(g) solubility in capillary water increases by 396% when pressure changes from 0.1 MPa to −100 MPa.

Soil water is predicted to be more oxygenated than free water, since the equilibrium constant of O_2(g) dissolution increases as soil water pressure decreases, which favors oxidation reactions (thermodynamic data from Mercury *et al.*, 2003). Nitrification reactions and oxidation of sulfides appear to be greatly favored in soil water. For instance, the ratio of $[NO_3^-]$ to

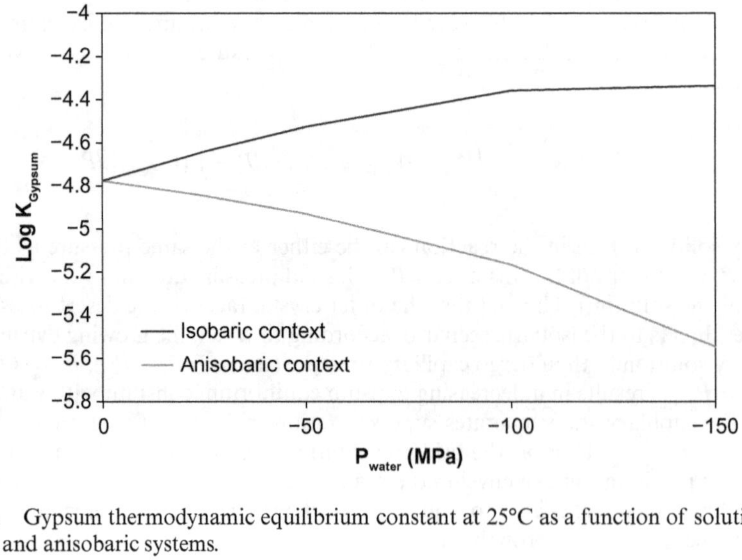

Figure 2.4. Gypsum thermodynamic equilibrium constant at 25°C as a function of solution pressure in isobaric and anisobaric systems.

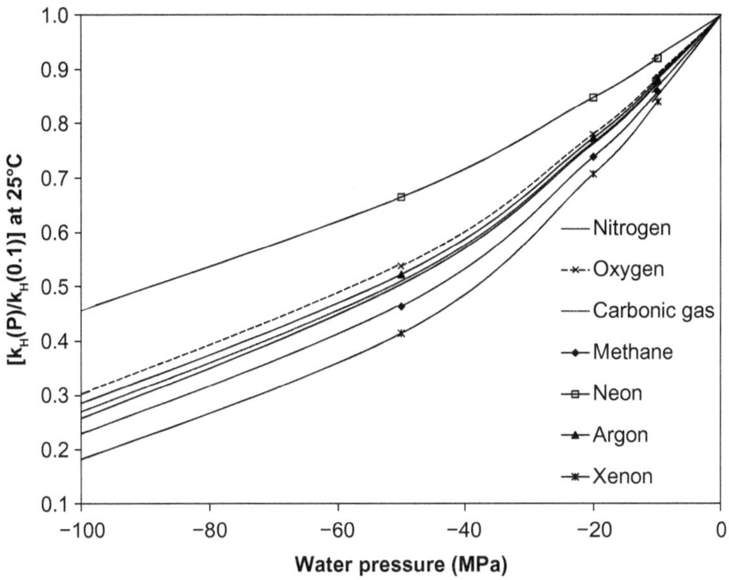

Figure 2.5. A ratio of the Henry's law constant in UZ water (at different capillary pressures) to the standard Henry's law constant for seven atmospheric gases at 25°C.

$[NH_4^+]$ increases by an order of magnitude when changing the water pressure from 0.1 MPa to −100 MPa (reaction: $NH_4^+ + 2O_2(g) \rightleftarrows NO_3^- + 2H^+ + H_2O$). Taking into account the influence of temperature as well, this ratio is multiplied by 10.8 at 0°C, by 9.5 at 25°C, and by 8 at 60°C, when water pressure decreases from 0.1 MPa to −100 MPa. This promotion by the "gas-in" effect is slightly buffered by a decrease of the redox potential of the O_2/H_2O couple, changing from 806 mV at 25°C and 0.1 MPa to 803 mV when pressure decreases to −100 MPa. Note that these calculations are performed with constant P_{O_2} in air, equal to 0.021 MPa, and at neutral *pH* under the considered pressure. The above oxidation reactions are mostly microbially driven, and the calculated values should be considered only as estimates. Nevertheless, they show the general tendency of acceleration of the oxidation processes in the UZ.

It is largely reported that concentrations of dissolved gases in the UZ far exceed those resulting from soil air—free soil solution equilibrium, for noble gases (e.g., Aeschbach-Hertig *et al.*, 2000), gas hydrates (e.g., Clenell *et al.*, 1999) and CO_2 (e.g., Fennemore *et al.*, 1998). Such enrichment can be partially explained by the capillary properties of the UZ.

2.5 HYDROGEOCHEMICAL MODELING IN THE UNSATURATED ZONE

A computer code called "THERMO-ZNS" has been developed (e.g., Lassin *et al.*, 2005) to calculate the relevant thermodynamic properties of water, aqueous species, minerals and gases in capillary systems. It is based on three existing computer codes: "IAPWS-97" (Wagner and Kruse, 1998), "Supcrt92" (Johnson *et al.*, 1992) and "Eqtest" (Span and Wagner, 2000). "Supcrt92" calculates standard thermodynamic properties in the SZ and supercritical domains. "Eqtest" calculates thermodynamic properties of water in the SZ and supercritical domains, using the most recent equation of state of water of Wagner and Pruss (2002). In "THERMO-ZNS" the domain of application is extended to the metastability field of capillary water according to the flowchart shown in Figure 2.6.

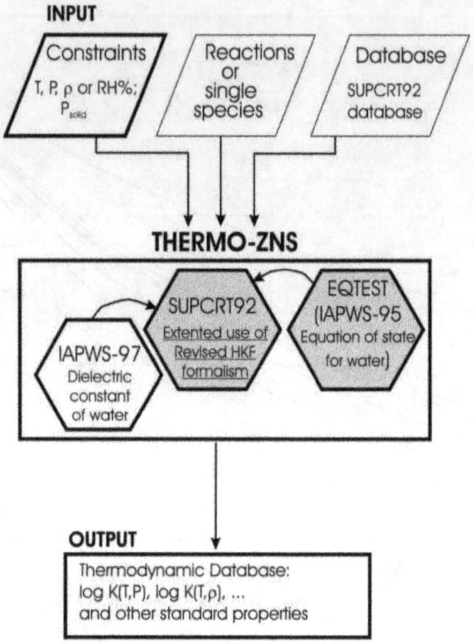

Figure 2.6. Flowchart of the "THERMO-ZNS" computer code. Grey zones designate existing computer codes used for building the THERMO-ZNS (see text). Bold frames indicate units modified from these codes.

New input data (compared to Supcrt92) are *RH*, and a relationship between pressures in different phases of the system (anisobaric/isobaric conditions). The resulting equilibrium constants calculated at (*T*, *P*) pairs of interest (*P* is a capillary pressure) can now be loaded into the PHREEQC database (Parkhurst and Appelo, 1999) which reads them as a specific case of the considered mineral. That allowed us to use the computation capabilities of PHREEQC for estimation of saturation indices in the UZ by including the correction of activity coefficients. We have also used the PHREEQC kinetic calculations which enabled us to propose a new way to account for the peculiar mineralogy of desert roses (see section 2.3.1, and Pettenati *et al.*, 2008). To go further, the modeling of the retention curves, characterizing the non-saturated water content and water conductivity, must be carried out up the highest suctions. This was done applying well-known methods of calculation (Pettenati *et al.*, 2008), paving the way for a reactive transport modeling in the UZ with consideration of its specific thermodynamics.

The first step we are now working on is modeling the 1D advection with PHREEQC through the dual column unit. This is a promising way to estimate the long-term influence of capillary effects on an unsaturated-saturated continuum. The effect of capillarity could be introduced into a 1D transport (PHREEQC) by calculating thermodynamic properties (THERMO-ZNS) of a pore solution subjected to a given matric potential, when this solution is moving in the column by piston effect for various durations. Studying evolution of both mineralogy in the column and output solutions (as a function of suctions imposed on the capillary water with time) enables us to better understand the influence of capillarity on the UZ-SZ geochemical continuum.

2.6 CONCLUSIONS

The present contribution describes the application of the metastable-water physics to hydrochemical modeling in surficial environments.

UZ environments are characterized by coexistence of (at least) two types of water, contrasted in terms of hydrodynamical and thermodynamic properties. The difference between the two types is related to the intensity of "capillarization", itself linked to local conditions, namely RH and pore radius. The appearance of capillary water is fundamental for hydrochemical modeling in the surficial domain, because the studied capillary properties imply that solid-gas-solution equilibria in SZ and UZ are different, and depend on aridity of climate (RH) and properties of porous media (sizes of pores and their distribution).

Except for peculiarities of the capillary state, the mass balance at the UZ scale depends on interrelation between the capillary volume and the percolating free-water volume. Considering that capillary water behavior can deviate from the Young-Laplace law for the hydrogeochemical modeling of soils, we have to consider the significant volume of capillary water (up to 100 L m^{-3} at very high suctions) located inside microporosity. Obviously, that introduces the modeling to the notion of lifetime, because such "non-Young-Laplace" capillary water is a metastable fluid, and we quoted recent results supporting that highly metstable water can live for a very long time in the nature. Furthermore, a long lifetime of the immobile capillary water allows it to reach the solid-solution equilibrium. Hydrogeochemical modeling needs to couple the almost immobile capillary "interacting" water with the mobile percolating water. The former forms the chemical signature, whereas the latter disseminates it.

REFERENCES

Adushkin, V.V., Andreev, S.N. & Popel, S.I.: Cavitation mechanism of formation of nano- and microsize particles of minerals in ore deposits. *Geol. Ore Deposits* 46:5 (2004), pp. 313–320.

Aeschbach-Hertig, W., Peeters, F., Beyerle, U. & Kipfer, R.: Palaeotemperature reconstruction from noble gases in groundwater taking into account equilibration with entrapped air. *Nature* 405 (2000), pp. 1040–1044.

Alvarenga, A.D., Grimsditch, M. & Bodnar, R.J.: Elastic properties of water under negative pressures. *J. Chem. Phys.* 98:11 (1993), pp. 8392–8396.

Andraski, B.: Soil-water movement under natural-site and waste-site conditions: A multiple-year study in the Mojave Desert, Nevada. *Water Ressour. Res.* 33 (1997), pp. 1901–1916.

Caupin, F. & Herbert, E.: Cavitation in water: a review. *C.R. Phys.* 6 (2006), pp. 1000–1017.

Chienthavorn, O. & Su-In, P.: Modified superheated water extraction of pesticides from spiked sediment and soil. *Anal. Bioanal. Chem.* 385 (2006), pp. 83–89.

Clennell, M.B., Hovland, M., Booth, J.S. Henry, P. & Winters, W.J.: Formation of natural gas hydrates in marine sediments. 1. Conceptual model of gas hydrate growth conditioned by host sediment properties. *J. Geophys. Res. B* 104 (1999), pp. 22,985–23,003.

Corte, A.E.: Vertical migration of particles in front of a moving freezing plane. *J. Geophys. Res.* 67 (1962), pp. 1085–1090.

Debenedetti, P.G.: *Metastable liquids*. Princeton University Press, Princeton, NJ, USA, 1996.

El Mekki, M., Ramboz, C., Perdereau, L., Shmulovich, K.I. & Mercury, L.: Lifetime of superheated water in a micrometric synthetic fluid inclusion. In: S.J. Rzoska & V. Mazur (eds): *Metastable systems under pressure*. NATO Science Series, Springer Verlag., Vienna, 2009.

Fennemore, G.G., Neller, W.C. & Davis, A.: Modeling of pyrite oxidation in arid environments. *Environ. Sci. Technol.* 32 (1998), pp. 2680–2687.

Flint, A.L., Campbell, G.S., Ellet, K.M. & Calissendorff, C.: Calibration and temperature correction of heat dissipation matric potential sensors. *Soil Sci. Soc. Am. J.* 66 (2002), pp. 1439–1445.

Friedman, G.M. & Sanders, J.E.: Origin and occurrence of dolostones. In: G.V. Chilingar (ed): *Developments in sedimentology*, 9A, Elsevier Amsterdam, the Netherlands, 1967, pp. 267–348.

Green, J.L., Durben, D.J., Wolf, G.H. & Angell, C.A.: Water and solutions at negative pressure: Raman spectroscopic study to minus 80 Megapascals. *Science* 249 (1990), pp. 649–652.

Henry, P., Thomas, M. & Clennell, M.B.: Formation of natural gas hydrates in marine sediments. 2. Thermodynamic calculations of stability conditions in porous sediments. *J. Geophys. Res. B*, 104 (1999), pp. 23,005–23,022.

Kastner, M.: An inclusion hourglass pattern in synthetic gypsum. *Am. Miner.* 55 (1970), pp. 2128–2130.

Kew, G. & Gilkes, R.: Classification, strength and water retention characteristics of lateritic regolith. *Geoderma* 136 (2006), pp. 184–198.

Klute, A.: Methods of soil analysis. *Agron. Monogr.* 9. ASA and SSSA: Madison, WI, USA, 1986.

Krumgaltz, B.S., Pogorelsky, R. & Pitzer, K.S.: Volumetric properties of single aqueous electrolytes from zero to saturation concentration at 298.15°K represented by Pitzer's ion-interaction equations. *J. Phys. Chem. Ref. Data* 25:2 (1996), pp. 663–689.

Lassin, A., Azaroual, M. & Mercury, L.: Geochemistry of unsaturated soil systems: aqueous speciation and solubility of minerals and gases in capillary solutions. *Geochim. Cosmochim. Acta* 69:22 (2005), pp. 5187–5201.

McCrea, A.F., Anand, R.R. & Gilkes, R.J.: Mineralogical and physical properties of lateritic pallid zone materials developed from granite and dolerite. *Geoderma* 47 (1990), pp. 33–57.

Mercury, L. & Tardy, Y.: Negative pressure and thermodynamic properties of capillary water. *C. R. Acad. Sci. Paris* 324:11 (1997a), pp. 863–873.

Mercury, L. & Tardy, Y.: Physicochemical features of water in capillaries and fog water droplets. *C. R. Acad. Sci. Paris* 325:12 (1997b), pp. 947–954.

Mercury, L. & Tardy, Y.: Negative pressure of stretched liquid water. Geochemistry of soil capillaries. *Geochim. Cosmochim. Acta* 65:20 (2001), pp. 3391–3408.

Mercury, L. & Tardy, Y.: Response to the comment by J.V. Walther. *Geochim. Cosmochim. Acta* 68 (2004), pp. 2775–2780.

Mercury, L., Azaroual, M., Zeyen, H. & Tardy, Y.: Thermodynamic properties of solutions in metastable systems under negative or positive pressures. *Geochim. Cosmochim. Acta* 67 (2003), pp. 1769–1785.

Mercury, L., Pinti, D.L. & Zeyen, H.: The effect of the negative pressure of capillary water on atmospheric noble gas solubility in groundwater and palaeotemperature reconstruction. *Earth & Planetary Sci. Lett.* 223 (2004), pp. 147–161.

Owen, B.B. & Brinkley, S.R.: Calculation of the effect of pressure upon ionic equilibrium in pure water and in salt solutions. *Chem. Rev.* 42 (1941), pp. 461–474.

Parkhurst, D.L. & Appelo, C.A.J.: User's Guide to PHREEQC (Version 2)—a computer program for speciation, batch-reaction, one dimensional transport, and inverse geochemical calculation. U.S. Geological Survey, Water-Resources Investigation. Report 99-4259, 1999.

Pettenati, M., Mercury, L. & Azaroual, M.: Capillary geochemistry in non-saturated zone of soils. Water content and geochemical signatures. *Applied Geochem.* 23 (2008), pp. 3799–3818.

Putnis, A., Prieto, M. & Fernandez-Diaz, L.: Fluid supersaturation and crystallization in porous media. *Geol. Mag.* 132 (1995), pp. 1–13.

Roedder, E.: Metastable superheated ice in liquid-water inclusions under high negative pressure. *Science* 155 (1967), pp. 1413–1417.

Rossi, C. & Nimmo, J.R.: Modeling of soil water retention from saturation to oven dryness. *Water Resour. Res.* 30 (1994), pp. 701–708.

Scanlon, B.R., Keese, K., Reedy, R.C., Šimůnek, J. & Andraski, B.J.: Variations in flow and transport in thick desert vadose zones in response to paleoclimatic forcing (0–90 kyr): field measurements, modeling, and uncertainties. *Water Resour. Res.* 39 (2003), p. 1179.

Scherer, G.W.: Crystallization in pores. *Cement Concrete Res.* 29 (1999), pp. 1347–1358.

Scholander, P.F., Hammel, H.T., Bradstreet, E.D. & Hemmingsen, E.A.: Sap pressure in vascular plants. Negative hydrostatic pressure can be measured in plants. *Science* 148 (1965), pp. 339–346.

Sha-Qingan, Pan-Zhengpu & Wang-Yao: Recent dedolomitization in the vadose zone. *Scientica Geologica Sinica* 4 (1979), pp. 378–383.

Shearman, D.J.: Displacement of sand grains in sandy gypsum crystals. *Geol. Mag.* 118 (1981), pp. 303–306.

Shmulovich, K., Mercury, L., Thiéry, R., Ramboz, C. & El Mekki, M.: Superheating ability of water and aqueous solutions. Experiments and geochemical consequences. *Geochim. Cosmochim. Acta* 73 (2009), pp. 2457–2470.

Shock, E.L. & Helgeson, H.C.: Calculation of the thermodynamic and transport properties of aqueous species at high pressures and temperatures: Correlation algorithms for ionic species and equation of state predictions to 5 Kb and 1000°C. *Geochim. Cosmochim. Acta* 52 (1982), pp. 2009–2036.

Smith R.M.: Superheated water: the ultimate green solvent for separation science. *Anal. Bioanal. Chem.* 385 (2006), pp. 419–421.

Stockwell, J., Smith, L., Jambor, J.L. & Beckie, R.: The relationship between fluid flow and mineral weathering in heterogeneous unsaturated porous media: a physical and geochemical characterization of a waste-rock pile. *Appl. Geochem.* 21 (2006), pp. 1347–1361.

Stumm, W. & Morgan, J.J.: *Aquatic chemistry*. Wiley-Interscience, NY, USA, 1996.

Tamari, S., Gaudu, J.-C. & Simonneau, T.: Tensiometric measurement and metastable state of water under tension. *Soil Sci.* 156 (1993), pp. 149–155.

Tanger IV, J.C. & Helgeson H.C.: Calculation of the thermodynamic and transport properties of aqueous species at high pressures and temperatures: revised equations of state for the standard partial molal properties of ions and electrolytes. *Am. J. Sci.* 288 (1988), pp. 19–98.

Tardy, Y. & Nahon, D.: Geochemistry of laterites, stability of Al-Goethite, Al-Hematite, and Fe^{3+}-Kaolinite in bauxites and ferricretes: an approach to the mechanism of concretion formation. *Am. J. Sci.* 285 (1985), pp. 865–903.

Tardy, Y. & Novikoff, A.: Activité de l'eau et déplacement des équilibres gibbsite-kaolinite dans les profils latéritiques. *C. R. Acad. Sci. Paris* 306 (1988), pp. 39–44.

Trolard, F. & Tardy, Y.: The stabilities of gibbsite, boehmite, aluminous goethites and aluminous hematites in bauxites, ferricretes and laterites as a function of water activity, temperature and particle size. *Geochim. Cosmochim. Acta* 51 (1987), pp. 945–957.

Trolard, F. & Tardy, Y.: A model of Fe^{3+}-kaolinite, Al^{3+}-goethite, Al^{3+}-hematite equilibria in laterites. *Clay Minerals* 24 (1989), pp. 1–21.

Tyree, M.T. & Zimmermann, M.H.: *Xylem structure and the ascent of sap*. Springer Verlag, 2002.

Van Genuchten, M. & Wierenga, P.J.: Mass transfert studies in sorbing porous media: I. Analytical solutions. *Soil Sci. Soc. Am. J.* 40 (1976), pp. 473–480.

Wagner, W. & Pruss, A.: The IAPWS Formulation 1995 for the Thermodynamic Properties of Ordinary Water Substance for General and Scientific Use. *J. Phys. Chem. Ref. Data* 31:2 (2002), pp. 387–535.

Wu, W. & Nancollas, G.H.: Determination of interfacial tension from crystallization and dissolution data: a comparison with other methods. *Adv. Colloid Interf. Sci.* 79 (1999), pp. 229–279.

Zheng, Q., Durben, D.J., Wolf, G.H. & Angell, C.A.: Liquids at large negative pressures: water at the homogeneous nucleation limit. *Science* 254 (1991), pp. 829–832.

Zilberbrand, M.: A nonelectrical mechanism of ion exclusion in thin water films in finely dispersed media. *J. Colloid Interf. Sci.* 192 (1997), pp. 471–474.

Zilberbrand, M.: On equilibrium constants for aqueous geochemical reactions in water unsaturated soils and sediments. *Aquatic Geochem.* 5 (1999), pp. 195–206.

CHAPTER 3

Governing equations and solution algorithms for geochemical modeling

Carlos Ayora, Maarten W. Saaltink & Jesus Carrera

"Science is facts; just as houses are made of stones, so is science made of facts; but a pile of stones is not a house and a collection of facts is not necessarily science."

Henri Poincaré (1902)*

3.1 THE FORMULATION OF REACTIONS

This chapter contains the basic concepts needed for the mathematical formulation of chemical reactions and calculation of species concentrations. Although they are recently included in some geochemical codes (Parkhurst and Appelo, 1999), isotope reactions and distributions are outside the scope of this chapter. While this is not strictly necessary, a vector and matrix notation is adopted here because it facilitates concise writing, which will be critical in subsequent chapters. Matrix notation also facilitates the use of linear algebra techniques, such as multiplication of matrices and the inverse, rank and kernel of a matrix. We hope that the chapter can be read without being familiar with these techniques. Still, algorithmically oriented readers may consult any of the various textbooks on this field (e.g., Meyer, 2000).

3.1.1 *Species, reactions and stoichiometric coefficients*

A chemical system consists of chemical reactions and species. A chemical species is defined as any chemical entity distinguishable from the rest by: (i) its chemical composition or (ii) the phase in which it is present. For instance, CO_2(g) in the gas phase is a different species than CO_2(aq) dissolved in water. A chemical reaction is the process by which some species (reactants) are consumed to yield other species (products). For the purposes of this chapter, chemical reactions are described by the stoichiometric coefficients that relate moles of each participating chemical species produced or consumed by the reaction. Mathematically, this can be represented by a "mass-balance" equation that reflects this linear combination of the species:

$$\sum_{j=1}^{j=n_s} S_{ij} q_j = 0 \quad i = 1, \ldots, n_r \tag{3.1}$$

where q_j is the chemical formula of the j-th species, and S_{ij} is its stoichiometric coefficient in reaction i and where n_s and n_r are the number of species and reactions, respectively. By convention the stoichiometric coefficients of the reactants are negative and those of products positive. In this book the convention of writing reactions as dissociation and dissolution is also adopted. It is convenient to write the chemical mass-balance equation (3.1) in a matrix-vector notation:

$$\mathbf{Sq} = \mathbf{0} \tag{3.2}$$

---

*Source: Henri Poincaré: La Science et l'hypothèse, París, Flammarion, 1902.

where **S** is the $n_r \times n_s$ stoichiometric matrix, whose rows contain the stoichiometric coefficients in each reaction, and **q** a vector of size n_s, containing the chemical formula of all species.

These concepts can be illustrated by means of three simple examples that will be used throughout this chapter.

Example 1a: Consider the dissolution-precipitation of anhydrite assuming that other species can be ignored. Then, the chemical system consists of three species (SO_4^{2-}, Ca^{2+} and $CaSO_4$(s)) and one reaction:

$$\text{(R1)} \quad CaSO_4(s) \rightleftarrows SO_4^{2-} + Ca^{2+}$$

Vector **q** is:

$$\mathbf{q} = \begin{pmatrix} CaSO_4(s) \\ SO_4^{2-} \\ Ca^{2+} \end{pmatrix}$$

and the stoichiometric matrix **S** is:

$$\mathbf{S} = \begin{pmatrix} & CaSO_4(s) & SO_4^{2-} & Ca^{2+} \\ R1 & -1 & 1 & 1 \end{pmatrix}$$

where it should be evident that, with these definitions, the reaction can be written either as (R1) above or as equation (3.2).

Example 1b: Consider now that calcite is added to the system, and that *pH* is so high that we do not need to worry about it or about carbonate species other than CO_3^{2-}. Then, two new species need to be considered (CO_3^{2-} and $CaCO_3$(s)) to account for the new reaction:

$$\text{(R2)} \quad CaCO_3(s) \rightleftarrows CO_3^{2-} + Ca^{2+}$$

Vector **q** and matrix **S** become:

$$\mathbf{q} = \begin{pmatrix} Ca^{2+} \\ SO_4^{2-} \\ CO_3^{2-} \\ CaSO_4(s) \\ CaCO_3(s) \end{pmatrix} \quad \mathbf{S} = \begin{pmatrix} & CaSO_4(s) & CaCO_3(s) & Ca^{2+} & SO_4^{2-} & CO_3^{2-} \\ R1 & -1 & 0 & 1 & 1 & 0 \\ R2 & 0 & -1 & 1 & 0 & 1 \end{pmatrix}$$

which are somewhat more complex than those of example 1a, but still allow writing the set of reactions (R1, R2) as equation (3.2).

Example 1c: We consider finally a chemical system of seven aqueous species (H_2O, Cl^-, CO_3^{2-}, H^+, OH^-, HCO_3^- and CO_2(aq)) and three reactions:

$$\begin{aligned} &\text{(R3)} \quad H_2O && \rightleftarrows \quad H^+ + OH^- \\ &\text{(R4)} \quad HCO_3^- && \rightleftarrows \quad H^+ + CO_3^{2-} \\ &\text{(R5)} \quad CO_2(aq) + H_2O && \rightleftarrows \quad H^+ + HCO_3^- \end{aligned}$$

Then, vector **q** of all species and the stoichiometric matrix **S** are:

$$\mathbf{q} = \begin{pmatrix} H_2O \\ Cl^- \\ CO_3^{2-} \\ H^+ \\ OH^- \\ HCO_3^- \\ CO_2(aq) \end{pmatrix} \qquad \mathbf{S} = \begin{pmatrix} & H_2O & Cl^- & CO_2(aq) & H^+ & OH^- & HCO_3^- & CO_3^{2-} \\ R3 & -1 & 0 & 0 & 1 & 1 & 0 & 0 \\ R4 & 0 & 0 & 0 & 1 & 0 & -1 & 1 \\ R5 & -1 & 0 & -1 & 1 & 0 & 1 & 0 \end{pmatrix}$$

Note that Cl^- does not participate in any reaction (i.e., it is called a conservative species), and, therefore, the Cl^- column contains only zeros.

What should emerge from these three examples is that all information about the chemical system is contained in the vector of participating species, **q**, and the stoichiometric matrix, **S**, so that the set of chemical reactions becomes described always by equation (3.2), regardless of the actual complexity of the problem. This facilitates encapsulating all reactions in a simple expression. We make extensive use of these concepts in the following sections.

3.1.2 *Equilibrium reactions in terms of the stoichiometric matrix*

Equilibrium reactions obey the mass action law. This law relates the activities of the chemical species for each reaction when the system is in thermodynamic equilibrium. It can be written as:

$$\prod_{j=1}^{j=n_s} a_j^{S_{ij}} = K_i \Rightarrow \sum_{j=1}^{j=n_s} S_{ij} \log a_j = \log K_i \quad i = 1, \ldots, n_r \tag{3.3}$$

where a_j is the activity of the j-th species and K_i is the equilibrium constant of the i-th reaction. Of course, equation (3.3) can also be written more concisely using a matrix-vector notation:

$$\mathbf{S} \log \mathbf{a} = \log K \tag{3.4}$$

There are alternative methods to compute chemical equilibrium based on Gibbs energy minimization (Harvie *et al.*, 1998; Kulik, 2000; Kulik *et al.*, 2000). These methods are very robust for highly non-linear systems. However, their coupling with reactive flow algorithms are more complicated and will not be described here.

In the case of example 1a, equation (3.4) simply becomes:

$$\begin{pmatrix} 1 & 1 & -1 \end{pmatrix} \begin{pmatrix} \log a_{SO_4^{2-}} \\ \log a_{Ca^{2+}} \\ \log a_{CaSO_4(s)} \end{pmatrix} = \log K_{CaSO_4(s)}$$

or

$$\log a_{SO_4^{2-}} + \log a_{Ca^{2+}} - \log a_{CaSO_4(s)} = \log K_{CaSO_4(s)}$$

If anhydrite can be considered a pure phase, its activity is 1 (see section 3.3), so that this equation simply becomes:

$$a_{SO_4^{2-}} \cdot a_{Ca^{2+}} = K_{CaSO_4(s)} = K_{anh}$$

where K_{anh} is the solubility product of anhydrite. This equation is represented graphically in Figure 3.1, which also displays the graphical representation of examples 1a and 1b. Notice that the anhydrite equilibrium equation describes a hyperbole when plotting activities or a straight line when plotting log-activities. In the case of example 1b, these lines can be projected in the direction of the species that does not participate in the reaction to become surfaces. For instance, the anhydrite equilibrium equation becomes a plane parallel to the log a_{CO_3} axis, because anhydrite equilibrium will be satisfied independently of the activity of carbonate.

It is clear that graphs become cumbersome and their use is limited. However, readers familiar with algebraic structures may recognize a vector space both for species and the logarithms of activities. Equilibrium reactions impose constrains that can be viewed as subspaces. The issue is developed by Friedly and Rubin (1992).

We now turn to example 1c. The equilibrium constants of reactions R3, R4 and R5 are $10^{-14.0}$, $10^{-10.3}$ and $10^{-6.3}$, respectively, at 25°C. So, the three mass action laws of this example can be written in matrix-vector notation, equation (3.4), as:

$$\begin{pmatrix} -1 & 0 & 0 & 1 & 1 & 0 & 0 \\ 0 & 0 & 0 & 1 & 0 & -1 & 1 \\ -1 & 0 & -1 & 1 & 0 & 1 & 0 \end{pmatrix} \cdot \begin{pmatrix} \log a_{H_2O} \\ \log a_{Cl^-} \\ \log a_{CO_2(aq)} \\ \log a_{H^+} \\ \log a_{OH^-} \\ \log a_{HCO_3^-} \\ \log a_{CO_3^{2-}} \end{pmatrix} = \begin{pmatrix} -14.0 \\ -10.3 \\ -6.3 \end{pmatrix}$$

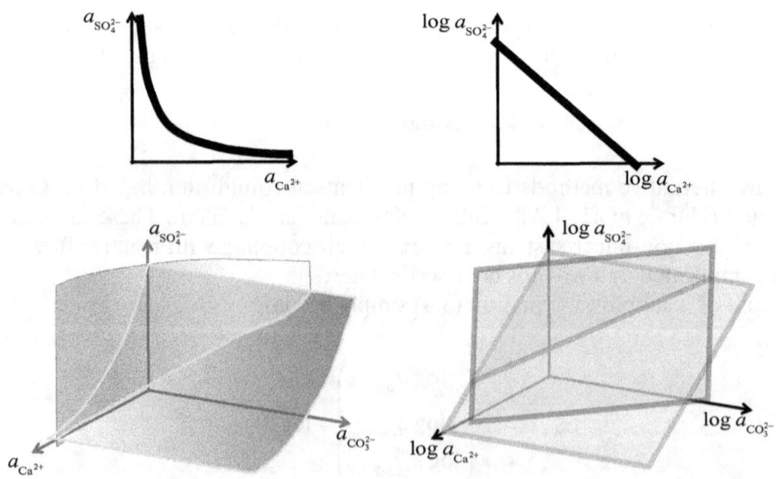

Figure 3.1. Equilibrium provides a constraint on concentrations, which must lie on equilibrium lines (or surfaces). As for examples 1a (above) and 1b (below), equilibrium constraints are non-linear functions of activities or concentrations (left), but linear functions of log activities (right). Therefore it is convenient to write equilibrium equations as in equation (3.4).

Note for instance that the first equation simply expresses $pH + pOH = 14$, assuming that the activity of water is 1 (dilute solutions).

The three reactions of the example are independent of each other. That is, no reaction can be constructed by combination of other reactions. The reaction

$$(\text{R4} + \text{R5}) \quad CO_2 + H_2O \rightleftarrows 2H^+ + CO_3^{2-}$$

is not independent of the three other reactions, because it can be constructed by summing reaction R4 and R5. It also means that, if the mass action laws for reactions R4 and R5 are met, then the mass law of reaction R4 + R5 is inevitably met as well. So, if thermodynamic equilibrium is assumed, it is not necessary to include this extra reaction in the definition of the chemical system. The number of independent reactions is equal to, what is called in linear algebra, the rank of stoichiometric matrix, **S**. This rank can be calculated by means of a procedure called *singular value decomposition* (Meyer, 2000). As a rule, we build the chemical system in such a way that all its equilibrium reactions are independent. Therefore, the rank of matrix, **S**, equals its number of rows or **S** is fully ranked. This condition needs not to be met for kinetic reactions as we shall see below.

3.1.3 *Primary and secondary species*

The mass action laws (3.4) represent a set of mathematical equations that can be used for geochemical calculations. They consist of n_e equations (one for each equilibrium reaction notice that we are now using the subindex e instead r) and n_s unknowns (the activities of vector **a**). Except for some rare cases, the number of reactions (n_e) is lower than the number of species (n_s). Therefore, we cannot calculate the activities of all species from just the mass action laws alone. However, if we knew the activities of ($n_s - n_e$) species (e.g., from mass balance calculations), we could calculate those of the other (n_e) species. As we shall see in subsequent sections and chapters, this is indeed a common way to proceed. We will term primary species the ($n_s - n_e$) known species and secondary the (n_e) remaining species. So, by definition, the activities of the secondary species can be calculated from the activities of the primary species through the mass action law. We can write the mass action laws in terms of primary and secondary species by splitting matrix **S** and vector **a** into two parts:

$$\mathbf{S} = (\mathbf{S}_1 \mid \mathbf{S}_2) \tag{3.5}$$

$$\mathbf{a} = \begin{pmatrix} \mathbf{a}_1 \\ \mathbf{a}_2 \end{pmatrix} \tag{3.6}$$

where $\mathbf{a}_1$ contains the activities of the ($n_s - n_e$) primary and $\mathbf{a}_2$ the (n_e) secondary species and where $\mathbf{S}_1$ contains the stoichiometric coefficients for the primary and $\mathbf{S}_2$ those for the secondary species. Substituting (3.5) and (3.6) into (3.4) gives:

$$\mathbf{S}_1 \log \mathbf{a}_1 + \mathbf{S}_2 \log \mathbf{a}_2 = \log \mathbf{K} \tag{3.7}$$

Equation (3.7) is often written in a form such that the activities of secondary species are an explicit function of primary species, that is, $\mathbf{a}_2 = f(\mathbf{a}_1)$. If matrix $\mathbf{S}_2$ happens to be a diagonal matrix, such an expression can be deduced easily. If not, it can also be obtained by multiplying equation (3.7) by the inverse of matrix $\mathbf{S}_2$, denoted by $\mathbf{S}_2^{-1}$. Multiplying the mass action law (eq. 3.7), or the definition of reactions (eq. 3.4) by any square non-singular matrix implies substituting the reactions in the original chemical system by a new set of reactions obtained from linear combinations of the original ones. Multiplying equation (3.7) by $\mathbf{S}_2^{-1}$ yields:

$$\log \mathbf{a}_2 = \mathbf{S}_1^* \log \mathbf{a}_1 + \log \mathbf{K}^* \tag{3.8}$$

with

$$\mathbf{S}_1^* = -(\mathbf{S}_2)^{-1}\,\mathbf{S}_1 \tag{3.9}$$

$$\log \mathbf{K}^* = (\mathbf{S}_2)^{-1} \log \mathbf{K} \tag{3.10}$$

In this way we obtain a new description of the same chemical system, which is now defined by equilibrium constants **K*** instead of **K** and of stoichiometric matrix for the primary species $\mathbf{S}_1^*$ instead of $\mathbf{S}_1$. In fact, we could also define a stoichiometric matrix for the secondary species ($\mathbf{S}_2^*$ instead of $\mathbf{S}_2$). However, by definition, $\mathbf{S}_2^*$ equals the identity matrix and, hence, can be left out of equation (3.8).

To illustrate these concepts, let us consider the examples. As mentioned above, minerals are often treated as pure phases, and consequently their activities are equal to 1 and do not have to be computed. Therefore, the columns corresponding to these minerals can be temporarily deleted. As a result, we are left with 3 aqueous species and 2 equilibrium reactions, which demands one primary species. We choose Ca^{2+}, which leads to:

$$\mathbf{a}_1 = \left(a_{Ca^{2+}}\right) \qquad \mathbf{S}_1 = \begin{pmatrix} & Ca^{2+} \\ R1 & 1 \\ R2 & 1 \end{pmatrix}$$

$$\mathbf{a}_2 = \begin{pmatrix} a_{SO_4^{2-}} \\ a_{CO_3^{2-}} \end{pmatrix} \qquad \mathbf{S}_2 = \begin{pmatrix} & SO_4^{2-} & CO_3^{2-} \\ R1 & 1 & 0 \\ R2 & 0 & 1 \end{pmatrix}$$

Note that $\mathbf{S}_2 = \mathbf{I}$. Therefore, $\mathbf{S}_1^* = -\mathbf{S}_1$, and equation (3.8) can be written directly as:

$$\begin{pmatrix} \log a_{SO_4^{2-}} \\ \log a_{CO_3^{2-}} \end{pmatrix} = \begin{pmatrix} -1 \\ -1 \end{pmatrix}\left(\log a_{Ca^{2+}}\right) + \begin{pmatrix} \log K_{CaSO_4(s)} \\ \log K_{CaCO_3(s)} \end{pmatrix}$$

Notice that this can be interpreted (Fig. 3.2) as the equations of two planes. Their intersection with the plane that defines the activity of Ca^{2+} yields the activities of the secondary species, sulfate and carbonate.

The situation is a bit more complex in example 1c. That system consists of seven species and three reactions. Therefore, there are four primary species and three secondary species. If we choose H_2O, Cl^-, $CO_2(aq)$ and H^+, to be primary and OH^-, HCO_3^- and CO_3^{2-} to be secondary, $\mathbf{a}_1$, $\mathbf{a}_2$, $\mathbf{S}_1$ and $\mathbf{S}_2$ become:

$$\mathbf{a}_1 = \begin{pmatrix} a_{H_2O} \\ a_{Cl^-} \\ a_{CO_2(aq)} \\ a_{H^+} \end{pmatrix} \qquad \mathbf{S}_1 = \begin{pmatrix} & H_2O & Cl^- & CO_2(aq) & H^+ \\ R3 & -1 & 0 & 0 & 1 \\ R4 & 0 & 0 & 0 & 1 \\ R5 & -1 & 0 & -1 & 1 \end{pmatrix}$$

$$\mathbf{a}_2 = \begin{pmatrix} a_{OH^-} \\ a_{HCO_3^-} \\ a_{CO_3^{2-}} \end{pmatrix} \qquad \mathbf{S}_2 = \begin{pmatrix} & OH^- & HCO_3^- & CO_3^{2-} \\ R3 & 1 & 0 & 0 \\ R4 & 0 & -1 & 1 \\ R5 & 0 & 1 & 0 \end{pmatrix}$$

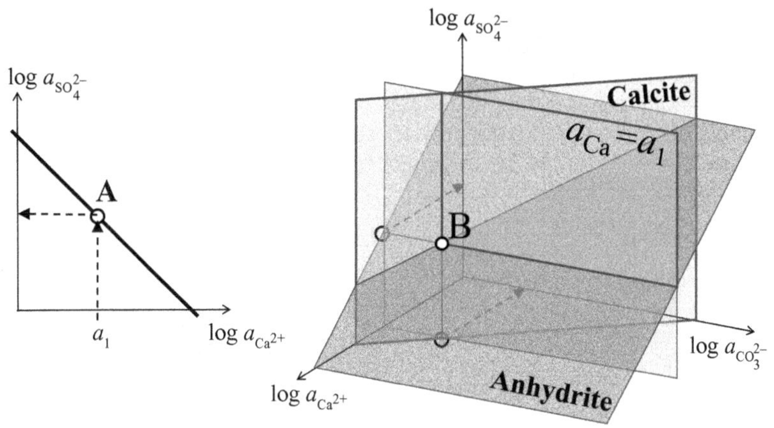

Figure 3.2. Ca^{2+} is the primary species adopted for examples 1a (left) and 1b (right). If its activity is known (a_1), the activities of the remaining (secondary) species can be easily (explicitly) calculated from the intersection with the equilibrium equations, which yields points A (example 1a, left) and B (example 1b, right).

Then, a new stoichiometric matrix, $\mathbf{S}_1^*$, and new equilibrium constants, $\mathbf{K}^*$, can be calculated by applying equations (3.9) and (3.10):

$$\mathbf{S}_1^* = -\begin{pmatrix} 1 & 0 & 0 \\ 0 & -1 & 1 \\ 0 & 1 & 0 \end{pmatrix}^{-1} \begin{pmatrix} -1 & 0 & 0 & 1 \\ 0 & 0 & 0 & 1 \\ -1 & 0 & -1 & 1 \end{pmatrix} = -\begin{pmatrix} 1 & 0 & 0 \\ 0 & 0 & 1 \\ 0 & 1 & 1 \end{pmatrix} \begin{pmatrix} -1 & 0 & 0 & 1 \\ 0 & 0 & 0 & 1 \\ -1 & 0 & -1 & 1 \end{pmatrix} = \begin{pmatrix} 1 & 0 & 0 & -1 \\ 1 & 0 & 1 & -1 \\ 1 & 0 & 1 & -2 \end{pmatrix}$$

$$\log \mathbf{K}^* = \begin{pmatrix} 1 & 0 & 0 \\ 0 & -1 & 1 \\ 0 & 1 & 0 \end{pmatrix}^{-1} \begin{pmatrix} -14.0 \\ -10.3 \\ -6.3 \end{pmatrix} = \begin{pmatrix} -14.0 \\ -6.3 \\ -16.6 \end{pmatrix}$$

So, the mass action laws can be rewritten as:

$$\begin{pmatrix} \log a_{OH^-} \\ \log a_{HCO_3^-} \\ \log a_{CO_3^{2-}} \end{pmatrix} = \begin{pmatrix} 1 & 0 & 0 & -1 \\ 1 & 0 & 1 & -1 \\ 1 & 0 & 1 & -2 \end{pmatrix} \begin{pmatrix} \log a_{H_2O} \\ \log a_{Cl^-} \\ \log a_{CO_2(aq)} \\ \log a_{H^+} \end{pmatrix} + \begin{pmatrix} -14.0 \\ -6.3 \\ -16.6 \end{pmatrix}$$

and by combining equations (3.7) and (3.8), the equilibrium reaction can be also written as:

$$\mathbf{S}_1^* \log \mathbf{a}_1 - \mathbf{S}_2^* \log \mathbf{a}_2 = -\log \mathbf{K}^* \tag{3.11}$$

Notice that after this transformation, the new stoichiometric matrix becomes:

$$\mathbf{S}^* = (\mathbf{S}_1^* | -\mathbf{S}_2^*) = \begin{pmatrix} & H_2O & Cl^- & CO_2(aq) & H^+ & OH^- & HCO_3^- & CO_3^{2-} \\ R3^* & 1 & 0 & 0 & -1 & -1 & 0 & 0 \\ R4^* & 1 & 0 & 1 & -1 & 0 & -1 & 0 \\ R5^* & 1 & 0 & 1 & -2 & 0 & 0 & -1 \end{pmatrix}$$

where R3* is identical to the original R3, but multiplied by −1, R4* is −R5, and R5* is −R4 −R5, which as discussed earlier makes perfect chemical sense. In short, what we have done is to arrange the equilibrium reactions to ensure that the secondary species are written in terms of monomolar dissociation reactions.

Different sets of primary and secondary species can be used. If in our example we choose HCO_3^- as primary species and H^+ as secondary species, $\mathbf{a}_1$, $\mathbf{a}_2$, $\mathbf{S}_1$ and $\mathbf{S}_2$ would become:

$$\mathbf{a}_1 = \begin{pmatrix} a_{H_2O} \\ a_{Cl^-} \\ a_{CO_3^{2-}} \\ a_{HCO_3^-} \end{pmatrix} \quad \mathbf{S}_1 = \begin{pmatrix} & H_2O & Cl^- & CO_3^{2-} & HCO_3^- \\ R3 & -1 & 0 & 0 & 0 \\ R4 & 0 & 0 & 1 & -1 \\ R5 & -1 & 0 & 0 & 1 \end{pmatrix}$$

$$\mathbf{a}_2 = \begin{pmatrix} a_{OH^-} \\ a_{CO_3^{2-}} \\ a_{CO_2(aq)} \end{pmatrix} \quad \mathbf{S}_2 = \begin{pmatrix} & OH^- & H^+ & CO_2(aq) \\ R3 & 1 & 1 & 0 \\ R4 & 0 & 1 & 0 \\ R5 & 0 & 1 & -1 \end{pmatrix}$$

Then the mass action laws could be rewritten as:

$$\begin{pmatrix} \log a_{OH^-} \\ \log a_{H^+} \\ \log a_{CO_2(aq)} \end{pmatrix} = \begin{pmatrix} -1 & 0 & -1 & 1 \\ 0 & 0 & 1 & -1 \\ 1 & 0 & 1 & -2 \end{pmatrix} \begin{pmatrix} \log a_{H_2O} \\ \log a_{Cl^-} \\ \log a_{CO_3^{2-}} \\ \log a_{HCO_3^-} \end{pmatrix} + \begin{pmatrix} -3.7 \\ -10.3 \\ -4.0 \end{pmatrix}$$

Not all sets of primary and secondary species are valid. One must ensure that matrix $\mathbf{S}_2$ of equation (3.7) is not singular, so that it can be inverted. If we choose HCO_3^-, CO_3^{2-} and $CO_2(aq)$ to be secondary, then matrix $\mathbf{S}_2$ becomes:

$$\mathbf{S}_2 = \begin{pmatrix} & HCO_3^- & CO_3^{2-} & CO_2(aq) \\ R3 & 0 & 0 & 0 \\ R4 & -1 & 1 & 0 \\ R5 & 1 & 0 & -1 \end{pmatrix}$$

As the first row of matrix $\mathbf{S}_2$ contains zeros, this is a singular matrix. Therefore, HCO_3^-, CO_3^{2-} and $CO_2(aq)$ cannot be a set of secondary species.

The chosen set of primary species is called the basis. In vector space nomenclature, a basis is a set of variables that allow defining any element of the system, which is precisely what primary species are. If $\mathbf{a}_1$ is known for a given water, then $\mathbf{a}_2$ is easily computed by means of equation (3.8), so that the activities are known for all considered species in that water.

3.1.4 *Components and component matrix*

In the previous section we have seen that by knowing the activities of primary species, one can calculate the activities of all other species. However, if we do not know them, we have more unknowns (the concentrations of n_s species) than equations (n_e mass action law equations). The way to solve this is to add mass balance equations, so that the number of unknowns is equal to the number of equations. The problem is that reactions affect the mass of reacting species (and therefore the mass balance equations), which causes the resulting equations to be complex. This is where components come into place. Components are defined here as the chemical entities that are not affected by equilibrium reactions. This condition facilitates expressing mass balances. Since equilibrium reactions establish relations among species, our definition of components is consistent with the definition in the Gibbs phase rule: the number of components is the number of chemically independent constituents of the system. The atomic constituents could serve as a set of components, but they are never used as such because they are rarely present in the aqueous phase. Besides, it would not be convenient because reactions are expressed in terms of chemical species.

To understand well the meaning of components, consider the mass balances of all species undergoing chemical reactions in a closed system (that is, without mass entering or leaving the system):

$$\frac{\partial \mathbf{c}}{\partial t} = \mathbf{S}^T \mathbf{r} \tag{3.12}$$

where **c** is a vector of concentrations of all species, **r** is a vector of the rates of all chemical reactions and $\mathbf{S}^T$ is the transpose of the stoichiometric matrix **S** (that is, columns and rows are exchanged). Equation (3.12) just expresses the change in concentration due to the chemical reactions. As a vector-matrix notation is used, it is actually a set of equations (one equation for each chemical species). Note, that the unknowns of (3.12) are the n_s concentrations (**c**) and the n_e reaction rates (**r**). So, in theory, we could calculate concentrations and reaction rates from the n_s mass balances (3.12) together with the n_e mass action laws (3.4) or (3.8). Now the number of unknowns equals the number of equations. Unfortunately, this system contains many unknowns and it is highly non-linear. It is much easier if reaction rates (**r**) are eliminated from (3.12). This can be done by multiplying (3.12) by a matrix orthogonal to matrix $\mathbf{S}^T$. This is a fully ranked matrix defined in such a way that:

$$\mathbf{U}\mathbf{S}^T = \mathbf{0} \tag{3.13}$$

where **U** is the $(n_s - n_r) \times n_s$ kernel matrix (or null space) of $\mathbf{S}^T$. The kernel is a subspace orthogonal to the valid (equilibrium) states of the system. That is, it indicates the directions, in the space of species, in which a chemical system cannot move (without violating the equilibrium constraints). As this matrix defines the components, we call it the component matrix. Multiplying (3.12) by **U** gives:

$$\frac{\partial \mathbf{U}\mathbf{c}}{\partial t} = \mathbf{0} \tag{3.14}$$

If we define a vector $\mathbf{u} = \mathbf{U}\mathbf{c}$, then $\partial \mathbf{u}/\partial t = 0$ and **u** is constant in time in case of a closed system. Vector **u** is the concentration of the components. This illustrates an important feature of components: their concentrations do not depend on reaction rates (**r**). It also gives a more physical meaning to the concept of components: chemical reactions do not destroy or create components, but only change the species of which they are part of.

Matrix **U** is not unique and there are various methods to calculate it. We describe several methods below.

3.1.4.1 *Method 1 (aqueous components)*

This is, by far, the most widely used method. It is based on Gauss-Jordan elimination (Steefel and MacQuarrie, 1996), which consists of defining the following component matrix, as follows:

$$\mathrm{U} = (\mathbf{I} \,|\, -\mathbf{S}_1^T(\mathbf{S}_2^T)^{-1}) = (\mathbf{I} \,|\, (\mathbf{S}_1^*)^T) \tag{3.15}$$

where **I** is the $(n_s - n_r) \times (n_s - n_r)$ identity matrix (that is, a matrix with 1's on the diagonal and 0's elsewhere). Matrices $\mathbf{S}_1^T$ and $\mathbf{S}_2^T$ are the transposed matrices of $\mathbf{S}_1$ and $\mathbf{S}_2$ of equation (3.5). So, we can use the same calculations as explained in the previous section. We can check that indeed equation (3.13) is met:

$$\mathbf{U}\mathbf{S}^T = \left(\mathbf{I} \mid -\mathbf{S}_1^T\left(\mathbf{S}_2^T\right)^{-1}\right)\begin{pmatrix}\mathbf{S}_1^T \\ \mathbf{S}_2^T\end{pmatrix} = \mathbf{S}_1^T - \mathbf{S}_1^T\left(\mathbf{S}_2^T\right)^{-1}\mathbf{S}_2^T = \mathbf{S}_1^T - \mathbf{S}_1^T = \mathbf{0} \tag{3.16}$$

The vector of concentrations of the components (**u**) can be written as:

$$\mathbf{u} = \mathbf{U}\mathbf{c} = \left(\mathbf{I} \mid \left(\mathbf{S}_1^*\right)^T\right)\begin{pmatrix}\mathbf{c}_1 \\ \mathbf{c}_2\end{pmatrix} = \mathbf{c}_1 + (\mathbf{S}_1^*)^T\mathbf{c}_2 \tag{3.17}$$

This equation allows us to calculate the component matrix of example 1c. Using the first set of primary species of the previous section gives:

$$\mathbf{U} = \left(\begin{array}{ccccc|ccc} & H_2O & Cl^- & CO_2(aq) & H^+ & OH^- & HCO_3^- & CO_3^{2-} \\ \Sigma H_2O & 1 & 0 & 0 & 0 & 1 & 1 & 1 \\ \Sigma Cl & 0 & 1 & 0 & 0 & 0 & 0 & 0 \\ \Sigma C & 0 & 0 & 1 & 0 & 0 & 1 & 1 \\ \Sigma H^+ & 0 & 0 & 0 & 1 & -1 & -1 & -2 \end{array}\right)$$

The rows of matrix **U** refer to the components and its columns to the species. For instance, we can consider the species HCO_3^- to be composed of the components H_2O, $CO_2(aq)$ and $-H^+$, because the column of HCO_3^- contains a one for the rows of H_2O, $CO_2(aq)$ and minus one for H^+. The primary species are composed of only one component. For instance, the species $CO_2(aq)$ is composed of only one component: $CO_2(aq)$. If we multiply this matrix by the vector of concentrations we obtain the concentration of the components (**u**):

$$\mathbf{u} = \mathbf{U}\mathbf{c} = \left(\begin{array}{cccc|ccc} 1 & 0 & 0 & 0 & 1 & 1 & 1 \\ 0 & 1 & 0 & 0 & 0 & 0 & 0 \\ 0 & 0 & 1 & 0 & 0 & 1 & 1 \\ 0 & 0 & 0 & 1 & -1 & -1 & -2 \end{array}\right)\begin{pmatrix} c_{H_2O} \\ c_{Cl^-} \\ c_{CO_2(aq)} \\ c_{H^+} \\ c_{OH^-} \\ c_{HCO_3^-} \\ c_{CO_3^{2-}} \end{pmatrix} = \begin{pmatrix} c_{H_2O} + c_{OH^-} + c_{HCO_3^-} + c_{CO_3^{2-}} \\ c_{Cl^-} \\ c_{CO_2(aq)} + c_{HCO_3^-} + c_{CO_3^{2-}} \\ c_{H^+} - c_{OH^-} - c_{HCO_3^-} - 2c_{CO_3^{2-}} \end{pmatrix}$$

The diagonal matrix in the front portion of the component matrix (i.e., the one coinciding with the primary species) ensures that the components are somehow associated to the primary species. This association is sometimes weak, but examining the components often provides some insight into the chemical system. For example, note that the third element of **u** (third component) coincides with the total inorganic carbon (*TIC*), which is a typical measurement of natural waters. It is clear that the reactions do not affect the total inorganic carbon. A decrease in HCO_3^- concentration due to reaction R4 will be compensated by an increase in CO_3^{2-} concentration. For the particular selection of primary species of the example, the value of the fourth component represents the total acidity of the solution (its capacity to buffer external addition of alkalinity), and can be positive or negative. The second component coincides with the chloride concentration, which is the total chloride in the system. The meaning of the first component is more puzzling. We could imagine that, in a system initially consisting of only the primary species (i.e., zero concentrations for the secondary species) some of the water could be used up in forming, together with the other primary species, OH^- (by dissociation, R3), HCO_3^- (by means of R4), or CO_3^{2-} (by means of R5). This points out that one does not need to seek any special meaning to components. Still, if primary species are chosen to be relevant elements of the system, the component concentrations will represent the total concentration of those species. That is the sum of the concentration of the primary species plus all the complexes that can be formed with it (hence the Σ symbol to represent components).

Of course, we can also use the second set of primary species of the previous section. This leads to:

$$\mathbf{U} = \left(\begin{array}{lcccc|ccc} & H_2O & Cl^- & CO_3^{2-} & HCO_3^- & OH^- & H^+ & CO_2(aq) \\ \text{Compl} & 1 & 0 & 0 & 0 & 1 & 0 & -1 \\ \text{Comp2} & 0 & 1 & 0 & 0 & 0 & 0 & 0 \\ \text{Comp3} & 0 & 0 & 1 & 0 & 1 & -1 & -1 \\ \text{Comp4} & 0 & 0 & 0 & 1 & -1 & 1 & 2 \end{array}\right)$$

and

$$\mathbf{u} = \begin{pmatrix} c_{H_2O} + c_{OH^-} - c_{CO_2(aq)} \\ c_{Cl^-} \\ c_{CO_3^{2-}} + c_{OH^-} - c_{H^+} - c_{CO_2(aq)} \\ c_{HCO_3^-} - c_{OH^-} + c_{H^+} + 2c_{CO_2(aq)} \end{pmatrix}$$

Note that the third element of **u** does not coincide anymore with the total inorganic carbon, and the fourth element has no direct translation to acidity-alkalinity concept. Nevertheless, although less intuitive from the geochemical point of view, both are a valid components in the sense that their value is not affected by chemical reactions.

Redox reactions can also be included in the component matrix. However, they require a special treatment; for example, redox reactions can be written as half-reactions including the electron as an additional chemical species (e^-). As the concentration of e^- is purely fictitious (there are no free electrons present in the aqueous phase), the component matrix should not consider this concentration, but the electron can still be used as a component. This has caused some confusion and errors as discussed by Lichtner (2004).

We will illustrate the application of the component matrix to redox reactions by a small example:

$$\text{(R6)} \quad O_2 + 4H^+ + 4e^- \rightleftarrows 2H_2O$$

$$\text{(R7)} \quad Fe(OH)_3(s) + 3H^+ + e^- \rightleftarrows Fe^{2+} + 3H_2O$$

There are six species (H_2O, H^+, e^-, Fe^{2+}, O_2, $Fe(OH)_3(s)$), two reactions and, therefore, four components. The stoichiometric matrix is:

$$\mathbf{S} = \begin{pmatrix} & H_2O & H^+ & e^- & Fe^{2+} & O_2 & Fe(OH)_3(s) \\ R6 & 2 & -4 & -4 & 0 & -1 & 0 \\ R7 & 3 & -3 & -1 & 1 & 0 & -1 \end{pmatrix}$$

If we choose O_2 and $Fe(OH)_3$ to be secondary and the rest primary, the component matrix is:

$$\mathbf{U} = \left(\begin{array}{c|cccc|cc} & H_2O & H^+ & e^- & Fe^{2+} & O_2 & Fe(OH)_3(s) \\ \Sigma H_2O & 1 & 0 & 0 & 0 & 2 & 3 \\ \Sigma H^+ & 0 & 1 & 0 & 0 & -4 & -3 \\ \Sigma e^- & 0 & 0 & 0 & 0 & -4 & -1 \\ \Sigma Fe & 0 & 0 & 0 & 1 & 0 & 1 \end{array}\right)$$

and the concentration of the components (**u**):

$$\mathbf{u} = \begin{pmatrix} c_{H_2O} + 2c_{O_2} + 3c_{Fe(OH)_3\,(s)} \\ c_{H^+} - 4c_{O_2} - 3c_{Fe(OH)_3\,(s)} \\ -4c_{O_2} - c_{Fe(OH)_3\,(s)} \\ c_{Fe^+} + c_{Fe(OH)_3\,(s)} \end{pmatrix}$$

Note that in the matrix **U** the element of column and row e^- contains a zero instead of a one. It is zero because the electron concentration is fictitious. Nevertheless, there is still a component e^- (third element of vector **u**) and the concentration of e^- is still a variable, because the concentrations of $Fe(OH)_3$ and O_2 are a function of e^- through the mass action laws. As in the case of H^+, the physical meaning of the third component would be the total reductive capacity of the solution (its capacity to buffer external addition of oxidants), and can be positive or negative.

An alternative is avoiding the electron. Instead of the two half-reactions R6 and R7 we can also write one reaction without the electron:

$$\text{(R8)}\quad Fe(OH)_3(s) + 2H^+ = Fe^{2+} + 0.25O_2 + 2.5H_2O$$

Now there are five species (H_2O, H^+, Fe^{2+}, O_2, $Fe(OH)_3(s)$), one reaction and, therefore, four components (just as before). The stoichiometric matrix is:

$$\mathbf{S} = \begin{pmatrix} & H_2O & H^+ & Fe^{2+} & O_2 & Fe(OH)_3(s) \\ R8 & 2.5 & -2 & 1 & 0.25 & -1 \end{pmatrix}$$

If we choose $Fe(OH)_3$ to be secondary and the rest primary, the component matrix will be:

$$\mathbf{U} = \left(\begin{array}{c|cccc|c} & H_2O & H^+ & O_2 & Fe^{2+} & Fe(OH)_3(s) \\ \Sigma H_2O & 1 & 0 & 0 & 0 & 2.5 \\ \Sigma H^+ & 0 & 1 & 0 & 0 & -2 \\ \Sigma O_2 & 0 & 0 & 1 & 0 & 0.25 \\ \Sigma Fe & 0 & 0 & 0 & 1 & 1 \end{array}\right)$$

and the concentration of the components (**u**):

$$\mathbf{u} = \begin{pmatrix} c_{H_2O} + 2.5c_{Fe(OH)_3\,(s)} \\ c_{H^+} - 2c_{Fe(OH)_3\,(s)} \\ c_{O_2} + 0.25c_{Fe(OH)_3\,(s)} \\ c_{Fe^+} + c_{Fe(OH)_3\,(s)} \end{pmatrix}$$

We should get the same result, as obtained previously, by using O_2 instead of e^- as a primary species.

3.1.4.2 *Method 2 (eliminate constant activity species)*

Method 1 is very convenient in the sense that it yields components that often have a well-defined meaning and, in reactive transport problems, are the same throughout the transport domain. This is very useful because the solute transport equation simply expresses the mass balance of solutes, therefore, components defined by Method 1 can be transported conservatively (we will return to this issue in Chapter 4). However, it is not convenient in cases, such as examples 1a or 1b, where non-aqueous species participate in the reactions. In such cases, several options exist. The most commonly used is to include those species in the chemical system. This is inconvenient, however, when the activity of those species is fixed (e.g., H_2O in example 1c, or a pure phase mineral in examples 1a or 1b) because one is including as unknown a species that is not really unknown (its activity is fixed!). There are two options to overcome this difficulty:

- Reduce the set of primary species (Saaltink *et al.*, 1998). The basic idea is to proceed as in Method 1, but then multiply the components matrix by an elimination matrix, **E.**
- Eliminate constant activity species from the outset (De Simoni *et al.*, 2005). In essence, this method does not include constant activity species in the vector **q** of species, and then proceeds like Method 1.

We illustrate these approaches by application to examples 1a and 1b.

Eliminating $CaSO_4(s)$ as a species yields:

$$\mathbf{q}=\begin{pmatrix} Ca^{2+} \\ SO_4^{2-} \end{pmatrix},\quad \mathbf{S}=\begin{pmatrix} & Ca^{2+} & SO_4^{2-} \\ R1 & 1 & 1 \end{pmatrix},\quad \mathbf{U}=\begin{pmatrix} & Ca^{2+} & SO_4^{2-} \\ U_{anh} & 1 & -1 \end{pmatrix}$$

where we have denoted $U_{anh} = Ca^{2+} - SO_4^{2-}$. The meaning of this component is somewhat puzzling because Ca^{2+} and SO_4^{2-} appear to be two independent species. Actually, they are not, as equilibrium with anhydrite provides a strong link between them. In fact, dissolution of anhydrite will increase in the same amount the number of moles of Ca^{2+} and SO_4^{2-}. Therefore, U_{anh} will not be altered by dissolution (or precipitation) of anhydrite, which was indeed the way we decided components in the first place. Note also that U_{anh} is orthogonal to the anhydrite equilibrium line (Fig. 3.3), which is indeed the property we demanded for **U.**

Example 1b is similar. Eliminating $CaSO_4(s)$ and $CaCO_3(s)$ from the chemical system (because we assume their activity to be 1) yields:

$$\mathbf{q}=\begin{pmatrix} Ca^{2+} \\ SO_4^{2-} \\ CO_3^{2-} \end{pmatrix},\quad \mathbf{S}=\begin{pmatrix} & Ca^{2+} & SO_4^{2-} & CO_3^{2-} \\ R1 & 1 & 1 & 0 \\ R2 & 1 & 0 & 1 \end{pmatrix},\quad \mathbf{U}=\begin{pmatrix} & Ca^{2+} & SO_4^{2-} & CO_3^{2-} \\ U_{ac} & 1 & -1 & -1 \end{pmatrix}$$

where now U_{ac} remains unaltered by dissolution precipitation of anhydrite and/or calcite (recall that we are assuming pH to be high, so that we do not worry about CO_3^{2-} evolving to HCO_3^- or CO_2).

3.1.4.3 *Other methods*

Molins *et al.* (2004) and Kräutle and Knabner (2005, 2007) describe other methods to compute the component matrix. Broadly speaking, the goal is to define components so that kinetic species do not affect all components. As we shall see in Chapter 4, this facilitates reactive transport computations.

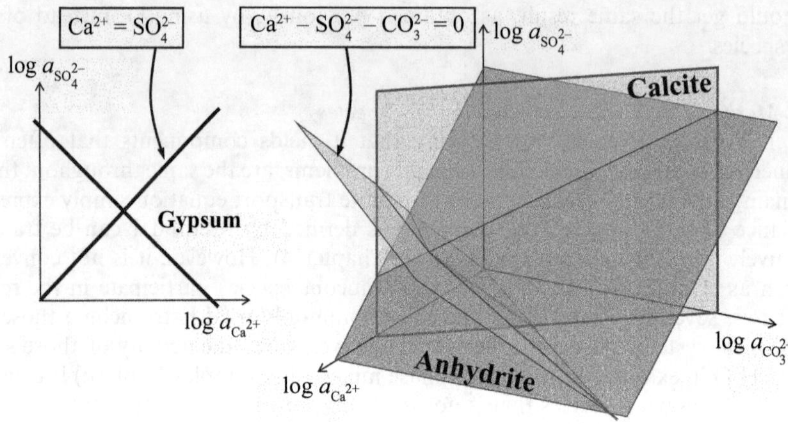

Figure 3.3. Component equations are orthogonal to equilibrium equations, so individual activities at equilibrium remain invariant. This is shown here for example 1a (left), where component $Ca^{2+} - SO_4^{2-}$ is orthogonal to anhydrite equilibrium (dissolution/precipitation of anhydrite leaves the component unchanged) and example 1b (right), where component $Ca^{2+} - SO_4^{2-} - CO_3^{2-}$ remains unchanged when calcite or anhydrite dissolve.

3.2 HOMOGENEOUS REACTIONS

Homogeneous reactions take place among chemical species belonging to the same phase, here restricted to species in the aqueous phase. They include aqueous complexation, acid-base and redox reactions. The simplest geochemical problem is to calculate the concentration of all aqueous species in a given solution (i.e., the speciation of the solution). The continuous motion and large number of dissolved ions in solution results in numerous collisions which are responsible for the formation of ion pairs and/or dissolved complexes. Since these reactions are almost instantaneous, they can be effectively considered as equilibrium reactions. Therefore, the mass action law described in equation (3.7) can be applied to the reactions among aqueous species.

The activity of i-th aqueous species, a_i, can be calculated as:

$$a_i = c_i \gamma_i \tag{3.18}$$

where c_i is its molal concentration and γ_i the activity coefficient. Here the activity coefficient is a measure of how the solution behavior is far from the ideality, being the reference state a 1 mol kg^{-1} water ideal solution of solute i. Then, according to equation (3.8), if n_e is the number of homogeneous reactions the vector of concentrations $\mathbf{c}_2$ of n_e aqueous species can be calculated from the concentration of the rest $n_s - n_e$ aqueous primary species:

$$\log \mathbf{c}_2 = \mathbf{S}_1^* \log \mathbf{c}_1 + \mathbf{S}_1^* \log \boldsymbol{\gamma}_1 + \log \mathbf{K}^* - \log \boldsymbol{\gamma}_2 \tag{3.19}$$

In these equations, $\mathbf{S}_1^*$ is known from the stoichiometry of the reactions, $\mathbf{K}^*$ is known, at fixed p and T, from thermodynamic databases and equation (3.10), and $\boldsymbol{\gamma}_1$ and $\boldsymbol{\gamma}_2$ are functions of $\mathbf{c}_1$ and $\mathbf{c}_2$ (see section 1.2 in Chapter 1 for details). Therefore, the aqueous speciation problem still contains n_s unknowns ($\mathbf{c}_1$ and $\mathbf{c}_2$), and n_e equations (mass action law equations).

The rest of the n_s–n_e equations required are supplied by the mass conservation of the components:

$$\mathbf{u} = \mathbf{c}_1 + (\mathbf{S}_1^*)^T \mathbf{c}_2 \tag{3.20}$$

In general, the components of the vector of total concentration of components, **u**, are values known from analytical techniques, such as the total Cl or TIC in the example following equation (3.17). There are, however, some exceptions. In the same example, the first element of **u** represents the total water of the system. There, the value of c_{H_2O} (i.e., the free water or mol H_2O/kg water) is 1000/18, much higher than the rest of terms of the component when only an aqueous solution is considered. Therefore, what is known is the concentration c_{H_2O} of the species instead the concentration of the first element of the component **u**, and we can leave out this equation. A similar case is the fourth element of the component **u** in the same example, where the analytical value is the concentration of the species c_{H^+} (more specifically its activity obtained from electrochemical methods): the value of the component in this case does not necessarily have any empirical meaning. A third case also occurs when half-reactions are used in redox problems. There, the analytical value is the concentration of the species c_{e^-} (its activity). For these three cases (H_2O, H^+ and e^-) the component equations can be removed.

Therefore, equations (3.19) and (3.20) form a system of n_S equations with n_S unknowns which can be solved. On solving for $\mathbf{c}_1$ and $\mathbf{c}_2$, the value of $\boldsymbol{\gamma}_1$ and $\boldsymbol{\gamma}_2$ are assumed known. Therefore, it is necessary to assume an initial value for $\boldsymbol{\gamma}$ and update it with the value of **c** in an iterative process (see the method of solution in section 3.5).

3.2.1 *Speciation calculations*

We have discussed extensively how to express reactions, the mass action law, or components. It should be noticed, however, that equilibrium was expressed in terms of activities (recall eq. 3.4 or 3.8), whereas components, and mass balances in general, are expressed in terms of concentrations (molalities). Moreover, chemical analyses are given in terms of concentrations. The objective of this section is to discuss how to complete the concentrations of all species from information about the chemical system. Definition of the chemical system has been discussed in earlier chapters and specific issues regarding these computations are discussed in section 3.2, for homogeneous reactions, and 3.3, for heterogeneous reactions. Depending on the type of available information, we will typically need to compute secondary species concentrations, $\mathbf{c}_2$, from primary ones, $\mathbf{c}_1$, or all species concentrations from components. The resulting systems of equations are non-linear. This section is devoted to the algorithms for solving such systems.

There are two types of iterative methods commonly used to solve non-linear systems: the Picard and Newton-Raphson methods. Both are outlined below. Given an initial guess of a solution, each method seeks a better approximation of the solution, and repeats the procedure, until a satisfactory solution is found.

The Picard method is also called the "sequential iteration" or "fixed point" method. It requires writing the system of equations in the form $\mathbf{x} = \mathbf{g}(\mathbf{x})$ where **x** is the vector of unknowns and **g** is the algorithm function used to compute **x**. An example is equation (3.19), where given $\mathbf{c}_1$ and $\mathbf{c}_2$ one can obtain the activity coefficients $\boldsymbol{\gamma}$ to obtain again $\mathbf{c}_2$. The Picard method proceeds as:

$$\mathbf{x}^{k+1} = g(\mathbf{x}^k) \tag{3.21}$$

where k is the iteration number, and the equation is repeated until convergence, (i.e., until $\mathbf{x}^{k+1} \approx \mathbf{x}^k$, that is, until a "fixed point" has been reached). The actual working of the system is shown in Figure 3.4. Examining this figure, it should be obvious that the algorithm converges

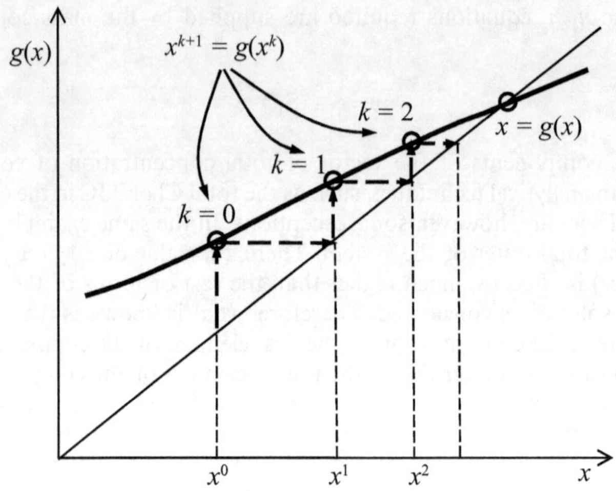

Figure 3.4. Schematic description of Picard method for solving non-linear (systems of) equations written in the form $x = g(x)$. The solution is found as the limit of the sequence $x^{k+1} = g(x^k)$. Convergence to the solution is fast if the slope of g is small.

if $dg/dx < 1$ (in general when g is not overly sensitive to the value of x). Otherwise, $\mathbf{x}^{k+1}$ will be further away from the solution than $\mathbf{x}^k$ and the procedure will diverge.

The Newton-Raphson method requires writing the equation system as $\mathbf{f}(\mathbf{x}) = 0$ ($\mathbf{x} - \mathbf{g}(x) = 0$ would do if the problem is defined as $\mathbf{x} = \mathbf{g}(\mathbf{x})$). The idea is to linearize the system of equations and to solve the linear approximation. That is:

$$\mathbf{f}^{k+1} = \mathbf{f}(\mathbf{x}^{k+1}) \approx \mathbf{f}(\mathbf{x}^k) + \frac{\partial \mathbf{f}}{\partial \mathbf{x}}\left(\mathbf{x}^{k+1} - \mathbf{x}^k\right) = 0 \tag{3.22}$$

The algorithm is quite powerful, but requires computing $\partial\mathbf{f}/\partial\mathbf{x}$ (called the Jacobian of **f**), which can be tedious. The system may also fail to converge if the initial guess ($\mathbf{x}^o$) is too far from the solution.

The application of both methods to speciation calculations is outlined below.

3.2.1.1 *Algorithm 1*

Compute $\mathbf{c}_2$ given $\mathbf{c}_1$ using Picard's method. This calculation is required for reactive transport computations based on the "direct substitution approach" that will be discussed in Chapter 4. The calculation basically uses equation (3.8), where the activities are written in terms of concentrations as $\mathbf{a} = \boldsymbol{\gamma}\mathbf{c}$:

$$\log \mathbf{c}_2 = \mathbf{S}_1^* \log \boldsymbol{\gamma}_1 \mathbf{c}_1 + \log K^* - \log \boldsymbol{\gamma}_2 \tag{3.23}$$

This equation would be explicit if the activity coefficients were known. As the dependence of activity coefficients on species concentrations is weak, the system is almost linear. In fact, recalling Figure 3.4, and thinking of equation as $\mathbf{c}_2 = g(\mathbf{c}_2)$, it appears that $\boldsymbol{g}$ is almost constant and Picard method should converge quickly. The algorithm becomes:

Step 0: Initialization. Set $\boldsymbol{c}_2^0$, $k = 0$
Step 1: Given $\boldsymbol{c}_2^k$ and $\boldsymbol{c}_1$, compute $\boldsymbol{\gamma}_1^k$ and $\boldsymbol{\gamma}_2^k$
Step 2: Use $\boldsymbol{\gamma}_1^k$, $\boldsymbol{\gamma}_2^k$ and $\boldsymbol{c}_1$ in equation to compute $\boldsymbol{c}_2^{k+1}$
Step 3: Converge check. If $\boldsymbol{c}_2^{k+1}$ is close to $\boldsymbol{c}_2^k$ (e.g., if $\left|c_{2i}^k - c_{2i}^{k+1}\right| < \varepsilon_i$, for all species i, where ε_i is a species dependent tolerance, then convergence has been reached and the algorithm stops. Otherwise, set $k = k + 1$ and return to step 1.

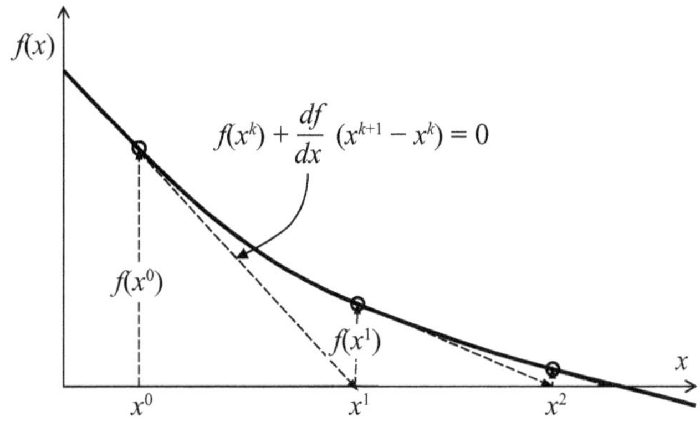

Figure 3.5. Schematic description of Newton-Raphson method for solving non-linear (systems of) equations written in the form $f(x) = 0$. The solution is found as the limit of the sequence x^{k+1}, obtained by setting to 0 the linear approximation of f at x^k. Convergence to the solution is fast if the curvature of f is small.

The algorithm is quite straight forwards and usually converges easily. The only open issues include how to initialize the algorithm. Unit activity coefficients could be a good starting. However, the algorithm should not be very sensitive to the initialization method and a good initialization will rarely do more than save a few iterations. The convergence criterion, ε_i is usually defined so as to ensure that the relative error is small. For example ε_i may be chosen as $\varepsilon_i = r(c_{2i}^k + c_{2i}^{k+1})/2$ where r is a relative tolerance (e.g., $r = 10^{-6}$).

3.2.1.2 *Algorithm 2*

Compute **c** given **u** using the Newton-Raphson method. This computation is required for reactive transport simulations based on sequential iteration methods that will be discussed in Chapter 4.

The system of equations to be solved includes the components definition and mass action law. That is:

$$f_c(\mathbf{c}_1) = \mathbf{U}_1\mathbf{c}_1 + \mathbf{U}_2\mathbf{c}_2 - \mathbf{u} = 0 \tag{3.24}$$

$$\log \mathbf{c}_2 = \mathbf{S}_1^* \log \boldsymbol{\gamma}_1 \mathbf{c}_1 + \log \mathbf{K}^* - \log \boldsymbol{\gamma}_2 \tag{3.25}$$

We may see these equations as a non-linear system of n_s unknowns ($n_1 = n_s - n_e$ primary species concentrations and n_e secondary species) and n_s equations ($n_s - n_e$ components definitions and n_e mass action law equations). However, given that $\mathbf{c}_2$ is almost explicit, this system is usually solved in terms of $\mathbf{c}_1$. That is, we use equation (3.25), formally, to solve $\mathbf{c}_2$ as a function of $\mathbf{c}_1$ and then solve equation (3.24) for $\mathbf{c}_1$. Proceeding iteratively, the latter can be expanded as:

$$\mathbf{f}_C(\mathbf{c}_1^{k+1}) \approx \mathbf{f}_c(\mathbf{c}_1^k) + \frac{\partial \mathbf{f}_c}{\partial \mathbf{c}_1}(\mathbf{c}^{k+1} - \mathbf{c}_1^k) = 0 \tag{3.26}$$

Then, the algorithm proceeds as follows:

Step 0: Initialization. Set $k = 0$ and initial guess for $\mathbf{c}_1^k$.

Step 1: Given $\mathbf{c}_1^k$, compute $\mathbf{c}_2^k$ by means of equation (3.25). We also need to compute $\partial\mathbf{c}_2/\partial\mathbf{c}_1$. This requires taking derivatives of equation (3.25) with respect to $\mathbf{c}_1$, which leads to the following system:

$$\left(\frac{1}{\mathbf{c}_2}\mathbf{I} + \frac{1}{\gamma_2}\frac{\partial\gamma_2}{\partial\mathbf{c}_2} - \frac{\mathbf{S}_1^*}{\gamma_1 c_1}\frac{\partial\gamma_1}{\partial\mathbf{c}_2}\right)\frac{\partial\mathbf{c}_2}{\partial\mathbf{c}_1} = \frac{\mathbf{S}_1^*}{\gamma_1\mathbf{c}_1}\left(\frac{\partial\gamma_1}{\partial\mathbf{c}_1} + \mathbf{I}\right) - \frac{1}{\gamma_2}\frac{\partial\gamma_2}{\partial\mathbf{c}_1} \tag{3.27}$$

Step 2: Given $\mathbf{c}_1^k$ and $\mathbf{c}_2^k$, compute $\mathbf{f}_c^k$ by means of equation (3.24). Since $\mathbf{f}_c^k$ is the error in equation (3.24), it is often called the residual.

Step 3: Given $\partial c_2/\partial c_1$ compute $\Delta\mathbf{c}_1 = (\mathbf{c}_1^{k+1} - \mathbf{c}_1^k)$ by solving the system, equivalent to equation (3.26):

$$\left(\mathbf{U}_1 + \mathbf{U}_2 \frac{\partial c_2}{\partial c_1}\right)\Delta\mathbf{c}_1 = -f_c^k \tag{3.28}$$

Step 4: Compute $\mathbf{c}_1^{k+1} = \mathbf{c}_1^k + \Delta\mathbf{c}_1$. This computation may require some extra checks to ensure a robust solution. For example, if $\mathbf{c}_{1i}^{k+1} \le 0$, then one may choose $\mathbf{c}_{1i}^{k+1} = \alpha_i c_{1i}^k$, where α_i is a reduction factor (for instance $\alpha_i = 0.1$). In fact, it is often safe to restrict the maximum change per iteration so that if $\mathbf{c}_{1i}^{k+1} \le \alpha_i c_{1i}^k$, then $\mathbf{c}_{1i}^{k+1} = \alpha_i c_{1i}^k$ and if $\mathbf{c}_{1i}^{k+1} \ge \mathbf{c}_{1i}^k/\alpha_i$, then $\mathbf{c}_{1i}^{k+1} = \mathbf{c}_{1i}^k/\alpha_i$.

Step 5: Convergence check. If $\mathbf{c}_1^{k+1}$ is close to $\mathbf{c}_1^k$ (recall comments on step 4 of algorithm 3.1) and or $f_c^{k+1} \approx 0$, then stop. Otherwise, set $k = k+1$ and return to step 1.

This algorithm is applied to example 1a, where we will keep assuming that all activity coefficients are equal to 1 for simplicity. The system can then be written as:

$$f_c(c_1) = c_1 - c_2 - u = 0$$
$$c_1 \cdot c_2 = K_{anh}$$

This system can be solved analytically by substituting the second equation, $c_2 = K_{anh}/c_1$, into the first. This leads to a second degree equation whose solution is:

$$c_1 = \frac{-u \pm \sqrt{u^2 + 4K_{anh}}}{2}$$

Notice that the + sign must be chosen to ensure a positive concentration. The algorithm, which can be easily programmed into a spreadsheet, proceeds as follows (we solve for $u = 1.37 \cdot 10^{-5}$ M).

iter.	$c_1(k=0) = 3.00\text{E}{-05}$		$c_1(k=0) = 0.02$		$c_1(k=0) = 3$	
no.	$c_1(k)$	resid.	$c_1(k)$	resid.	$c_1(k)$	resid.
0	3.00E–5	1.67E–2	2.00E–2	–2.00E–2	3.00E+0	–3.00E+0
1	5.99E–5	8.30E–3	3.64E–5	1.37E–2	–1.34E–5	–3.75E–2
2	1.19E–4	4.09E–3	7.25E–5	6.82E–3	–2.67E–5	–1.87E–2
3	2.31E–4	1.93E–3	1.43E–4	3.34E–3	–5.34E–5	–9.35E–3
4	4.16E–4	7.76E–4	2.75E–4	1.53E–3	–1.06E–4	–4.62E–3
5	6.15E–4	1.87E–4	4.76E–4	5.63E–4	–2.08E–4	–2.21E–3
6	6.95E–4	1.23E–5	6.51E–4	1.04E–4	–3.84E–4	–9.34E–4
7	7.01E–4	5.39E–8	6.99E–4	3.87E–6	–5.97E–4	–2.57E–4
8	7.01E–4	1.04E–12	7.01E–4	5.34E–9	–7.03E–4	–2.28E–5
9	7.01E–4	–4.57E–20	7.01E–4	1.02E–14	–7.15E–4	–1.81E–7

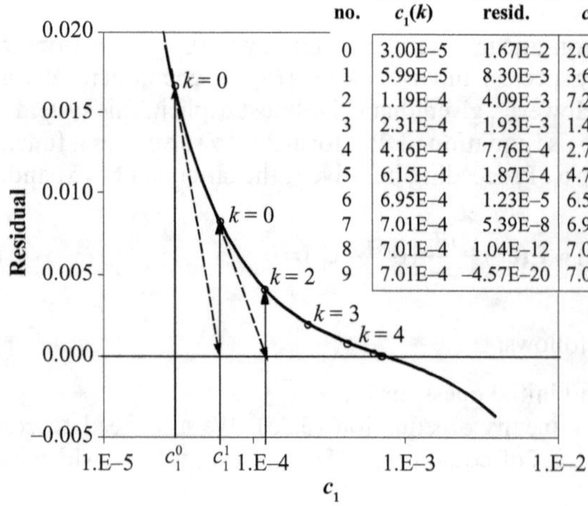

Figure 3.6. Application of Newton-Raphson method to the speciation problem of Example 1a given $u = 1.37\text{E}{-5}$ for an initial guess of $c_1 = 3\text{E}{-5}$ (the table displays the iterations for other initial values).

Step 0: We choose $c_1 = 0.003$
Step 1: $c_2 = K_{anh}/c_1 = 0.5 \cdot 10^{-6}/3 \cdot 10^{-5} = 1.67 \cdot 10^{-2}$
$dc_2/\partial c_1 = -K_{anh}/c_1^2 = -0.5 \cdot 10^{-6}/(3 \cdot 10^{-5}) = -557$
Step 2: $f_c^o = c_1 - c_2 - u = 3.10^{-5} - 1.67 \cdot 10^{-2} - 1.37 \cdot 10^{-5} \simeq -1.67 \cdot 10^{-2}$
Step 3: $\left(1 - \frac{dc_2}{\partial c_1}\right)\Delta c_1 = -f_c^o$, so that $(1-(-557))\Delta c_1 = 1.67 \cdot 10^{-2}$
Therefore, $\Delta c_1 = 1.67 \cdot 10^{-2}/558 = 2.99 \cdot 10^{-5}$
Step 4: $\mathbf{c}_1^{k+1} = \mathbf{c}_1^k + \Delta\mathbf{c}_1 = 3.10^{-5} + 2.99 \cdot 10^{-5} = 5.99 \cdot 10^{-5}$
Step 5: Convergence check. Since $\mathbf{c}_1^1 = 5.99 \cdot 10^{-5}$ is different from $\mathbf{c}_1^0 = 3.10^{-5}$, we have to repeat the calculation from step 1. Notice that now $k = 1$.

Iterations are summarized in Figure 3.6 for several values of the initial concentration. It is worth pointing that, unlike in algorithm 3.1, convergence is sensitive to the initial guess. Therefore, here it is important to get a good initial guess. Notice also that when $\mathbf{c}_1^0 = 3$, the algorithm converges to a negative concentration. This negative concentration is also a solution to the problem, but it is not adequate meaningful one. This stresses the importance of having intermediate checks in the algorithm.

3.3 HETEROGENEOUS REACTIONS

Heterogeneous reactions involve species in several phases. The geochemical problem is now extended to calculate the concentration of the chemical species in each phase. We will describe here the procedure to solve the speciation when an aqueous solution coexists with a solid surface, a bulk solid and a gas phase. As in the case of the aqueous species, reactions among the species of the different phases are assumed here to be at equilibrium.

3.3.1 *Surface complexation reactions*

Many minerals such as metal oxides, hydroxides and layered silicates (Al_2O_3, TiO_2, FeOOH, SiO_2, kaolinite, etc.) exhibit electrically charged surfaces in the presence of natural waters. These surfaces contain ionizable functional groups (ex.: silanol groups in hydrated silica: Si-OH), being responsible for chemical reactions at the surface. We will represent these surface groups or sites as *X*OH.

The sorption of a solute on the solid surfaces is described as a chemical reaction between the aqueous species and specific surface sites (surface complexation). These surface reactions include protonation and deprotonation of the surface, and anion and cation complexation reactions with the protonated or deprotonated surfaces. For example, deprotonation and desorption of a metal M^{z+} could be represented as:

$$XO^- = XOH - H^+$$
$$XOM^{Z_+-1} = XOH - H^+ + M^{Z_+}$$

Usually, surface complexation reactions reach equilibrium within hours. Therefore, similar to aqueous complexation, they can be described by the mass action law (eq. 3.7). However, in doing so, the activity of surface species must be evaluated. As a first approach, we assume ideal behavior of the empty sites *X*OH and the surface complexes. That is, we assume that a particular surface complex is not influenced by the rest of sites (empty or full) in the neighborhood. Then, γ_2 is the unity, and the equilibrium constant is usually referred to as the intrinsic constant (Dzombak and Morel, 1990).

Therefore, the aqueous+surface phase speciation problem still contains n_s unknowns ($\mathbf{c}_1$ and $\mathbf{c}_2$), and n_e mass action law equations (3.7). The rest of $n_s - n_e$ equations required are supplied by the mass conservation of the components (3.17). To ensure that the set of primary species is linearly independent, at least one of them must be a surface species. Therefore, one component of **u** must also include the surface species. In surface speciation problems the value of this component is the total number of sorption sites, and is an experimental datum.

We will illustrate the application of the component matrix of a surface complexation by a small example with only one reaction of water dissociation in the aqueous phase:

(R9) $H_2O = H^+ + OH^-$
(R10) $XOH_2^+ = XOH + H^+$
(R11) $XO^- = XOH - H^+$
(R12) $XOCd^+ = XOH - H^+ + Cd^{2+}$

There are eight species (H_2O, H^+, OH^-, Cd^{2+} XOH, XO^-, XOH_2^+, $XOCd^+$), four reactions and, therefore, four components. The stoichiometric matrix is:

$$\mathbf{S} = \begin{pmatrix} & H_2O & H^+ & OH^- & Cd^{2+} & XOH & XO^- & XOH_2^+ & XOCd^+ \\ R9 & -1 & 1 & 1 & 0 & 0 & 0 & 0 & 0 \\ R10 & 0 & 1 & 0 & 0 & 1 & 0 & -1 & 0 \\ R11 & 0 & -1 & 0 & 0 & 1 & -1 & 0 & 0 \\ R12 & 0 & -1 & 0 & 1 & 1 & 0 & 0 & -1 \end{pmatrix}$$

If we choose OH^-, XO^-, XOH_2^+, $XOCd^+$ to be secondary and the rest primary, the component matrix is:

$$\mathbf{U} = \begin{pmatrix} & H_2O & H^+ & Cd^{2+} & XOH & OH^- & XO^- & XOH_2^+ & XOCd^+ \\ \Sigma H_2O & 1 & 0 & 0 & 0 & 1 & 0 & 0 & 0 \\ \Sigma H^+ & 0 & 1 & 0 & 0 & -1 & -1 & 1 & 1 \\ \Sigma Cd & 0 & 0 & 1 & 0 & 0 & 0 & 0 & 1 \\ \Sigma XOH & 0 & 0 & 0 & 1 & 0 & 1 & 1 & 1 \end{pmatrix}$$

and the concentration of the components (**u**):

$$\mathbf{u} = \begin{pmatrix} c_{H_2O} + c_{OH^-} \\ c_{H^+} - c_{OH^-} - c_{XO^-} + c_{XOH_2^+} + c_{XOCd^+} \\ c_{Cd^{2+}} + c_{XOCd^+} \\ c_{XOH} + c_{XO^-} + c_{XOH_2^+} + c_{XOCd^+} \end{pmatrix}$$

The four mass action laws of reactions R9 to R12, and the four mass conservation equations of the components (with the exception of c_{H_2O} and c_{H^+} discussed above), allow to solve the speciation problem. The value of K_{int}, of the reactions, the *pH*, the total inventory of Cd of the system (u_3) or in solution ($c_{Cd^{2+}}$), and the total sorption sites (u_4) have to be known to solve the problem.

In aqueous complexation reactions the electric charge is assumed to be homogeneously distributed in the solution. However, surface complexes attach to a fixed surface and a local electrostatic field is created. To account for the non-ideal behavior of the surface species, we have to consider an additional energetic term accounting for the work needed for the aqueous species to travel across the surface electric field:

$$\Delta G_{ads} = \Delta G_{\text{int}} + \Delta G_{coul} = \Delta G_{\text{int}} + \Delta z F \psi \quad (3.29)$$

where ΔG is the free energy change of the overall adsorption reaction, ΔG_{int} and ΔG_{coul} are the free energy change due to chemical bonding and to the electrostatic work, respectively, Δz is the charge loss of the surface complexation reaction, F the Faraday's constant (96,485 C mol^{-1}) and ψ is the mean surface potential [V]. In terms of equilibrium constants:

$$K_{ads} = K_{\text{int}} e^{-\frac{\Delta z F \psi}{RT}} \tag{3.30}$$

where R is the gas constant (8.314 J mol^{-1} K^{-1}) and T the absolute temperature [K], K_{ads} the equilibrium constant of the overall adsorption process and K_{int} does not depend on the surface charge. In the following calculations, the thermodynamic activities of the surface complexes are substituted by their concentrations. This is based on the assumption that all non-ideal behavior of the surface species is accounted for by the electrostatic term. Therefore, the reference state to calculate the thermodynamic activity would be 1 mol sorbed species /kg water with no surface electrostatic field.

By considering the electrostatic term in the equilibrium formulation another unknown, the mean surface potential, ψ is introduced. A new equation to find the solution of the problem is, therefore, required. This equation relates the potential to the surface charge through a capacitance. Several theoretical models describe the relationship between the mean surface potential and the charge density. Some of these models are: the constant capacitance model, the diffuse layer model and the triple layer model. The option for a particular model depends on the amount of experimental data needed to obtain the equilibrium constant of the reactions. We discuss first the triple layer model because, from the modeling point of view, it includes the rest as particular cases.

This model was developed by Davies and Leckie (1978) to be applied to adsorption of metals on oxide surfaces. All sorbed ions are ascribed to one of two surface layers, 0 and β, whereas a diffuse layer, d, is formed with counter ions in solution (Fig. 3.7). Some surface reactions, such as protonation and deprotonation, take place at the immediate surface of the mineral or 0 layer. This layer has a total surface charge σ_0 and a potential ψ_0. Other ions in solution are binded to protonated/deprotonated groups of opposite sign and create a layer β with a charge σ_β and a potential ψ_β. The surface charge of layers 0 and β is calculated from the total surface species adsorbed on the layer:

$$\sigma_0 = \frac{F}{SA}\sum_{i=1}^{N_0} z_i c_i \tag{3.31}$$

$$\sigma_\beta = \frac{F}{SA}\sum_{i=1}^{N_\beta} z_i c_i \tag{3.32}$$

where F is the Faraday constant [C mol^{-1}], S is the solid concentration [g kg^{-1} water], and A is the surface area of the solid per unit of mass [m^2 g^{-1}], z_i and c_i are the charge and the concentration of the i-th adsorbed species [mol kg^{-1} water], respectively; and N_0 and N_β are the number of species adsorbed in layers 0 and β, respectively.

In order to neutralize the surface charge, a zone with a lower density of delocalized ions of the same charge is hypothesized to form in the aqueous phase close to the solid surface. The surface charge and the potential of this diffuse layer of counterions are related by Guy-Chapman theory:

$$\sigma_d = (8RT\varepsilon\varepsilon_0 I)^{1/2}\sinh(F\psi_\delta/2RT) \tag{3.33}$$

where ε is the dielectric constant of water, ε_0 the permitivity of free space (8.854 × 10^{-13} C V^{-1} dm^{-1}), and I the ionic strength of the solution [mol kg^{-1} water].

The 0 and β layers are separated by a region of capacitance C_1, and the layers β and d are separated by a region of capacitance C_2. The surface charge of the three layers is also

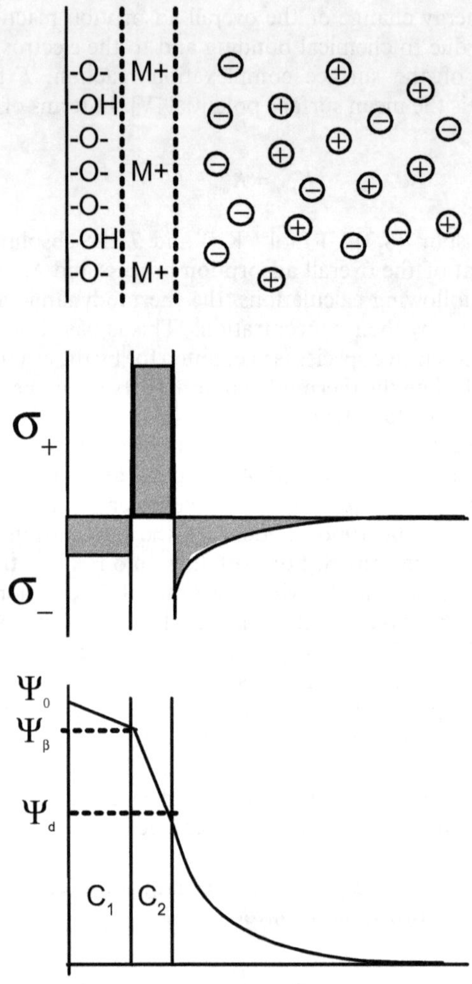

Figure 3.7. Conceptual representation of the charge distribution and electrical potential at a planar surface (modified from Davies and Leckie, 1978).

related to potentials by electrostatic capacitances, and the total surface charge must also be null:

$$\sigma_0 = C_1(\psi_0 - \psi_\beta) \tag{3.34}$$

$$\sigma_\beta = C_1(\psi_\beta - \psi_0) + C_2(\psi_\beta - \psi_d) \tag{3.35}$$

$$\sigma_0 + \sigma_\beta + \sigma_d = 0 \tag{3.36}$$

The values of C_1 and C_2 [F m^{-2}] are known and normally obtained by fitting experimental data.

In the triple layer model, three new unknowns (ψ_0, ψ_β and ψ_d) are present, and, therefore, three new equations are needed to solve the problem. These equations are obtained by substituting equations (3.31) to (3.33) into equations (3.34) to (3.36).

The electrostatic correction defined in equation (3.30) can be taken into account by defining a vector of three new species containing the potentials ψ. In order to automatically build the mass action law equations, the new species are conveniently defined as $e^{-F\psi/RT}$ (Westall, 1980). Then the mass action law would result in:

$$S\log\mathbf{a} = \log\mathbf{K}_{\text{int}} + \mathbf{S}_{\psi}\log e^{-F\psi/RT} \tag{3.37}$$

where $\mathbf{S}_{\psi}$ is a matrix of dimension $r \times 3$ formed by the charge variation Δz of the surface complexation reaction. A new set of r mass action law equations can be automatically built:

$$(S\,|\,S_{\nu})\log\begin{pmatrix}\mathbf{a}\\ e^{-F\psi/RT}\end{pmatrix} = \log\mathbf{K}_{\text{int}} \tag{3.38}$$

To complete the system of equations, 3 new elements should be added to the non-electrostatic component matrix **U**. These elements are related to the total component of the three dummy primary species $e^{-F\psi/RT}$. The physical meaning of the two first total components is the electrical charge of each surface after conversion to mol kg^{-1} units, and the meaning of the third component is the total charge balance (3.38):

$$\begin{aligned}
u_0 &= \sum_{i=1}^{N_0} z_i c_i = \frac{SA}{F} C_1(\psi_0 - \psi_\beta)\\
u_\beta &= \sum_{i=1}^{N_\beta} z_i c_i = \frac{SA}{F}\left[C_1(\psi_\beta - \psi_0) + C_2(\psi_\beta - \psi_d)\right]\\
u_d &= \sigma_0 + \sigma_\beta + \sigma_d = C_1(\psi_0 - \psi_\beta) + C_1(\psi_\beta - \psi_0) + C_2(\psi_\beta - \psi_d) + (8RT\varepsilon\varepsilon_0 I)^{\frac{1}{2}}\sinh(F\psi_d/2RT)
\end{aligned} \tag{3.39}$$

The application of the electrostatic correction to the cadmium sorption example described above (reactions R10 to R12) would lead to the K_{ads} constants:

$$\text{(R10)}\quad XOH_2^+ \rightleftarrows XOH + H^+ \qquad K_{ads} = \frac{c_{XOH_2^+}}{c_{XOH}\,a_{H^+}}\, e^{-\frac{(+1)F\psi_0}{RT}}$$

$$\text{(R11)}\quad XO^- \rightleftarrows XOH - H^+ \qquad K_{ads} = \frac{c_{XO^-}\,a_{H^+}}{c_{XOH}}\, e^{-\frac{(-1)F\psi_0}{RT}}$$

$$\text{(R12)}\quad XOCd^+ \rightleftarrows XOH - H^+ + Cd^{2+} \qquad K_{ads} = \frac{c_{XOCd^+}\,a_{H^+}}{c_{XOH}\,a_{Cd^{2+}}}\, e^{-\frac{(-1)F\psi_0}{RT}-\frac{(+2)F\psi_\beta}{RT}}$$

There are eight species (H_2O, H^+, OH^-, Cd^{2+} XOH, XO^-, XOH_2^+, $XOCd^+$), four reactions and, therefore, four components. The stoichiometric matrix is:

$$\mathbf{S} = \begin{pmatrix}
 & H_2O & H^+ & OH^- & Cd^{2+} & XOH & XO^- & XOH_2^+ & XOCd^+ & e^{-F\psi_0/RT} & e^{-F\psi_\beta/RT} & e^{-F\psi_d/RT}\\
R9 & -1 & 1 & 1 & 0 & 0 & 0 & 0 & 0 & 0 & 0 & 0\\
R10 & 0 & 1 & 0 & 0 & 1 & 0 & -1 & 0 & 1 & 0 & 0\\
R11 & 0 & -1 & 0 & 0 & 1 & -1 & 0 & 0 & -1 & 0 & 0\\
R12 & 0 & -1 & 0 & 1 & 1 & 0 & 0 & -1 & -1 & 2 & 0
\end{pmatrix}$$

If we maintain the same secondary species as the previous example (OH^-, XO^-, XOH_2^+, $XOCd^+$), the component matrix is:

$$
\mathbf{U}=\left(\begin{array}{l|ccccccccccc}
 & H_2O & H^+ & Cd^{2+} & XOH & e^{-F\psi_0/RT} & e^{-F\psi_\beta/RT} & e^{-F\psi_d/RT} & OH^- & XO^- & XOH_2^+ & XOCd^+ \\
\Sigma H_2O & 1 & 0 & 0 & 0 & 0 & 0 & 0 & -1 & 0 & 0 & 0 \\
\Sigma H^+ & 0 & 1 & 0 & 0 & 0 & 0 & 0 & -1 & -1 & 1 & -1 \\
\Sigma Cd & 0 & 0 & 1 & 0 & 0 & 0 & 0 & 0 & 0 & 0 & 1 \\
\Sigma XOH & 0 & 0 & 0 & 1 & 0 & 0 & 0 & 0 & 1 & 1 & 1 \\
\Sigma e^{-F\psi_0/RT} & 0 & 0 & 0 & 0 & 0 & 0 & 0 & 0 & -1 & 1 & -1 \\
\Sigma e^{-F\psi_\beta/RT} & 0 & 0 & 0 & 0 & 0 & 0 & 0 & 0 & 0 & 0 & 2 \\
\Sigma e^{-F\psi_d/RT} & 0 & 0 & 0 & 0 & 0 & 0 & 0 & 0 & 0 & 0 & 0
\end{array}\right)
$$

and the concentration of the components (**u**):

$$
\mathbf{u}=\begin{pmatrix}
c_{H_2O}+c_{OH^-} \\
c_{H^+}-c_{OH^-}-c_{XO^-}+c_{XOH_2^+}+c_{XOCd^+} \\
c_{Cd^{2+}}+c_{XOCd^+} \\
c_{XOH}+c_{XO^-}+c_{XOH_2^+}+c_{XOCd^+} \\
-c_{XO^-}+c_{XOH_2^+}-c_{XOCd^+} \\
2c_{XOCd^+} \\
0
\end{pmatrix}
$$

The speciation problem can be solved using the four mass action laws of reactions R9 to R12 and the seven mass conservation equations of the components just defined.

The constant capacitance model has been introduced among others by Schindler *et al.* (1976). The diffuse double layer model has been fully described by Dzombak and Morel (1990) and applied to adsorption of metals on iron oxide surfaces. In these two models, the solid-water interface is composed by two layers: a layer of surface-bound complexes and a diffuse layer of counter ions in solution. All the sorbed ions are assigned to the unique surface layer, whose surface charge is calculated according to the addition of equations (3.31) and (3.32). Similarly to the triple-layer model, the relationship between the surface charge and the potential can be calculated from the Guy-Chapman theory described in equation (3.33). The calculation is very similar to that described for the triple layer model merging the primary species and components of surfaces 0 and β in only one surface.

3.3.2 *Cation exchange reactions*

Cation exchange takes place when an exchangeable cation located on the surface and interlayers of minerals is exchanged with a cation in the solution. The main difference with the surface complexation models previously described lies on the fact that no free sorption sites are assumed, and the capture of cations from the solution is always coupled with the release of cations from the surface. Moreover, since cations are exchanged, protonated and deprotonated surfaces play no role in exchange reactions, which are not *pH*-dependent, except for very low *pH* where protons can compete with the rest of cations for the exchange position.

In building the mass action law of exchange reactions, the activities of the ions in solution, are calculated according to any of the aqueous species formalisms (Debye-Hückel, Pitzer, etc.). However, there are different conventions to describe the non-ideal behavior of the ions attached to the exchangeable sites. These conventions become relevant when treating heterovalent exchange, i.e., one divalent cation is exchanged for two monovalent cations.

According to the *Gaines-Thomas convention*, the activity of an exchanged cation is expressed as a function of the total exchangeable cations. The expression for cation exchange reactions is:

$$\frac{1}{z_i}I^{z_i}+\frac{1}{z_j}(X_{z_j}-J)=\frac{1}{z_i}(X_{z_i}-I)+\frac{1}{z_j}J^{z_j}$$

where z is the electric charge of the cation. The expression $(X_z - I)$ indicates that each exchangeable cation occupies z exchangeable sites. According to this convention the activity of an exchangeable cation is equated to its equivalent fraction. Similar to activity in solid solutions, the standard state (activity equal 1), is an exchanger solid with all the sites occupied by the cation of interest:

$$a_{(X_{z_i}-I)}=\beta_I=\frac{[X_{z_i}-I]_{eq}}{\sum_{j=1}^{N_Y}[X_{z_j}-J]_{eq}}=\frac{z_i[X_{z_i}-I]_M}{CEC} \tag{3.40}$$

where $[X_{zi} - I]_{eq}$ and $[X_{zi} - I]_M$ are the number of equivalents and moles of the exchangeable cation per kg of water respectively; β_I is the equivalent fraction, CEC is the cation exchange capacity (in equivalents per kg of water), and N_Y the total number of exchangeable cations.

In the *Vanselow convention*, the activity of an exchanged cation is also expressed as a function of the total exchangeable cations. However, the activity of an exchangeable cation is equated to its molar fraction, χ:

$$a_{(X_{z_i}-I)}=\chi_{(X_{z_i}-I)}=\frac{[X_{z_i}-I]_M}{\sum_{j=1}^{N_Y}[X_{z_j}-J]_M} \tag{3.41}$$

According to the *Gapon convention*, the activity of an exchanged cation is expressed as a function of the total exchangeable sites. Then expression for cation exchange reactions is:

$$\frac{1}{z_i}I^{z_i}+(X-J_{\frac{1}{z_j}})=(X-I_{\frac{1}{z_j}})+\frac{1}{z_j}J^{z_j}$$

and the activity of an exchangeable cation is equated to its molar fraction, χ. Since the Gapon convention is based on single exchanger sites with charge –1, the molar and equivalent fractions are coincident:

$$a_{(X-I_{\frac{1}{z_j}})}=\chi_{(X-I_{\frac{1}{z_j}})}=\frac{[X-I_{\frac{1}{z_j}}]_M}{\sum_{j=1}^{N_Y}[X-J_{\frac{1}{z_j}}]_M}=\frac{[X-I_{\frac{1}{z_j}}]_{eq}}{CEC} \tag{3.42}$$

Since an atom of I cannot split into z_i parts, each of it occupying a single exchange site X, the Gapon equation may not have a direct molecular meaning. The expression $(X - I_{1/zi})$, however, reflects the chances of finding the cation I in an exchange site, which is coherent with the theory of regular mixtures (Appelo, 1994).

The speciation calculation of a system including cation exchange reactions is very similar to that explained for non-electrostatic surface complexation. The differences are that the empty sites (*X*OH) and protonation and deprotonation reactions are not needed. Another important difference is that an ideal behavior is assumed for the exchange sites in the solid phase, i.e., they do not depend on the composition of this phase. However, experiments show that this is not so, and therefore, *K* is not a true equilibrium constant and is currently called exchange coefficient or conditional equilibrium constant. Corrections to transform molar or equivalent fractions into thermodynamic activities are required, and are mostly based on principles similar to those used to describe solid-solution thermodynamics (Fletcher and Sposito, 1989).

We will illustrate the application of the cation exchange reactions with the Gaines-Thomas formulation with a small example with only one reaction of water dissociation in the aqueous phase:

$$
\begin{array}{lll}
\text{(R13)} & H_2O & \rightleftarrows H^+ + OH^- \\
\text{(R14)} & K^+ + X-Na & \rightleftarrows X-K + Na^+ \\
\text{(R15)} & \tfrac{1}{2}Ca^{2+} + X-Na & \rightleftarrows \tfrac{1}{2}X_2-Ca + Na^+
\end{array}
$$

There are nine species (H_2O, H^+, OH^-, K^+, $X-Na$, $X-K$, Na^+, Ca^{2+}, X_2-Ca), three reactions and, therefore, six components. The stoichiometric matrix is:

$$
\mathbf{S} = \begin{pmatrix}
 & H_2O & H^+ & OH^- & K^+ & X\text{-Na} & X\text{-K} & Na^+ & Ca^{2+} & X_2\text{-Ca} \\
R13 & -1 & 1 & 1 & 0 & 0 & 0 & 0 & 0 & 0 \\
R14 & 0 & 0 & 0 & -1 & -1 & 1 & 1 & 0 & 0 \\
R15 & 0 & 0 & 0 & 0 & -1 & 0 & 1 & -\frac{1}{2} & \frac{1}{2}
\end{pmatrix}
$$

If we choose H^+, $X-K$ and X_2-Ca to be secondary and the rest primary, the component matrix will be:

$$
\mathbf{U} = \begin{pmatrix}
 & H_20 & OH^- & K^+ & X\text{-Na} & Na^+ & Ca^{2+} & H^+ & X\text{-K} & X_2\text{-Ca} \\
\Sigma H_20 & 1 & 0 & 0 & 0 & 0 & 0 & 1 & 0 & 0 \\
\Sigma OH^- & 0 & 1 & 0 & 0 & 0 & 0 & -1 & 0 & 0 \\
\Sigma K & 0 & 0 & 1 & 0 & 0 & 0 & 0 & 1 & 0 \\
\Sigma X & 0 & 0 & 0 & 1 & 0 & 0 & 0 & 1 & 2 \\
\Sigma Na & 0 & 0 & 0 & 0 & 1 & 0 & 0 & -1 & -2 \\
\Sigma Ca & 0 & 0 & 0 & 0 & 0 & 1 & 0 & 0 & 1
\end{pmatrix}
$$

and the concentration of the components (**u**):

$$
\mathbf{u} = \begin{pmatrix}
c_{H_2O} + c_{H^+} \\
c_{OH^-} - c_{H^+} \\
c_{K^+} + c_{X\text{-K}} \\
c_{X\text{-Na}} + c_{X\text{-K}} + 2c_{X_2\text{-Ca}} \\
c_{Na^+} - c_{X\text{-K}} - 2c_{X_2\text{-Ca}} \\
c_{Ca^{2+}} + c_{X_2\text{-Ca}}
\end{pmatrix}
$$

The three mass action laws of reactions R13 to R15, and the six mass conservation equations of the components (with the exception of c_{H_2O} and c_{H^+} discussed above), allow to solving the speciation problem. The values of log *K* of the reactions, *pH* and *CEC* (u_4) have to be known to solve the problem. With respect to the rest of elements (Na, K and Ca), there are two possible speciation cases, depending on what is known: (i) the total inventories (u_3, u_5 and u_6) or (ii) the concentrations in solution (K^+, Na^+ and Ca^{2+}).

3.3.3 *Reactions with a solid phase*

For solid/aqueous reactions, the reactive species belong to different phases. In building the mass action law (3.7), the activities of the ions in solution, are calculated as described above. In most geochemical calculations, minerals are assumed to be pure phases, and therefore, their activity is one. However, for a general case the activity of species in a solid phase is calculated according to:

$$a_i = X_i \lambda_i \tag{3.43}$$

where X_i is the molar fraction of the i-th species in the solid phase, and λ_i its activity coefficient accounting for the non-ideality of the solid solution. The reference state would be a pure solid phase made up of species i. X_i is calculated from chemical analysis, and λ_i is calculated through a thermodynamic model. For example, for a binary solid solution described by a two parameter Margules function (Anderson, 2005, p. 311):

$$\begin{aligned} RT \ln \lambda_1 &= (2w_{G2} - w_{G1})X_2^2 + 2(w_{G1} - w_{G2})X_2^3 \\ RT \ln \lambda_2 &= (2w_{G1} - w_{G2})X_1^2 + 2(w_{G2} - w_{G1})X_1^3 \end{aligned} \tag{3.44}$$

where w_{G1} and w_{G2} are parameters depending on P and T. For in depth descriptions of solid-solution aqueous-solution equilibrium calculations, see Glynn (1991, 2000).

3.3.4 *Reactions with a gas phase*

The activity of a species in a gas is calculated as a ratio between the fugacity of the species in the gas and the fugacity of the species in a reference state (Anderson, 2005, p. 206). The reference state is arbitrary. The most common reference state is the pure ideal gas species at 1 bar and 283.15 K. In ideal gas mixtures, fugacity is equivalent to partial pressure, which in turn can be calculated as the product of the total pressure and the molar fraction of the species in the gas phase. However, in non-ideal mixtures, fugacity requires a correction of the partial pressure accounting for the interaction of the gas particles. Then, the activity of species in a gas phase can be calculated according to:

$$a_i = \frac{X_i P \Gamma_i}{1} \tag{3.45}$$

where X_i is the molar fraction of the i-th species in the gas phase, P is the total pressure and Γ_i its fugacity coefficient accounting for the non-ideality of the gas mixture. The one in the denominator is just to remind that activity is dimensionless. X_i is calculated from chemical analysis, and Γ_i is calculated from the equation of state of the gas phase.

For example, for non-ideal mixtures of H_2O and CO_2, described by a Redlich-Kwong equation of state, the fugacity coefficient of CO_2 can be calculated as (Spycher *et al.*, 2003):

$$\begin{aligned} \ln \Gamma_{CO_2}^{mix} = {} & \ln\left(\frac{v}{v - b_{mix}}\right) + \left(\frac{b_{CO_2}}{v - b_{mix}}\right) - \frac{2(\chi_{H_2O} a_{H_2O-CO_2} + \chi_{CO_2} a_{CO_2})}{RT^{1.5} b_{mix}} \ln\left(\frac{v + b_{mix}}{v}\right) \\ & + \frac{a_{mix} b_{CO_2}}{RT^{1.5} b_{mix}^2}\left[\ln\left(\frac{v+b}{v}\right) + \frac{b_{mix}}{v + b_{mix}}\right] - \ln\left(\frac{pv}{RT}\right) \end{aligned} \tag{3.46}$$

where v is the molar volume, and a and b are parameters of the equation of state.

The important point in equations (3.43) and (3.45) is that the value of λ_l and Γ_I depend in turn on the concentration of the species in the solid and gas phases (X_i), which are unknowns in the problems. Therefore, as already mentioned for γ_i in the aqueous speciation problem, the value of λ_i and Γ_i are initially estimated and are updated in an iterative process (see section 3.5).

We will illustrate the application of the reactions with a solid and a gas phase with the equilibrium of water with Mg-calcite and the atmospheric CO_2:

(R16) $H_2O \rightleftarrows H^+ + OH^-$
(R17) $CO_2(aq) + H_2O \rightleftarrows HCO_3^- + H^+$
(R18) $HCO_3^- \rightleftarrows CO_3^{2-} + H^+$
(R19) $CaHCO_3^+ \rightleftarrows HCO_3^- + Ca^{2+}$
(R20) $MgHCO_3^+ \rightleftarrows HCO_3^- + Mg^{2+}$
(R21) $CO_2(g) \rightleftarrows CO_2(aq)$
(R22) $CaCO_3(s) \rightleftarrows CO_3^{2-} + Ca^{2+}$
(R23) $MgCO_3(s) \rightleftarrows CO_3^{2-} + Mg^{2+}$

There are thirteen species (H_2O, H^+, OH^-, $CO_2(aq)$, HCO_3^-, CO_3^{2-}, $CaHCO_3^+$, $MgHCO_3^+$, Ca^{2+}, Mg^{2+}, $CO_2(g)$, $CaCO_3(s)$, $MgCO_3(s)$), eight reactions, and, therefore, five components. The stoichiometric matrix is:

S =

	H_2O	H^+	OH^-	CO_2 (aq)	HCO_3^-	CO_3^{2-}	$CaHCO_3^-$	$MgHCO_3^-$	Ca^{2+}	Mg^{2+}	CO_2 (g)	$CaCO_3$ (s)	$MgCO_3$ (s)
R16	−1	1	1	0	0	0	0	0	0	0	0	0	0
R17	−1	1	0	−1	1	0	0	0	0	0	0	0	0
R18	0	1	0	0	−1	1	0	0	0	0	0	0	0
R19	0	0	0	0	1	0	−1	0	1	0	0	0	0
R20	0	0	0	0	1	0	0	−1	0	1	0	0	0
R21	0	0	0	1	0	0	0	0	0	0	−1	0	0
R22	0	0	0	0	0	1	0	0	1	0	0	−1	0
R23	0	0	0	0	0	1	0	0	0	1	0	0	−1

If we select H_2O, H^+, $CO_2(aq)$, Ca^{2+} and Mg^{2+} to be primary and the rest secondary, the vector of concentration of the components (**u**) is:

$$\mathbf{u} = \begin{pmatrix} c_{H_2O} + c_{OH^-} + c_{HCO_3^-} + c_{CO_3^{2-}} + c_{CaHCO_3^+} + c_{MgHCO_3^+} + c_{CaCO_3(s)} + c_{MgCO_3(s)} \\ c_{H^+} - c_{OH^-} - c_{HCO_3^-} - 2c_{CO_3^{2-}} - c_{CaHCO_3^+} - c_{MgHCO_3^+} - 2c_{CaCO_3(s)} - 2c_{MgCO_3(s)} \\ c_{CO_2(aq)} + c_{HCO_3^-} + c_{CO_3^{2-}} + c_{CaHCO_3^+} + c_{MgHCO_3^+} + c_{CO_2(g)} + c_{CaCO_3(s)} + c_{MgCO_3(s)} \\ c_{Ca^{2+}} + c_{CaHCO_3^+} + c_{CaCO_3(s)} \\ c_{Mg^{2+}} + c_{MgHCO_3^+} + c_{MgCO_3(s)} \end{pmatrix}$$

The eight mass action laws of reactions R16 to R23, and the five mass conservation equations of the components, allow solving the speciation problem. The values of log K of the reactions, as well as c_{H_2O} and c_{H^+} (1000/18 and 10^{-pH}, respectively) are usually known. For the rest of equations, total inorganic carbon (u_3), and total Ca and Mg (u_4 and u_5) are also analytical data.

Another common speciation problem is the case where the amount of solute is not known, but the solution is assumed in equilibrium with a solid phase or with a given partial pressure of gas. In the case of equilibrium with the Mg-calcite, the mass conservation equation of two components (u_3, u_4 or u_5) cannot be used, and the mass amount of species in the solid phase ($c_{CaCO_3(s)}$ and $c_{MgCO_3(s)}$) cannot be calculated. We have removed two equations and two unknowns, and the system can be solved. The mass action equations are still valid, since equilibrium was assumed between the solution and the solid phase. In the case of equilibrium with a partial pressure of CO_2, then one mass conservation equation can be removed (the one for which u_3, u_4 or u_5 is not known), and the value of $c_{CO_2(g)}$ can be calculated from the partial pressure.

3.4 REACTION PATHS

In the previous sections 3.1 to 3.3, we have described how to calculate the concentration of species in a geochemical system in equilibrium. We are now concerned about processes or reaction paths: the successive equilibrium states of the geochemical system caused by an external perturbation. Since we have already discussed how to calculate the speciation at equilibrium, the problem is how to introduce the external perturbations. These can be of different types. There are three main types of perturbations to a system: variations in temperature and pressure, variations in the activity of a species, and/or mass transfers to/from the system.

From the modeling point of view, variations in temperature (and/or pressure), only affect the value of the equilibrium constant K in the mass action equations (3.7), and do not affect the mass balance of components (3.17). This means that changes in temperature may change the concentration of the different species of the system, but not the total amount of components. The temperature variation is introduced externally to the system:

$$T = T^0 + \xi(T^f - T^0) \tag{3.47}$$

where T^0 and T^f are the initial and final temperatures, and ξ is a function which varies from zero to one and describes the reaction progress.

Rather than emulate a natural process, the evolution of a system after the variation of the activity of a particular species, such as *pH* or *pe* is commonly used to show the predominance *pH-pe* field of the species. Then, the modification is only applied to the activity of one species:

$$a_i = a_i^0 + \xi a_i^r \tag{3.48}$$

where ξ is a function which varies from zero to one and describes how the addition takes place. As an example, the increment of the $a_{CO_2(g)}$ to twice that of the previous example (through the increment in the partial pressure of CO_2), in ten steps, will modify the third element of the vector of components, and for the first step will be:

$$a_{CO_2(g)} = a^0_{CO_2(g)} + 0.1 a^0_{CO_2(g)}$$

Then, according to reaction R21, the amount of $c_{CO_2(aq)}$ will increase, but the variation of the rest of species, including those in the solid phase, cannot be predicted in a straight forwards manner and requires speciation calculations.

In the general case, any mass transfer to/from the system is always applied to the mass of the components. We have selected several cases: addition/removal of reactants, water mixing and evaporation. The addition/removal of a reactant is calculated by adding the vector of reactant $\mathbf{u}^r$ to the initial vector of components $\mathbf{u}^0$:

$$\mathbf{u} = \mathbf{u}^0 + \xi \mathbf{u}^r \tag{3.49}$$

where ξ is a function which varies from zero to one. As an example, the addition of m mol kg^{-1} water of dolomite in 50 steps to the previous example will modify the five elements of the vector of components according to the stoichiometry:

$$CaMg(CO_3)_2 \rightleftarrows Ca^{2+} + Mg^{2+} + 2CO_2(aq) + 2H_2O - 4H^+$$

and for the first step will be:

$$\begin{pmatrix} u_1 \\ u_2 \\ u_3 \\ u_4 \\ u_5 \end{pmatrix} = \begin{pmatrix} u_1^0 \\ u_2^0 \\ u_3^0 \\ u_4^0 \\ u_5^0 \end{pmatrix} + \frac{m}{50}\begin{pmatrix} 2 \\ -4 \\ 2 \\ 1 \\ 1 \end{pmatrix}$$

Water mixing can be calculated as a weighted addition of the component vector of the end-member waters:

$$\mathbf{u} = \sum \chi_i \mathbf{u}_i \tag{3.50}$$

where χ_i is the mixing ratio of end members $(\sum \chi_i = 1)$. A relevant case of water mixing is to calculate the pH of a mixture. The first step is to calculate the value of the component (u_2 in the example of the dolomite problem) of the end-members by solving the speciation problem, then calculate the mixed component through equation (3.50), and finally, calculate the pH value ($c_{H^+}\gamma_{H^+}$) of the mixture by solving a new speciation problem of this mixture water.

The water mixing formulation also holds for the acid/base titration. As an example, the addition of 25 mL of a solution 0.1 m HCl to 975 mL of the solution of the Mg-calcite problem will be calculated as:

$$\begin{pmatrix} u_1 \\ u_2 \\ u_3 \\ u_4 \\ u_5 \end{pmatrix} = 0.975\begin{pmatrix} u_1^0 \\ u_2^0 \\ u_3^0 \\ u_4^0 \\ u_5^0 \end{pmatrix} + 0.025\begin{pmatrix} 0 \\ 0.1 \\ 0 \\ 0 \\ 0 \end{pmatrix}$$

where the value of u_2 of the acid solution ($u_2 = c_{H^+} - c_{OH^-}$) has been assumed as c_{H^+}, and a sixth component accounting for Cl has not been added.

It should be evident that, unlike components, species concentration will not vary linearly with the mixing ratio because they are affected by equilibrium reactions. In fact, it may be of interest to compute the reaction rate (i.e., the number of moles of reactants that have evolved into products). These result from the mass balance of the secondary species associated to each reaction. To illustrate these concepts, let us consider the mixing of two waters, A and B, both in equilibrium with anhydrite in example 1a. Let χ be the mixing ratio, then:

$$u^M = \chi u^A + (1-\chi)u^B \tag{3.51}$$

where the super-index identifies the mixture. As it is shown Chapter 4 (section 4.5.4), in a binary system such as example 1a, the secondary species concentration is:

$$c_2^M = \frac{u^M + \sqrt{(u^M)^2 + 4k}}{2} \tag{3.52}$$

It is not difficult to check that c_2^M is smaller than that corresponding to non-reactive mixing

$$c_2^{NR} = \chi c_2^A + (1-\chi)c_2^B \tag{3.53}$$

so that some anhydrite $(c_2^{NR} - c_2^M)$ must have precipitated. The evolution is evident in Figure 3.8, where it is apparent that non-reactive mixing leads to waters that are supersaturated with respect to anhydrite. This effect is termed algebraic, because it is the non-linear nature of equilibrium (algebraic equations) what causes the difference. In general, however, there may be other causes of mixing disequilibrium: differences in temperature (which causes variations in equilibrium constants), in ionic strength (which causes variations in activity coefficients, or in *pH* and/or *pe* (which may cause variations in speciation). As a result, mixing of two aqueous solutions may lead to super or sub saturation with respect to minerals that were at equilibrium with the end-member solutions resulting in interesting geological features (Plummer, 1975; Corbella *et al.*, 2004).

Evaporation is a third interesting case of mass transfer. There, the removal of solvent can be simulated simply by increasing the other components. Geochemists typically simulate evaporation in a step-wise manner, so that for a particular step, *n*, of evaporation:

$$\mathbf{u}_{n+1} = \xi \mathbf{u}_n \tag{3.54}$$

where ξ is a factor higher than one that can be calculated as:

$$\log \xi = \frac{1}{N} \log C \tag{3.55}$$

where N is the number of steps of evaporation arbitrarily fixed and C is the final concentration factor of the solution ($u_N = C\,u_0$).

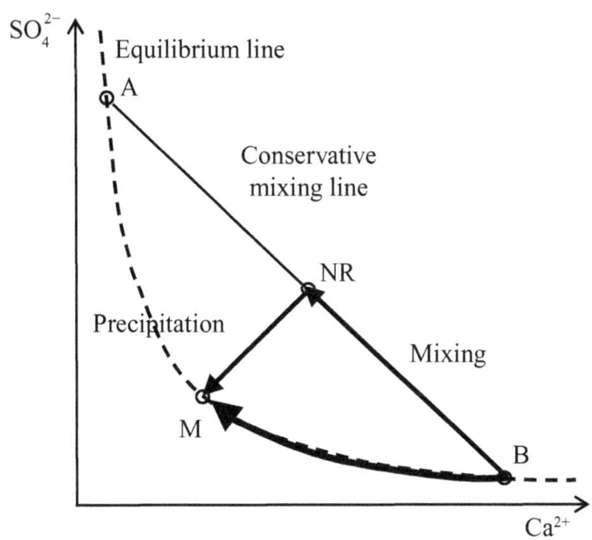

Figure 3.8. Ideal mixing of two end members (A and B) that are in equilibrium with anhydrite. If mixing is non-reactive, the resulting mixture will be oversaturated (point NR). Returning to equilibrium (point M) requires precipitation (algebraic effect).

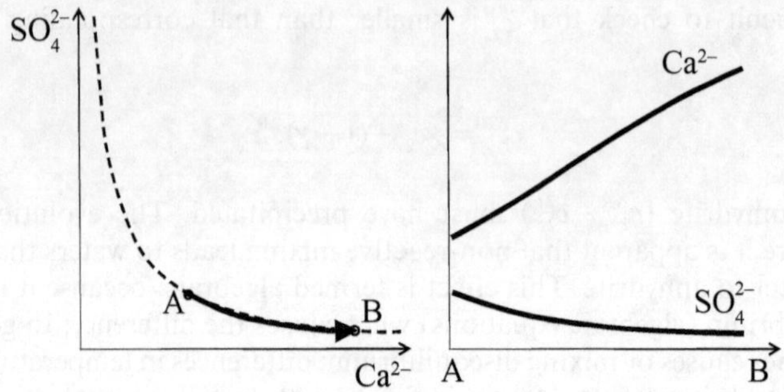

Figure 3.9. Ideal evaporation of water A. The solution evolves along the anhydrite equilibrium line, leading to a "geochemical divide" path (the largest concentration, Ca^{2+}, tends to increase, whereas the smallest, SO_4^{2-}, tends to decrease). This occurs because anhydrite removal predominantly affects the minor species, SO_4^{2-}, while evaporative concentration predominantly affects the major species, Ca^{2+}.

In mass balance based formulations (i.e., reactive transport or evaporation brines), the governing equations are:

$$\frac{dm_w}{dt} = -E \tag{3.56}$$

$$\frac{dm_w u}{dt} = 0 \tag{3.57}$$

where m_w is the mass of water, E is the evaporation rate and u is any (conservative) component. The solution of these equations is:

$$m_w(t) = m_w^o - Et \tag{3.58}$$

$$u = \frac{m_w^o u^o}{m_w^o - Et} \tag{3.59}$$

where m_w^0 and $u^0$are the initial mass and concentration of solute. Through speciation calculations one can obtain the concentrations of all participating species. Figure 3.9 illustrates the case for example 1a, where it is apparent that u (difference between SO_4^{2-} and Ca^{2+}) increases with time, causing the species concentrations to become increasingly apart (along a geochemical divide path: the initially largest species increases whereas the smallest one decreases).

3.5 FORMULATION OF KINETIC REACTIONS

Thermodynamic equilibrium only defines the minimum energy state of the system, but says nothing about the progress of the system towards such a state. Some geochemical processes (such as redox reactions or dissolution/precipitation of minerals) are known to progress so slowly that they may never reach equilibrium. Kinetics offers a way to analyze the chemical evolution of a reaction and allows incorporating explicitly the time variable. Unlike the general formulation of reactions offered by thermodynamic theory, rate laws are dependent on the specific species and on the mechanistic processes involved in each reaction.

Kinetic reactions are characterized by their reaction rate ($\mathbf{r}_k$), which is defined as the amount of reactants evolving to products per unit time. Here, we simply state that the reaction rate is a function of the concentration of all species:

$$\mathbf{r}_k = \mathbf{r}_k(\mathbf{c}) \tag{3.60}$$

The expression of equation (3.60) can be simple or complex depending on the phenomenon represented. The simplest case is that where r_k is constant, i.e., is not dependent on the concentration of species (this is termed a zero-order rate law). Another simple case occurs when there is a linear dependence of the rate on concentration (this is a first-order rate law):

$$\mathbf{r}_k = \mathbf{k} \cdot \mathbf{c} \tag{3.61}$$

where **k** is a vector of constant values. An example of this type of kinetics is the oxidation of Fe(II) to Fe(III) under atmospheric conditions (Singer and Stumm, 1970):

$$r_{\mathrm{Fe(II)}} = -\frac{d[\mathrm{Fe(II)}]}{dt} = k[\mathrm{Fe(II)}] \tag{3.62}$$

where k also integrates the atmospheric oxygen partial pressure.

More complex kinetic expressions describe microorganism-mediated reactions, such as the Monod-type expressions for the oxidation of organic matter due to a successive set of electron acceptors, each one inhibiting the action of other, less thermodynamically efficient, electron-acceptors (Wang and Van Cappellen, 1996):

$$r_i = k_{cat}\, c_i \frac{c_{cat}}{\kappa_{cat} + c_{cat}} \prod_{j=1}^{N_C-2} \frac{\kappa_j^{in}}{\kappa_j^{in} + c_j} \tag{3.63}$$

where the sub-index i represents organic matter, the sub-index cat represents the catalyst or active electron acceptor, and the sub-index j the rest of electron acceptor energetically favored acting as inhibitors, and κ represents the concentration of catalyst or inhibitor occurring when the reaction rate is at a maximum (Fig. 3.10).

Thus, for example, the oxidation of organic matter by nitrate would be represented as:

$$r_{\mathrm{CH_2O}} = k_{\mathrm{NO_3}} c_{\mathrm{CH_2O}} \frac{c_{\mathrm{NO_3}}}{\kappa_{\mathrm{NO_3}} + c_{\mathrm{NO_3}}} \frac{\kappa_{\mathrm{O_2}}^{in}}{\kappa_{\mathrm{O_2}}^{in} + c_{\mathrm{O_2}}}$$

where dissolved O_2 acts as an inhibitor. For high O_2 concentrations, the last term of equation (3.63) is low and oxidation by nitrate is inhibited. For low O_2 concentrations, the value of CO_2 is negligible with respect to $\kappa_{O_2}^{in}$, the last term value is close to 1, and NO_3 acts as an oxidant. For low NO_3 concentrations, the second fraction takes a value lower than 1, indicating that NO_3 is the limiting factor for the reaction to progress. For high values of NO_3, the second fraction value is close to 1, and the rate depends linearly on the organic matter concentration, as in the first order discussed above.

Another complex expression describes mineral dissolution rates (Steefel and Lasaga, 1994):

$$r_k = \sigma_k \zeta_k e^{-\frac{E_{a,k}}{RT}} \sum_{j=1}^{N_k} k_{kj} \prod_{i=1}^{N_c+N_x} a_i^{p_{kji}} \left(\Omega_k^{\theta_{kj}} - 1\right)^{\eta_{kj}} \tag{3.64}$$

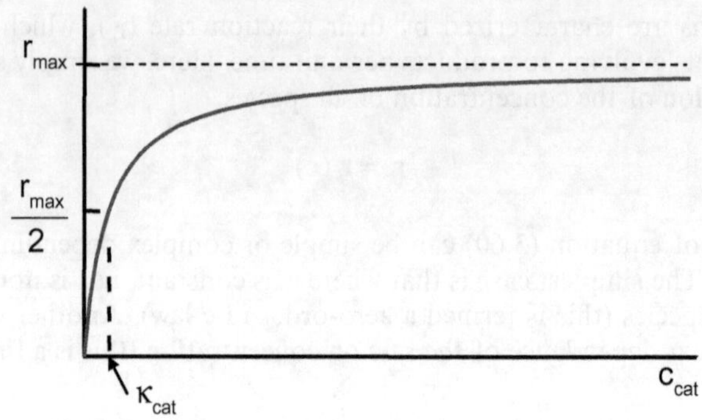

Figure 3.10. Variation of the reaction rate with the concentration of catalyst, according to the Monod kinetic expression.

where r_k is the mineral dissolution rate (moles of mineral per unit of volume and unit time), $E_{a,k}$ is the apparent activation energy of the overall reaction process; N_k is the number of terms of the experimental expression, and k_{mk} an experimental constant; $a_i^{p_{kji}}$ accounts for the catalytic effect of some species (particularly of H^+), where the value of p_i is determined experimentally; Ω_m is saturation; the parameters θ and η must be determined from experiments; usually, but not always, they are commonly assumed equal to 1. The term inside the parenthesis, called the far-from-equilibrium function), decreases the reaction rate in a non-linear way, as the solution approaches to equilibrium. Factor ζ_k takes on values of +1 or −1 depending on whether Ω_k is larger or smaller than 1 (precipitation or dissolution), respectively. At equilibrium, $\Omega_k = 1$ and therefore $r_k = 0$. σ_k is the reactive surface of the mineral per unit of volume.

For example, according to data compiled by Blum and Stillings (1995), the variation of the dissolution rate of albite with *pH*, can be represented by the addition of three terms:

$$r_k = \sigma_k \left[7.5 \cdot 10^{-10} \cdot (a_{H^+})^{0.5} + 1.0 \cdot 10^{-12} + 1.2 \cdot 10^{-14} \cdot (a_{H^+})^{-0.3}\right] (\Omega_k - 1)$$

For a reactive surface of 1 $m^2\,L^{-1}$, the dissolution rate is represented in Figure 3.11.

From the point of view of their governing equations, the stoichiometry of slow (kinetic) reactions is contained in the kinetic stoichiometric matrix, $\mathbf{S}_k$, of dimension $n_k \times n_s$ (n_k being the number of kinetic reactions, and n_s the number of species). As discussed above, there are a number of n_e equilibrium reactions among the species, which are described by the equilibrium stoichiometric matrix, $\mathbf{S}_e$. All this allows one to write the global stoichiometric matrix of the chemical system as:

$$\mathbf{S} = \begin{pmatrix} \mathbf{S}_e \\ \mathbf{S}_k \end{pmatrix} \tag{3.65}$$

The above matrices are illustrated by means of a simple example consisting of three equilibrium reactions:

(R24) $FeSO_4^\circ \rightleftarrows Fe^{2+} + SO_4^{2-}$
(R25) $H_2O \rightleftarrows H^+ + OH^-$
(R26) $Fe(OH)_3(s) \rightleftarrows Fe^{3+} + 3OH^-$

and two kinetic reactions:

(R27) $Fe^{3+} \rightleftarrows Fe^{2+} + H^+ - 0.5H_2O + 0.25O_2(aq)$
(R28) $FeS_2(s) \rightleftarrows Fe^{2+} + 2SO_4^{2-} + 2H^+ - H_2O - 3.5O_2(aq)$

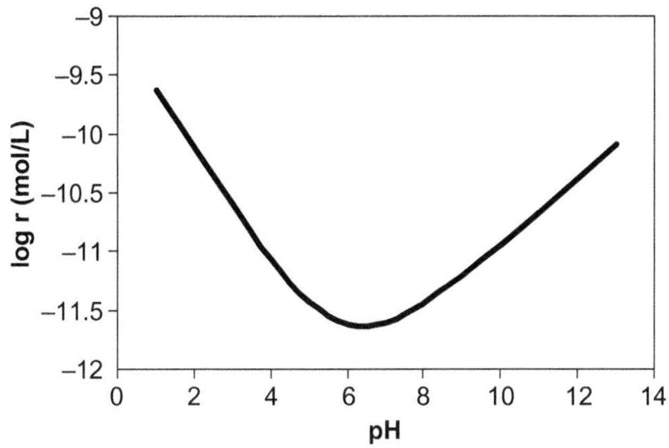

Figure 3.11. Representation of a dissolution expression for albite according to Blum and Stillings (1995).

The global stoichiometric matrix **S**, which includes $\mathbf{S}_e$ and $\mathbf{S}_k$, is:

$$\mathbf{S}=\left(\begin{array}{c|cccccccccc} & H^+ & OH^- & O_2(aq) & SO_4^{2-} & Fe^{2+} & Fe^{3+} & FeS_2(s) & FeSO_4^0 & H_2O & Fe(OH)_3(s) \\ FeSO_4^o & 0 & 0 & 0 & 1 & 1 & 0 & 0 & -1 & 0 & 0 \\ H_2O & 1 & 1 & 0 & 0 & 0 & 0 & 0 & 0 & -1 & 0 \\ Fe(OH)_3(s) & 0 & 3 & 0 & 0 & 0 & 1 & 0 & 0 & 0 & -1 \\ Fe^{3+} & 0.5 & -0.5 & 0.25 & 0 & 1 & -1 & 0 & 0 & 0 & 0 \\ FeS_2(s) & 1 & -1 & -3.5 & 2 & 1 & 0 & -1 & 0 & 0 & 0 \end{array}\right)$$

Then, a component matrix **U**, which allows to build the mass balance equation is required to solve the system of equations. We postulate that the new component matrix is the product of two matrices:

$$\mathbf{U}=\mathbf{U}_k\,\mathbf{U}_e \tag{3.66}$$

where $\mathbf{U}_e$ and $\mathbf{U}_k$ are the matrices describing the components for equilibrium and kinetic reactions, respectively.

For equilibrium reactions, it is possible to write the activities of n_e species (denoted secondary), in terms of the activities of the remaining $n_s - n_e$ primary species (see eqs. 3.8 and 3.19 as example). The procedure to obtain the matrix of equilibrium components is similar to that mentioned in equation (3.15):

$$\mathbf{U}_e=\left(\mathbf{I}\middle|-\mathbf{S}_{e1}^T\left(\mathbf{S}_{e2}^T\right)^{-1}\right)=\left(\mathbf{I}\,|\,(\mathbf{S}_{e1}^*)^T\right) \tag{3.67}$$

where $\mathbf{I}\,(n_s-n_e\times n_s-n_e)$ and $(\mathbf{S}_{e1}^*)^T\,(n_s-n_e-n_k\times n_e)$ are submatrices of U_e.

It is obvious that **U** defines the components that are conservative in equilibrium reactions, since it fulfills the condition of equation (3.13):

$$\mathbf{U}\mathbf{S}_e^T=\mathbf{U}_k\,\mathbf{U}_e\mathbf{S}_e^T=\mathbf{U}_k\,0=0 \tag{3.68}$$

The kinetic reactions are assumed to be linearly independent of each other. This assumption, however, should not be considered restrictive. Considering linearly dependent kinetic reactions would only modify the linearly independent reaction rates by adding to them the linearly dependent rates weighted by the coefficients that define such linear dependence (Lichtner, 1996).

Therefore, since the kinetic reactions are independent from equilibrium reactions, one can choose some of the vectors of the equilibrium basis to belong to the intersection between the equilibrium and kinetic constraints (hence, of dimension $n_s - n_e - n_k$). These vectors indicate those components that remain unaltered by either equilibrium or kinetic reactions. In short, such components are conservative. In terms of **U,** this means that it can be defined such that:

$$\mathbf{U}\mathbf{S}_k^T = \begin{pmatrix} \mathbf{U}_c \\ \mathbf{U}_r \end{pmatrix} \mathbf{S}_k^T = \begin{pmatrix} 0 \\ \mathbf{U}_r\mathbf{S}_k^T \end{pmatrix} \tag{3.69}$$

where $\mathbf{U}_c$ $(n_s - n_e - n_k \times n_s)$ and $\mathbf{U}_r$ $(n_k \times n_s)$ are submatrices of U, where sub-indices c and r stand for conservative and reactive (kinetically non-conservative), respectively.

On the other hand, the kinetic component matrix, $\mathbf{U}_k$, is built as follows:

$$\mathbf{U}_k = \begin{pmatrix} \mathbf{I}_1 & \mathbf{S}_{k1}^T \\ 0 & \mathbf{I}_2 \end{pmatrix} \tag{3.70}$$

where $\mathbf{I}_1$ $(n_s - n_e - n_k \times n_s - n_e - n_k)$, $\mathbf{S}_{k1}^T$ $(n_s - n_e - n_k \times n_k)$ and $\mathbf{I}_2$ $(n_k \times n_k)$ are submatrices of $\mathbf{U}_k$ $(n_s - n_e \times n_s - n_e)$. We can see that the product of the component matrix **U** times $\mathbf{U}_k$ fulfils the equation (3.69), which we split between conservative and reactive components:

$$\mathbf{U}\cdot\mathbf{S}_k^T = \mathbf{U}_k\cdot\mathbf{U}_e\cdot\mathbf{S}_k^T = \mathbf{U}_k\cdot\begin{pmatrix}\mathbf{I} & \mathbf{S}_{e1}^T\end{pmatrix}\begin{pmatrix}\mathbf{S}_{k1}^T \\ -\mathbf{I} \\ \mathbf{0}\end{pmatrix} = \mathbf{U}_k\begin{pmatrix}\mathbf{S}_{k1}^T \\ -\mathbf{I}_1\end{pmatrix} = \begin{pmatrix}\mathbf{I}_1 & \mathbf{S}_{k1}^T \\ \mathbf{0} & \mathbf{I}_2\end{pmatrix}\cdot\begin{pmatrix}\mathbf{S}_{k1}^T \\ -\mathbf{I}_1\end{pmatrix} = \begin{pmatrix}\mathbf{0} \\ -\mathbf{I}_1\end{pmatrix} \tag{3.71}$$

These definitions allow us to classify components according to the types of species defining them:

- **conservative**: their mass is invariant through chemical reactions (i.e., the top $n_s - n_e - n_k$ components in (3.70) or (3.71)). They can be mobile conservative, if all their species are aqueous, and mixed, if some of their species are immobile.
- **non-conservative**: their mass can vary through chemical reactions. The can be mobile kinetic, if all their species are aqueous; immobile, if all their species are immobile; and mixed.

As we shall see in the reactive transport section of Chapter 4, the this classification is useful because it can allow the transport of conservative components to be decoupled from the others.

To sum up, in the example, there are 10 species and 3 equilibrium reactions, therefore 7 components. We can select five aqueous species, H^+, OH^-, $O_2(aq)$, SO_4^{2-} and Fe^{2+}, and the two kinetic species Fe^{3+} and FeS_2 to complete the seven primary species. The rest of species, $FeSO_4^0$, H_2O and $Fe(OH)_3(s)$ can be expressed as a function of the 7 species selected as primary. According to equation (3.67), the component matrix will be:

$\mathbf{U}_e =$

	H^+	OH^-	O_2	SO_4^{2-}	Fe^{2+}	Fe^{3+}	$FeS_2(s)$	$FeSO_4^0$	H_2O	$Fe(OH)_3(s)$
ΣH^+	1	0	0	0	0	0	0	0	1	0
ΣOH^-	0	1	0	0	0	0	0	0	1	3
ΣO_2	0	0	1	0	0	0	0	0	0	0
ΣSO_4^{2-}	0	0	0	1	0	0	0	1	0	0
Fe(II)	0	0	0	0	1	0	0	1	0	0
Fe(III)	0	0	0	0	0	1	0	0	0	1
$FeS_2(s)$	0	0	0	0	0	0	1	0	0	0

Observe that each row is a component and each column is a species. We label each row with the total concentration obtained from each component.

$\mathbf{U}_k$ is defined by means of equation (3.70) as follows:

$$\mathbf{U}_k = \begin{pmatrix} & H^+ & OH^- & O_2 & SO_4^{2-} & Fe^{2+} & Fe^{3+} & FeS_2(s) \\ \Sigma H^+ & 1 & 0 & 0 & 0 & 0 & 0.5 & 1 \\ \Sigma OH^- & 0 & 1 & 0 & 0 & 0 & -0.5 & -1 \\ \Sigma O_2 & 0 & 0 & 1 & 0 & 0 & 0.25 & -3.5 \\ \Sigma SO_4^{2-} & 0 & 0 & 0 & 1 & 0 & 0 & 2 \\ Fe_T & 0 & 0 & 0 & 0 & 1 & 1 & 1 \\ Fe(III) & 0 & 0 & 0 & 0 & 0 & 1 & 0 \\ FeS_2(s) & 0 & 0 & 0 & 0 & 0 & 0 & 1 \end{pmatrix}$$

The product matrix is:

$$\mathbf{U} = \mathbf{U}_k \cdot \mathbf{U}_e = \begin{pmatrix} & H^+ & OH^- & O_2 & SO_4^{2-} & Fe^{2+} & Fe^{3+} & FeS_2(s) & FeSO_4 & H_2O & Fe(OH)_3(s) \\ \Sigma H^+ & 1 & 0 & 0 & 0 & 0 & 0.5 & 1 & 0 & 1 & 0.5 \\ \Sigma OH^- & 0 & 1 & 0 & 0 & 0 & -0.5 & -1 & 0 & 1 & 2.5 \\ \Sigma O_2 & 0 & 0 & 1 & 0 & 0 & 0.25 & 3.5 & 0 & 0 & 0.25 \\ \Sigma SO_4^{2-} & 0 & 0 & 0 & 1 & 0 & 0 & 2 & 1 & 0 & 0 \\ Fe_T & 0 & 0 & 0 & 0 & 1 & 1 & 1 & 1 & 0 & 1 \\ Fe(III) & 0 & 0 & 0 & 0 & 0 & 1 & 0 & 0 & 0 & 1 \\ FeS_2(s) & 0 & 0 & 0 & 0 & 0 & 0 & 1 & 0 & 0 & 0 \end{pmatrix}$$

In the definition of an equilibrium component (matrix $\mathbf{U}_e$), we include naturally only those species that are in equilibrium with the primary species. For Fe(II) these species are Fe^{2+} (primary) and $FeSO_4^0$ (secondary). In matrix $\mathbf{U}_e$, the component related to primary species Fe^{2+} was the ferrous iron in the system, Fe(II), and that related to Fe^{3+} was the ferric iron, Fe(III). Since a kinetic reaction takes place between Fe^{2+} and Fe^{3+}, neither Fe(II) nor Fe(III) are conservative components. If we take into account the kinetic reaction in building the component matrix **U** (i.e., by using matrix $\mathbf{U}_k$) the new component whose primary species is Fe^{2+} is total iron, including both Fe(II) and Fe(III). Total iron is conservative. This is why the product $\mathbf{U} \cdot \mathbf{S}_k^T$ for this component is zero.

Matrix U above yields the following components: one mobile conservative component (ΣH^+), three mixed conservative components (ΣOH^-, ΣO_2, ΣSO_4^{2-}, and Fe_T), one immobile non conservative component (FeS_2(s)), and one mixed non-conservative component (Fe(III)). This definition of components would be very convenient for working with static (non-flowing) systems because all, but two, components are conservative. We would only have to evaluate the evolution of the two non-conservative components. However, four of the conservative components are mixed. They involve mobile and immobile species. This is not convenient for flowing systems, as discussed in the reactive transport section of Chapter 4.

REFERENCES

Anderson, G.M.: *Thermodynamics in natural systems.* 2nd edition, Cambridge University Press, 2005.

Appelo, C.A.J.: Cation and proton exchange, pH variations and carbonate reactions in a freshening aquifer. *Water Resour. Res.* 30 (1994), pp. 2793–2805.

Blum, A.E. & Stillings, L.L.: Feldspar dissolution kinetics. *Rev. Mineral.* 31 (1995), pp. 291–341.

Corbella, M., Ayora, C. & Cardellach, E.: Hydrothermal mixing, carbonate dissolution and sulfide precipitation in Mississippi Valley-type deposits. *Mineral. Deposita* 39 (2004), pp. 344–357.

Davies, J.A. & Leckie, J.O.: Surface ionization and complexation at oxide/water interface. II. Surface properties of amorphous iron oxyhydroxide and adsorption of metal ions. *J. Colloid. Interface Sci.* 67 (1978), pp. 90–107.

De Simoni, M., Carrera, J., Sanchez-Vila, X. & Guadagnini, A.: A procedure for the solution of multicomponent reactive transport problems. *Water Resour. Res.* (2005), W11410.

Dzombak, D.A. & Morel, F.M.M.: *Surface complexation modelling.* Wiley Interscience, New York, 1990.

Fletcher, P. & Sposito, G.: The chemical modeling of clay/electrolyte interactions for montmorillonite. *Clay Minerals* 34 (1989), pp. 375–391.

Friedly, J.C. & Rubin, J.: Solute transport with multiple equilibrium-controlled or kinetically controlled chemical reactions. *Water Resour. Res.* 28:6 (1992), pp. 1935–1953.

Glynn, P.D.: MBSSAS: a code for the computation of Margules parameters and equilibrium relations in binary solid-solution aqueous-solution systems. *Computers & Geosciences* 17 (1991), pp. 907–966.

Glynn, P.D.: Solid-solution solubilities and thermodynamics: sulfates, carbonate and halides. In: C.N. Alpers, J.L. Jambor & D.K. and Nordstrom (eds): Sulfate minerals-crystallography, geochemistry and environmental significance. *Reviews in Mineralogy and Geochemistry*, 40, 2000, pp. 481–511.

Harvie, C.E., Greenberg, J.P. & Weare, J.H.: A chemical-equilibrium algorithm for highly nonideal multiphase systems—free-energy minimization. *Geochim. Cosmochim. Acta* 51 (1987), pp. 1045–1057.

Kräutle, S. & Knabner, P.: A new numerical reduction scheme for fully coupled multicomponent transport-reaction problems in porous media. *Water Resour. Res.* 41 (2005), W09414.

Kräutle, S. & Knabner, P.: A reduction scheme for coupled multicomponent transport-reaction problems in porous media: Generalization to problems with heterogeneous equilibrium reactions. *Water Resour. Res.* 43 (2007), W03429.

Kulik, D.A.: Thermodynamic properties of surface species at the mineral-water interface under hydrothermal conditions: A Gibbs energy minimization single-site 2pKA triple-layer model of rutile in NaCl electrolyte to 250°C. *Geochim. Cosmochim. Acta* 64 (2000), pp. 3161–3179. (errata: 2001, v. 65, p. 2027).

Kulik, D.A., Kersten, M., Heiser, U. & Neumann, T.: Application of Gibbs energy minimization to model early-diagenetic solid-solution aqueous-solution equilibria involving authigenic rhodochrosites in anoxic Baltic Sea sediments. *Aquat. Geochem.* 6 (2000), pp. 147–199.

Lichtner, P.C.: Continuous formulation of multicomponent-multiphase reactive transport, In: P.C. Lichtner, C.I. Steefel & E.H. Oeklers (eds): Reactive transport in porous media. *Reviews in Mineralogy* 34, 1996, pp. 83–129.

Lichtner, P.C.: Comment on "Coupled reactive mass transport and fluid flow: issues in model verification" by Freedman and Ibaraki. *Adv. Water Res.* 27:2 (2004), pp. 203–204.

Meyer, C.D.: *Matrix analysis and applied linear algebra.* SIAM, 2000; ISBN 0-89871-454-0, Available online: http://matrixanalysis.com.

Molins, S., Carrera, J., Ayora, C. & Saaltink, M.W.: A formulation for decoupling components in reactive transport problems. *Water Resour. Res.* (2004), W10301.

Parkhurst, D.L. & Appelo, C.A.J.: PHREEQC (Version 2)—A Computer Program for Speciation, Batch-Reaction, One-Dimensional Transport, and Inverse Geochemical Calculations. *Water-Resources Investigations Report* 99-4259, 1999.

Plummer, L.N.: Mixing of seawater with calcium carbonate groundwater. *Geol. Soc. Amer. Memoir* 142, 1975, pp. 219–236.

Saaltink, M.W., Ayora, C. & Carrera, J.: A mathematical formulation for reactive transport that eliminates mineral concentrations. *Water Resour. Res.* 34 (1998), pp. 1649–1656.

Schlinder, P.W., Furst, B., Dick, R. & Wolf, P.U.: Ligand properties of the surface silanol groups I. Surface complex formation with Fe^{3+}, Cu^{2+}, Cd^{2+} and Pb^{2+}. *J. Colloid. Interface Sci.* 55 (1976), pp. 469–475.

Spycher, N., Pruess, K. & Ennis-King, J.: CO_2-H_2O mixtures in the geological sequestration of CO_2. I. Assessment and calculation of mutual solubilities from 12 to 100°C and up to 600 bar. *Geochim. Cosmochim. Acta* 67 (2003), pp. 3015–3031.

Singer, P.C. & Stumm, W.: Acidic mine drainage: rate-determining step. *Science* 167 (1970), pp. 1121–1122.

Steefel, C.L. & Lasaga, A.C.: A coupled model for transport of multiple chemical species and kinetic precipitation/dissolution reactions with application to reactive flow in single phase hydrothermal systems. *Amer. J. Sci.* 294 (1994), pp. 529–592.

Steefel, C.I. & MacQuarrie, K.T.B.: Approaches to modeling reactive transport. *Rev. Mineral.* 34 (1996), pp. 83–129.

Wang, Y.F. & VanCappellen, P.: A multicomponent reactive transport model of early diagenesis: Application to redox cycling in coastal marine sediments. *Geochim. Cosmochim. Acta* 60 (1996), pp. 2993–3014.

Westall, J.: Chemical equilibrium including adsorption on charged surfaces. In: M. Kanaugh and J.O. *Advanced Chemical Series* 189, American Chemical Society, 1980, pp. 33–44.

CHAPTER 4

Fluid flow, solute and heat transport equations

Maarten W. Saaltink, Alexander Yakirevich, Jesus Carrera & Carlos Ayora

> *"What is referred to as "reactive transport modeling" is then an important set of interpretive tools for unraveling complex interactions between coupled processes and the effects of multiple space and time scales in the Earth. But reactive transport modeling can also be viewed as a research approach, a way of organizing and evaluating the effects of coupled dynamic geochemical, microbiological, and physical processes in the Earth sciences."*
>
> Carl I. Steefel *et al.* (2005)*

4.1 INTRODUCTION

Most descriptions of migration processes in the subsurface are based on the continuum mechanics approach to porous media. In this approach, the medium is viewed as consisting of solid matrix and void spaces, occupied by one or more fluids (e.g., water, air, oil, etc.), which represent different phases at the *microscopic level*. The values of state variables, and of material parameters or coefficients of the phase, can be assigned to every point within the domain. In the continuum representation of a porous medium, the state variables and properties describing the system, which are discontinuous at the pore scale, are replaced by the variables and properties that are continuous at the *macroscopic level* (Bear, 1972). Thus, the porous medium is replaced by a model represented as overlapping continua of solid and fluid phases. The value of any variable for each point in this continuum space is obtained by averaging the actual physical property over a representative elementary volume (REV). These averaged variables (e.g., porosity, density, pressure, temperature, concentration etc.) are referred to as macroscopic values of the considered physical properties. The macroscopic balance equations are derived using spatial averaging methods (Bear and Bachmat, 1990; Hassanizadeh and Gray, 1979a, 1979b, 1980; Whitaker, 1999; and Sorek *et al.*, 2005, 2010). This chapter presents the derivations of the basic equations needed to describe fluid flow, solute and heat transport in porous media, assuming a non-deformable solid matrix.

The continuum mechanics approach is not always appropriate. Spatial and temporal variability of physical and chemical properties may render the resulting equations inappropriate, so that alternative approaches may be required. These are reviewed in section 4.5.

A list of the notation used in this chapter is given at the end as appendix.

4.2 GROUNDWATER FLOW EQUATIONS

Groundwater flow equations describe the motion of one or more fluid phases. This requires equations for (i) mass conservation of each fluid phase, (ii) momentum balance, typically in the form of Darcy's law, and (iii) constitutive relations (equations of state) for fluid density and viscosity, as well as physical properties (e.g., relative permeability).

---

*Source: Carl I. Steefel, Donald J. DePaolo, Peter C. Lichtner: Reactive transport modeling: An essential tool and a new research approach for the Earth sciences. Earth and Planetary Science Letters 240 (2005), pp. 539–558.

4.2.1 *Single phase flow*

4.2.1.1 *The conservation mass for the fluid*

The conservation of fluid mass within a specified volume V_0 can be expressed as:

$$\begin{Bmatrix}\text{Rate of accumulation}\\ \text{of the fluid mass}\end{Bmatrix}=\begin{Bmatrix}\text{Mass flux across}\\ \text{boundary }\Gamma_0\text{ of }V_0\end{Bmatrix}+\begin{Bmatrix}\text{Rate of supply or}\\ \text{removal of mass in }V_0\end{Bmatrix} \tag{4.1}$$

The mass of fluid per unit volume of porous medium is given by $m_f = \varphi\rho_f$, where φ denotes the porosity and ρ_f denotes the fluid density. The mass flux across a unit surface area of the domain boundary is $\vec{F}_{M1} = \rho_f\vec{v}_f$, where $\vec{v}_f$ denotes a volumetric discharge per unit area. Denoting the volumetric rate of supply or removal of fluid per unit volume within V_0 as q_f, the integral form of the equation (4.1) reads:

$$\frac{d}{dt}\int_{V_0}\varphi\rho_f dV = -\int_{\Gamma_0}\vec{F}_{M1}\cdot\vec{n}d\Gamma + \int_{V_0}\rho_f q_f dV \tag{4.2}$$

Following the Gauss theorem:

$$\int_{\Gamma_0}\vec{F}_{M1}\cdot\vec{n}d\Gamma = \int_{V_0}\vec{\nabla}\cdot\vec{F}_{M1}dV \tag{4.3}$$

and substituting (4.3) into (4.2), results in:

$$\int_{V_0}\left[\frac{\partial(\varphi\rho_f)}{\partial t}+\vec{\nabla}\cdot\vec{F}_{M1}-\rho_f q_f\right]dV = 0 \tag{4.4}$$

Because the volume V_0 is arbitrary, the integrand itself must equal zero, which leads to the differential form of the equation of the fluid mass conservation:

$$\frac{\partial(\varphi\rho_f)}{\partial t}+\vec{\nabla}\cdot(\rho_f\vec{v}_f)=\rho_f q_f \tag{4.5}$$

which is also known as the equation of continuity for the fluid. Assuming constant fluid density and porosity, and no internal sink/source ($q_f = 0$), (4.5) reduces to:

$$\vec{\nabla}\cdot\vec{v}_f = 0 \tag{4.6}$$

It must be stressed that the three terms in equation (4.5) represent exactly the three terms in the "usual language" form of mass balance, equation (4.1). The only difference lies on the fact that (4.1) expressed the mass balance for an arbitrary volume V_0 whereas (4.5) expressed the balance per unit volume of porous medium at every point. Obviously, these two balances are equivalent, a principle that we used in deriving (4.5) from (4.4).

For deformable porous media, the principle of conservation of mass in porous rocks and the mass balance equations are presented by Bundschuh and Suárez Arriaga (2010) taking into account poroelastic properties of the rocks.

4.2.1.2 *The momentum mass balance equations for the fluid*

The motion of matter is typically derived from the conservation of momentum, which leads to equations relating the forces involved in the movement. In continuum fluid mechanics, the Navier-Stokes equation represents the momentum balance equation. This equation can be applied to describe flow of a viscous, Newtonian fluid through the pore network. However, this approach requires a detailed description of porous space geometry, and is therefore impractical for describing fluid motion at the macroscopic and larger scales. Darcy's law, and other approximations of fluid momentum (e.g., Brinkman, Forchheimer), represent momentum conservation in porous media in a much simpler way than the Navier-Stokes equation. Therefore, Darcy's law is typically used for modeling fluid motion in porous media.

Darcy's law

The engineer Henry Darcy from city Dijon, in the south of France, conducted experiments of water flow in a vertical, homogeneous, saturated sand filter of constant cross-sectional area. Based on the results of these experiments Darcy (1956) found that the volumetric flow rate (Q_V, volume of water flowing per unit time) through the sand column was proportional to the cross-sectional area of the column (A), proportional to the difference in water level elevations in the inflow and outflow reservoirs (h_1 and h_2, respectively), and inversely proportional to the column's length (L):

$$Q_V = -K_H A \frac{h_2 - h_1}{L} \tag{4.7}$$

where K_H denotes the hydraulic conductivity, which characterizes the resistance of porous material to viscous flow. The hydraulic conductivity depends on both solid matrix properties, (the distribution of pore and grain sizes and shapes and the overall porosity) and fluid properties (kinematic viscosity and density).

The specific discharge (v_f, or the Darcy flux) is defined as the flow rate per unit area of porous medium normal to the flow direction:

$$v_f = Q_V / A \tag{4.8}$$

The vector expression for Darcy flux of a single-phase fluid in three-dimensional space is:

$$\vec{v}_f = -\frac{\mathbf{K}}{\mu_f}\left(\vec{\nabla} p - \rho_f \vec{g}\right) \tag{4.9}$$

where **K** denotes the second rank tensor of absolute permeability (also called intrinsic permeability), μ_f is the fluid kinematic viscosity, p is the fluid pressure, and $\vec{g}$ is the gravity acceleration. For isotropic porous materials, permeability varies between 10^{-19} m^2 in clay formations and minimally fractured granite rocks and 10^{-7} m^2 in gravel deposits and highly fractured or karstic rocks.

The Darcy flux expresses the fluid discharge per unit cross-sectional area of porous medium, while at the microscopic level only the void portion of the area is available for fluid flow. The remaining portion is occupied by solid matrix. Assuming that the average areal porosity equals the volumetric one, the average velocity of the fluid in porous medium is given by:

$$u_f = v_f / \varphi \tag{4.10}$$

Henry Darcy derived his fundamental law through sand column experiments. However, during the last decades, several researchers have derived Darcy's law by transforming the microscopic momentum balance equation (Navier-Stokes) into a macroscopic equation. Bear and Cheng (2009) discuss three major techniques (volume averaging, homogenization, mixture theory) which can be applied to obtain Darcy's law under certain assumptions. The assumptions are: (i) inertial forces are negligible relative to the viscous ones; (ii) the solid phase is non-deformable and stationary; (iii) the body force is due to gravity only;

(iv) the effect of momentum transfer within the fluid, as a result of velocity gradients, is negligible, in comparison with the drag produced at the fluid-solid interface.

Non-Darcian momentum balance equations
Darcy law which is valid for laminar flow in porous media has same limitations. In fine grained materials, a lower limit of Darcy's law is indicated by existence of a minimum threshold hydraulic gradient (~30 for dense clays) required to initiate flow especially in clay formations (Bear, 1979). Explanations for the deviations of fluid flow from Darcy low in fine grained materials include: very small pores and double layer effects, non-Newtonian behavior of the fluid in very thin capillary spaces, electroosmotic counterflow. The seepage velocity, v_f, in Darcy's equation (4.9) is linear with respect to the pressure gradient. Column experiments, similar to those conducted by Darcy, indicate that the linear relationship holds when v_f is sufficiently small, which means that the Reynolds number, based on a representative grain or pore diameter (d_g):

$$R_e = \frac{v_f d_g}{\mu_f} \tag{4.11}$$

is equal or less than 1. Increasing v_f(for $R_e < 1-10$, which defines upper limit of Darcy's law) leads to a smooth transition to nonlinear drags, while the viscous forces that resist the flow remain predominant, i.e., Darcy's law is still valid (Bear, 1979). Further increasing the fluid velocity invalidates the linear relation as drag due to solid obstacles becomes comparable with the surface drag due to the friction (Nield and Bejan, 2006). In this case, the appropriate generalization of Darcy's law, when inertia effects at the microscopic level are included, takes the form (Joseph *et al.*, 1982):

$$\vec{v}_f = -\frac{\mathbf{K}}{\mu_f}(\vec{\nabla}p - \rho_f\vec{g}) - \frac{C_F\rho_f\mathbf{K}^{1/2}}{\mu_f}|\vec{v}_f|\vec{v}_f \tag{4.12}$$

where C_F is a dimensionless form drag constant. Nield and Bejan (2006) present a detailed discussion of the parameter C_F. Equation (4.12) is known as *Forchheimer* momentum balance equation (Forchheimer, 1901). This equation can also be theoretically derived by volume averaging the Navier-Stokes equation (Whitaker, 1996; Sorek *et al.*, 2005).

When both microscopic and macroscopic effects are negligible, while the dissipation of energy within the fluid is accounted for, the Brinkman (Brinkman, 1948) momentum mass balance equation is used to describe water flow in porous media:

$$\vec{v}_f = -\frac{\mathbf{K}}{\mu_f}(\vec{\nabla}p - \rho_f\vec{g}) - \frac{\mathbf{K}\mu'}{\mu_f}\nabla^2\vec{v}_f \tag{4.13}$$

where $\mu' = \mu[1+2.5(1-\varphi)]$ is the effective viscosity. Equation (4.13) includes the Laplacian term in the right hand side (r.h.s.), which is important when a no-slip boundary condition at the fluid-solid interface must be satisfied (Nield and Bejan, 2006). Bear and Bachmat (1990) theoretically proved that $\mu'/\mu_f = 1/\varphi\tau_f$, where τ_f is the tortuosity of the porous medium. Discussing applicability of the Brinkman equation, Nield and Bejan (2006) conclude that it is valid for large values of porosity: $\varphi > 0.8$ (Rubinstein, 1986) or even $\varphi > 0.95$ (Durlofsky and Brady, 1987), while naturally occurring porous media have porosities less than 0.6.

The semiheuristic momentum balance equation (Kaviani, 1994), which is an equivalent of the Navier-Stokes equation, was obtained by volume averaging and is used for the description of flow through porous media (Vafai and Tien, 1981):

$$\frac{\rho_f}{\varphi}\left(\frac{\partial\vec{v}_f}{\partial t} + \frac{1}{\varphi}\vec{v}_f\cdot\nabla\vec{v}_f\right) = -\frac{\mu}{\mathbf{K}}\vec{v}_f - \nabla p + \rho_f\vec{g} + \frac{\mu}{\varphi}\nabla^2\vec{v}_f - \frac{C_E}{\mathbf{K}^{1/2}}\rho_f|\vec{v}_f|\vec{v}_f \tag{4.14}$$

where C_E denotes the dimensionless Ergun coefficient accounting for deviation from the Stokes flow (Kaviani, 1995), which is found to vary with porosity and structure of the porous medium.

The left hand side (l.h.s.) of (4.14) represents the macroscopic inertial force. The first term on the r.h.s. of (4.14) represents the macroscopic viscous shear stress or Darcy term. The second and third terms on the r.h.s. of (4.14) represent the forces acting on the fluid, due to pressure gradient and gravity, respectively. The forth term on the r.h.s. of (4.14) represents the macroscopic viscous shear stress between layers of the fluid or Brinkman viscous term. The last term on the r.h.s. of equation (4.14) expresses the microscopic inertial force transferred through the fluid-solid interface. Depending on flow conditions, one of these terms is much smaller with respect to the remaining ones and, therefore, may be omitted from the momentum balance equation.

All momentum balance equations discussed here were formulated by neglecting the solid matrix velocity, which is a reasonable assumption when solving many practical problems.

4.2.1.3 *Flow equations*

General flow equation

Solving the mass balance equation together momentum balance equation provides the description of water flow in porous medium. Assuming that the water motion is described by Darcy's law, and combining equations (4.5) and (4.9) yields the flow equation:

$$\frac{\partial(\varphi\rho_f)}{\partial t}=\vec{\nabla}\cdot\left[\frac{\rho_f\mathbf{K}}{\mu_f}\left(\vec{\nabla}p-\rho_f\vec{g}\right)\right]+\rho q_f \tag{4.15}$$

In general, the density of a fluid depends on pressure, p, temperature, T, and solute concentration in the fluid, c_m, i.e.:

$$\rho_f=\rho_f\left(p,T,c_m\right) \tag{4.16}$$

A common form of (4.16) is:

$$\rho_f=\rho_{f0}\exp\left[\beta_p(p-p_0)-\beta_T\left(T-T_{f0}\right)+\beta_c\left(c_m-c_{m0}\right)\right] \tag{4.17}$$

which is often reduced to its linear form:

$$\rho_f=\rho_{f0}\left[1+\beta_p(p-p_0)-\beta_T(T-T_0)+\beta_c\left(c_m-c_{m0}\right)\right] \tag{4.18}$$

where ρ_{f0} is the density at the reference pressure (p_0), temperature (T_0) and concentration (c_{m0}) and β_p, β_T, β_c are empirical coefficients. The medium's porosity can also change as a function of pressure (Bear, 1979) and concentration of precipitated minerals, c_d, (Katz *et al.*, 2011):

$$\varphi=\varphi(p,c_d) \tag{4.19}$$

In view of (4.17)–(4.19), the l.h.s of (4.15) representing the rate of accumulation of fluid mass can be written as:

$$\begin{aligned}\frac{\partial(\varphi\rho_f)}{\partial t}&=\frac{\partial(\varphi\rho_f)}{\partial p}\frac{\partial p}{\partial t}+\frac{\partial(\varphi\rho_f)}{\partial T}\frac{\partial T}{\partial t}+\frac{\partial(\varphi\rho_f)}{\partial c_m}\frac{\partial c_m}{\partial t}+\frac{\partial(\varphi\rho_f)}{\partial c_d}\frac{\partial c_d}{\partial t}\\&=\frac{\partial(\varphi\rho_f)}{\partial t}=\frac{\partial(\varphi\rho_f)}{\partial p}\frac{\partial p}{\partial t}+\varphi\frac{\partial\rho_f}{\partial T}\frac{\partial T}{\partial t}+\varphi\frac{\partial\rho_f}{\partial c_m}\frac{\partial c_m}{\partial t}+\rho_f\frac{\partial\varphi}{\partial c_d}\frac{\partial c_d}{\partial t}\end{aligned} \tag{4.20}$$

For saturated conditions specific pressure storativity of the aquifer (S_{or}) is defined by (Bear, 1979): $\partial(\varphi\rho_f)/\partial p \equiv \rho_f S_{or}$. From (4.18) we obtain $\partial\rho_f/\partial T = \beta_T$ and $\partial\rho_f/\partial c_m = \beta_c$. Using these relations in (4.20), and substituting (4.20) into (4.15) yields the general fluid flow equation:

$$\rho_f S_{or}\frac{\partial p}{\partial t} + \varphi\beta_T\frac{\partial T}{\partial t} + \varphi\beta_c\frac{\partial c_m}{\partial t} + \rho_f\frac{\partial\varphi}{\partial c_d}\frac{\partial c_d}{\partial t} = \vec{\nabla}\cdot\left[\frac{\rho_f\mathbf{K}}{\mu_f}\left(\vec{\nabla}p - \rho_f\vec{g}\right)\right] + \rho q_f \tag{4.21}$$

When considering non-isothermal flow and solute transport problem, the primary variables in the equation (4.21) are p, T, c_m, and c_d. We note that this equation is coupled with the energy and solute transport equations and solved together with state equations (4.15), and (4.18). The equations for fluid viscosity as a function of temperature and solute concentration, $\mu_f = \mu_f\left(p_f, T, c_m\right)$, and for permeability as a function of porosity $\mathbf{K} = \mathbf{K}(\varphi)$ must be added (see, e.g., Chadam *et al.*, 1986; Ortoleva *et al.*, 1987; Chen and Liu, 2002).

Equation (4.21) accounts for buoyancy driven fluid flux, which needs to be coupled to the transport of thermal energy and/or chemical substances (Diersch and Kolditz, 2002). This equation can be used for simulating thermohaline (double-diffusive) convection in porous media problems (e.g., Rubin, 1976; Rubin and Roth, 1983; Oldenburg and Pruess, 1995, 1998; Diersch and Kolditz, 1998; Nield and Bejan, 2006), which is an important phenomenon in geothermal aquifers. For thermohaline convection to occur, three conditions must be met (Deirsch and Kolditz, 2002): (i) there should be a vertical gradient in two or more properties affecting the fluid density (e.g., concentrations of chemical species, temperature), (ii) the resulting gradients in the fluid density must have opposing signs, and (ii) the diffusivities of the properties must be different.

Oberbeck–Boussinesq approximation

Since coefficients β_p, β_T, β_c in (4.18) are generally small, the flow equation (4.21) can be simplified by using the so-called Oberbeck–Boussinesq (OB) approximation, after termed as Boussinesq approximation (Combarnous and Bories, 1975; Deirsch and Kolditz, 2002). The OB approximation states that spatial variations of fluid density can be neglected except for the buoyancy term $\rho_f\vec{g}$, which is retained in the momentum balance equation. Some additional assumptions are usually stated: (i) the porous medium is saturated, nondeformable and there are no fluid sinks/sources; (ii) the thermal characteristics of porous medium are assumed to be constant and independent on temperature; (iii) seepage velocities and their gradients are small. With these assumptions equation (4.21) reduces to:

$$\vec{\nabla}\cdot\left[\frac{\rho_f\mathbf{K}}{\mu_f}\left(\vec{\nabla}p - \rho_f\vec{g}\right)\right] = 0 \tag{4.22}$$

Deirsch and Kolditz (2002) pointed out that the OB approximation becomes invalid for large density variations, e.g., at high-concentration brines and/or high temperature gradients.

Variable saturation flow

In models of variably saturated flow, only a fraction of the total volume is filled by the wetting fluid (water in the present context):

$$\theta_f = S_f\varphi \tag{4.23}$$

where θ_f is volumetric fluid content, and S_f is fluid saturation. When $S_f = 1$, the porous medium is fully saturated with fluid, and when $S_f < 1$, the porous medium is referred to as being unsaturated. The fluid pressure is specified relative to atmospheric pressure. In unsaturated conditions ($S_f < 1$) the wetting fluid pressure is less than that of atmospheric because of fluid-solid interface tension effects and, therefore, it is negative, $p < 0$ (capillary pressure $p_c = -p$). In saturated conditions ($S_f = 1$), the fluid pressure is greater or equal to the atmospheric pressure, i.e., $p \geq 0$. The air is considered as a constant pressure (i.e., virtually immobile) gaseous phase, therefore the variable saturation model is referred to as single phase flow. The constitutive relation between capillary pressure and fluid saturation for a given porous medium, is called the retention curve (Fig. 4.1) and can be measured in laboratory or field experiments, or can be estimated by using pedotransfer functions (Worsten *et al.*, 2001; Pachepsky *et al.*, 2006). A number of functions are used to fit experimental retention data. The most widely used of these is Van Genuchten's equation, which reads:

$$S_f = S_{fr} + \left(1 - S_{fr}\right)\left[1 + \left(\alpha_{VG} p\right)^{n_{VG}}\right]^{-m_{VG}} \tag{4.24}$$

where S_{fr} is the residual saturation below which saturation does not drop and the fluid becomes immobile, α_{VG} and n_{VG} are fitting parameters, and m_{VG} is usually taken as equal to $1 - 1/n_{VG}$.

The Darcy momentum balance equation for the variable saturation porous medium has the form:

$$\vec{v}_f = -\frac{k_{rf}\mathbf{K}}{\mu_f}\left(\vec{\nabla} p - \rho_f \vec{g}\right) \tag{4.25}$$

where k_{rf} is the relative permeability to fluid flow (Fig. 4.1), which is independent of flow direction and drops with a decrease in saturation. It can be estimated using the Van Genuchten-Mualem equation (van Genuchten, 1980);

$$k_{rf} = \bar{S}_f^{0.5}\left[1 - \left(1 - \bar{S}_f^{1/m_{VG}}\right)^{m_{VG}}\right]^2 \tag{4.26}$$

where $\bar{S}_f = \left(S_f - S_{fr}\right)/\left(1 - S_{fr}\right)$ is the effective saturation.

The mass balance equation for variable saturation flow is:

$$\frac{\partial\left(\varphi S_f \rho_f\right)}{\partial t} + \vec{\nabla}\cdot\left(\rho_f \vec{v}_f\right) = \rho_f q_f \tag{4.27}$$

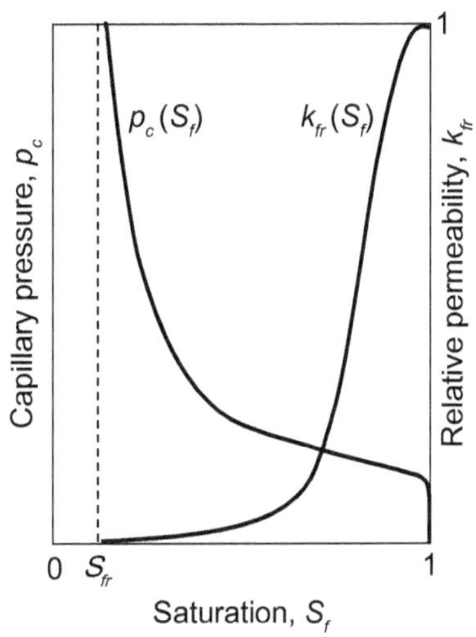

Figure 4.1. Typical retention and relative permeability curves.

Following Voss and Provost (2008), we assume that neither temperature nor concentration influence fluid saturation: changes in porosity for fully saturated media are due to pressure variations only. Actually, both temperature and solute concentration can affect the retention curve through changes in wettability and surface tension. This dependence is neglected here for the sake of simplicity. Expanding the mass storage term in the l.h.s. of (4.28) we obtain:

$$\begin{aligned}\frac{\partial(\varphi S_f\rho_f)}{\partial t} &= \frac{\partial(\varphi S_f\rho_f)}{\partial p}\frac{\partial p}{\partial t} + \frac{\partial(\varphi S_f\rho_f)}{\partial T}\frac{\partial T_f}{\partial t} + \frac{\partial(\varphi S_f\rho_f)}{\partial c_m}\frac{\partial c_m}{\partial t} \\ &= \left[S_f\frac{\partial(\varphi\rho_f)}{\partial p} + \varphi\rho_f\frac{\partial S_f}{\partial p}\right]\frac{\partial p}{\partial t} + \varphi S_f\left(\frac{\partial\rho_f}{\partial T}\frac{\partial T_f}{\partial t} + \frac{\partial\rho_f}{\partial c_m}\frac{\partial c_m}{\partial t}\right) \\ &= \left[S_f\rho_f S_{or} + \varphi\rho_f\frac{\partial S_f}{\partial p}\right]\frac{\partial p}{\partial t} + \varphi S_f\left(\beta_T\frac{\partial T_f}{\partial t} + \beta_c\frac{\partial c_m}{\partial t}\right)\end{aligned} \tag{4.28}$$

where water capacity $\partial S_f/\partial p$ is obtained by differentiating the retention curve. For example, function (4.24) yields (Voss and Provost, 2008):

$$\frac{\partial S_f}{\partial p} = \frac{\alpha_{VG}(n_{VG}-1)(1-S_{fr})(\alpha_{VG}p_c)^{n_{VG}-1}}{\left[1+(\alpha_{VG}p)^{n_{VG}}\right]^{(2n_{VG}-1)/n_{VG}}} \tag{4.29}$$

By substituting (4.24) and (4.27) into (4.26) we obtain:

$$\begin{aligned}&\left[S_f\rho_f S_{or} + \varphi\rho_f\frac{\partial S_f}{\partial p}\right]\frac{\partial p}{\partial t} + \varphi S_f\left(\beta_T\frac{\partial T}{\partial t} + \beta_c\frac{\partial c_m}{\partial t}\right) \\ &= \vec{\nabla}\cdot\left[\frac{\rho_f k_{rf}\mathbf{K}}{\mu_f}\left(\vec{\nabla}p - \rho_f\vec{g}\right)\right] + \rho_f q_f\end{aligned} \tag{4.30}$$

Note, that the specific pressure storativity, S_{or}, is sometimes neglected in the unsaturated zone where pressure is negative; in contrast, the relative water capacity, $\partial S_f/\partial p$, is equal to zero in the saturated zone, when the fluid pressure is positive or zero. Actually, equation (4.30) represents an extension of the Richards (1931) equation used to simulate isothermal flow in variably saturated conditions. In addition, (4.30) accounts for the effect of fluid density due to gradients of temperature and solute concentration.

4.2.2 *Multiphase flow*

The governing equations for multiphase flow are obtained from the principles of mass and momentum conservation. Major assumptions considered for multiphase flow modeling are (Xu and Pruess, 2001): (i) flow in all fluid phases occurs under pressure, viscous, and gravity forces; (ii) interactions between flowing phases are represented by characteristic curves for capillary pressure and relative permeability.

4.2.2.1 *Multiphase system*

Let us consider a porous medium with voids containing I_f immiscible fluids (e.g., water, oil, gas, etc.). Each fluid occupies some portion of the pore volume S_i, and each phase consists of one thermodynamic component only.

The macroscopic mass balance equation for each phase:

$$\frac{\partial(\varphi S_i \rho_i)}{\partial t} + \vec{\nabla} \cdot (\rho_i \vec{v}_i) = \rho_i q_i \tag{4.31}$$

where for each i phase, ρ_i is fluid density and q_i is volumetric rate of phase production per unit volume of porous medium (sink/source term).

The macroscopic momentum balance equation is obtained assuming that momentum exchange across the fluid-fluid interfaces is much smaller than across the fluid-solid interface (Bear and Bachmat, 1990). The resulting averaged momentum balance equation for each phase is similar to the equation applicable when the porous space is completely filled by one phase. When only a portion of the volume is occupied by a phase, the resistance to the flow depends on the phase saturation, which varies in space and with time; and the permeability relative to the phase becomes a function of saturation. It is commonly accepted that Darcy's equation is a good approximation for the flow of a fluid phase in a multiphase system:

$$\vec{v}_i = -\frac{k_{ri}\mathbf{K}}{\mu_i}\left(\vec{\nabla} p_i - \rho_i \vec{g}\right) \tag{4.32}$$

where k_{ri} and p_i are the relative permeability and pressure of the i phase, respectively. Substituting (4.32) into (4.31) yields the flow equation for the i-th phase:

$$\frac{\partial(\varphi S_i \rho_i)}{\partial t} = \vec{\nabla} \cdot \left[\frac{k_{ri}\mathbf{K}\rho_i}{\mu_i}\left(\vec{\nabla} p_i - \rho_i \vec{g}\right)\right] + \rho_i q_i \tag{4.33}$$

The capillary pressure (p_c) relations must be determined (one for each coupled fluids):

$$p_i - p_j = p_c\left(S_i, S_j\right) \tag{4.34}$$

where indexes i and j identify the fluid phases.

Bear and Bachmat (1990) note that only ($I_f - 1$) of the equations (4.34) are independent of each other.

In addition, the sum of all saturations must be equal to unit:

$$\sum_i S_i = 1 \tag{4.35}$$

All in all there are $2I_f$ unknowns (p_i and S_i) and $2I_f$ equations (4.33)–(4.35). To this system, the equations for calculating the relative permeabilities as a function of the phase saturations must also be added.

For two phase systems, typical relative permeabilities are shown in Figure 4.2. The relative permeabilities can be calculated using Bourdine's equations (Bear and Bachmat, 1990) for the wetting phase,

$$k_{rw} = \bar{S}_w^2 \int_{S_{rw}}^{S_w} \frac{ds}{p_c^2(s)} \Bigg/ \int_{S_{rw}}^{1} \frac{ds}{p_c^2(s)} \tag{4.36}$$

and for the non-wetting phase,

$$k_{rw} = \left(1 - \bar{S}_w\right)^2 \int_{\bar{S}_w}^{1} \frac{ds}{p_c^2(s)} \Bigg/ \int_{0}^{1} \frac{ds}{p_c^2(s)} \tag{4.37}$$

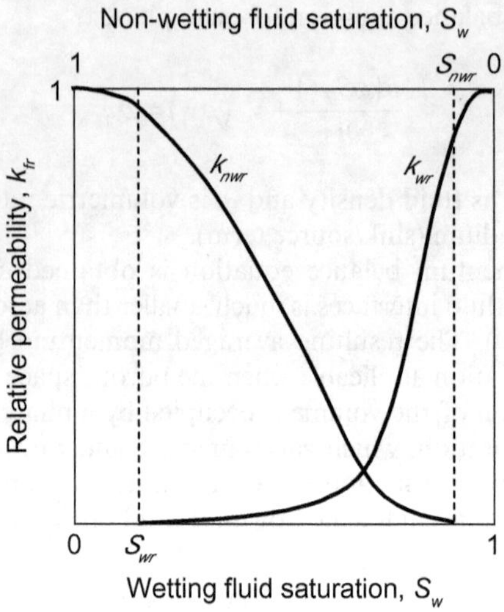

Figure 4.2. Typical relative permeabilities for two phase system.

where S_w, $\bar{S}_w$ are the saturation and effective saturation of the wetting phase, respectively; $p_c(S_w)$ is the retention curve.

For a two-phase system, if the saturation of the wetting fluid is less than its residual saturation, the wetting fluid will be held immobile by capillary forces, while the non-wetting fluid can flow. Likewise, if the saturation of the non-wetting fluid is less than its residual saturation, the non-wetting fluid becomes immobile, but the wetting fluid can migrate. Bundschuh and Suárez Arriaga (2010) develop and present relations describing relative permeabilities as a function of saturation for two- and three-phase systems.

4.3 TRANSPORT OF CONSERVATIVE SOLUTES

4.3.1 *Advection, diffusion and dispersion*

The transport of conservative solutes in porous media is governed by three major processes: advection, diffusion and mechanical dispersion. Solute migration is defined by quantifying solute mass fluxes (mass of the solute passing through a unit area of porous medium, normal to flow direction, per unit time) for each of these processes.

4.3.1.1 *Advection*

Advection governs the translation of solutes along trajectories aligned with the fluid flow direction. We consider transport of a dissolved solute of average concentration C (expressed as mass of solute per unit volume of the fluid) within a fluid that occupies the entire void space or part of it, at a volumetric fraction θ_f. The advective flux of the solute is defined by:

$$\vec{q}_{adv} = \theta_f \vec{u}_f C \tag{4.38}$$

where $\vec{u}_f = \vec{v}_f / \theta_f$ is the average pore (intrinsic) velocity of the fluid.

For a fluid of variable density, the macroscopic advective flux in the porous medium is:

$$\vec{q}_{adv} = \rho_f \theta_f \vec{u}_f c_m \tag{4.39}$$

where c_m is the mass concentration (i.e., mass of solute per unit mass of fluid).

4.3.1.2 *Diffusion*

Molecular diffusion of a solute (often called simply diffusion) reflects the random motion of solute molecules within the fluid. Diffusive fluxes are driven by gradients of chemical and electronic potentials (Anderson and Graf, 1976; Anderson, 1981). Diffusion can occur in gases, liquids, and solids, and it represents the net flux of molecules from a region of higher concentration to one of lower concentration. The result of diffusion is a gradual mixing of solutes. In a phase with uniform temperature and absent of external net forces acting on the particles, the diffusion process will eventually result in complete mixing and uniform concentration of particles.

At the microscopic level, the solute flux, q^*, due to molecular diffusion in the fluid is typically described by Fick's laws:

$$q^* = -D^* \frac{dC}{dx} \tag{4.40}$$

where D^* is the molecular diffusion coefficient, which is a function of pressure, temperature and concentration. In general, D^* grows with temperature and decreases with pressure (Bundschuh and Suárez Arriaga, 2010).

Adolf Eugen Fick introduced this law in 1855 to describe the diffusion of a gas across a fluid membrane. Later, Fick's law was widely applied to many physical problems associated with diffusion processes. The negative sign in (4.41) indicates that the solute mass diffuses in the direction of lower concentration. Diffusion occurs in both stagnant and flowing fluids. The diffusion coefficients for most ions in water are of order of 10^{-8}–10^{-10} m^2 s^{-1}, and in general, they increase with increasing temperature and decrease with increasing pressure (Poling *et al.*, 2000). However, the pressure dependence is weak and often neglected.

In three dimensions, the vector of the diffusion flux is:

$$\vec{q}^* = -D^* \vec{\nabla} C \tag{4.41}$$

The coefficient of molecular diffusion in porous medium is lower than that in pure aqueous solutions because of the longer pathway between grains, and the presence of the interface between the fluid and the solid grains. The macroscopic diffusive flux, $\vec{q}_D$, is obtained by averaging the microscopic diffusion fluxes over a REV (Bear and Cheng, 2009):

$$\vec{q}_D = -\theta_f D^* \boldsymbol{\tau}_f \left(\theta_f, \varphi\right) \cdot \nabla C = -\theta_f \mathbf{D} \cdot \nabla C \tag{4.42}$$

where $\mathbf{D} = D^* \boldsymbol{\tau}_f$ is the molecular diffusion coefficient (a second rank symmetric tensor) in porous medium, $\boldsymbol{\tau}_f$ is the tortuosity of porous medium (Bear and Bachmat, 1990), which is a second rank symmetric tensor. In an isotropic porous medium, the components of the tortuosity tensor $\tau_{fij} = \tau_s \delta_{ij}$, where $\tau_s (<1)$ is a scalar tortuosity, and δ_{ij} is the Kronecker delta (Bear and Cheng, 2009). The tortuosity depends on volumetric fluid fraction, for example (Millington, 1959):

$$\tau = \frac{\theta_f^{7/3}}{\varphi^2} \tag{4.43}$$

In fully saturated granular media, the tortuosity factor τ_f ranges between 0.3 and 0.7. For a fluid of variable density, the macroscopic diffusive flux in a porous medium is:

$$\vec{q}_D = -\theta_f \rho_f \mathbf{D} \cdot \nabla c_m \tag{4.44}$$

4.3.1.3 *Dispersion*

Solute transport in porous media is often studied by performing tracer tests in the laboratory or field. A sketch of a simple laboratory experiment to demonstrate the phenomenon of dispersion is presented in Figure 4.3a. At time $t = 0$, liquid solution containing a conservative tracer (e.g., chloride or bromide) of concentration C_0 starts discharging at a constant flow rate through a column of homogenous porous medium *initially* filled with a fluid with a zero concentration of tracer. Figure 4.3b shows the distribution of the tracer's relative concentration ($C_{ef}(x)/C_0$) within the column at a specific time $t^* < L_c/u_f$, where L_c is the column length and u_f is the intrinsic fluid velocity. The breakthrough curve (Fig. 4.3c) provides the relative concentration of the effluent measured at the column outlet and plotted as a function of time.

Transport by advection only yields a sharp front of tracer concentration (Figs. 4.3b,c, dotted line), where the initial pore fluid is completely replaced by the invading solute front (so-called "piston" or "plug" flow). Considering transport by both advection and diffusion results in a narrow transition zone (Figs. 4.3b,c, dashed line) close to the concentration front, where $0 < C_{ef}(t)/C_0 < 1$. In reality, the breakthrough curve has an S-shape form (Figs. 4.3b,c, solid line), with a relatively wide transition zone, where the tracer concentration changes from 0 to 1, showing the effect of tracer's spread near the concentration front (defined as a point where relative concentration is equal to 0.5).

An example of solute transport in two dimensions is presented in Figure 4.4. At time $t = 0$, a tracer is instantly injected into a two-dimensional uniform flow field in a saturated porous medium. Field observations reveal that as time evolves, the tracer's center of mass moves in

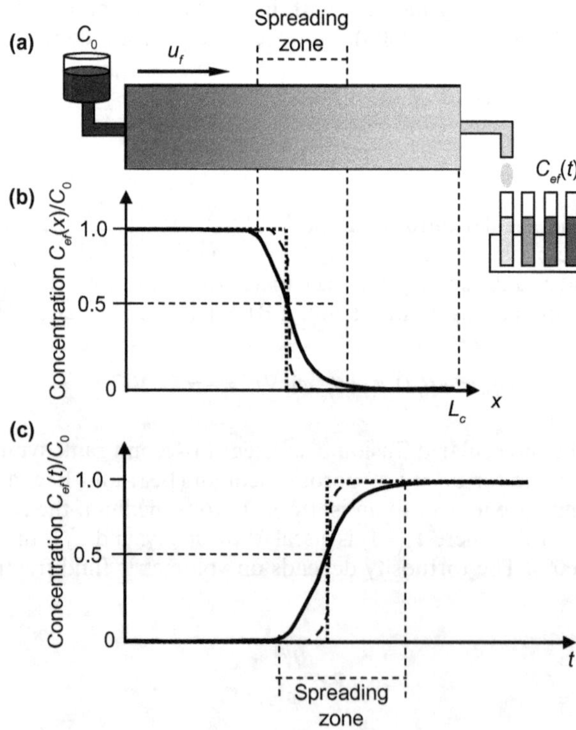

Figure 4.3. One-dimensional transport of a tracer through a column of homogenous porous medium: a) experimental sketch; b) distribution of tracer concentration within the column at specific time $t^* < L_c/u_f$; c) breakthrough curve of the tracer. Dotted line—transport by advection, dashed line—transport by advection and diffusion, solid line—transport by advection and dispersion.

the direction of flow with an average velocity. Simultaneously, the tracer spreads from its center of mass where its concentration is at maximum (Fig. 4.4). The area occupied by the tracer has the shape of an ellipse, growing with time in both the uniform flow direction (longitudinally) and in a direction perpendicular (transversely) to it (Fig. 4.4a). The value of the concentration peak decreases with time, and spreading of the tracer is higher in the direction of water flow in comparison to that of the perpendicular direction (Figs. 4.4b,c).

The process of spreading in a porous medium described above is called *hydrodynamic dispersion* (Bear, 1972). Dispersion causes a zone of mixing to develop between a fluid of one composition that is adjacent to, or being displaced by, a fluid of another composition. The mixing is produced by local velocity variations, which is macroscopically similar to molecular diffusion (Bear, 1972). It is an unsteady, irreversible process, because reversing the direction of the uniform flow field will not yield the initial tracer distribution.

The description above equates spreading and mixing. As the volume occupied by the solute plume increases, its concentration must decrease, so that the total mass of solute remains unchanged, as illustrated by Figure 4.4. Actually, the rate of spreading (increase in plume volume) does not need to equate the rate of mixing (decrease in concentrations). Instead, the growing plume may not fill all the pore space, so that high concentrations, suggesting limited mixing, may still be present. We shall return to this issue in section 4.6. For the time being, we continue with the formalism implicit in Figure 4.5 that equates mixing and spreading.

At the pore scale, variations in fluid velocity are due to several factors (Fig. 4.5): i) variations in pore sizes creating variations in flow velocities; ii) tortuous path lengths; and iii) local viscous friction at the solid-fluid interface and between fluid layers. Molecular diffusion of the solute also occurs. The first three factors cause the spreading phenomenon referred to as *mechanical dispersion*. Molecular diffusion caused by the random motion of molecules

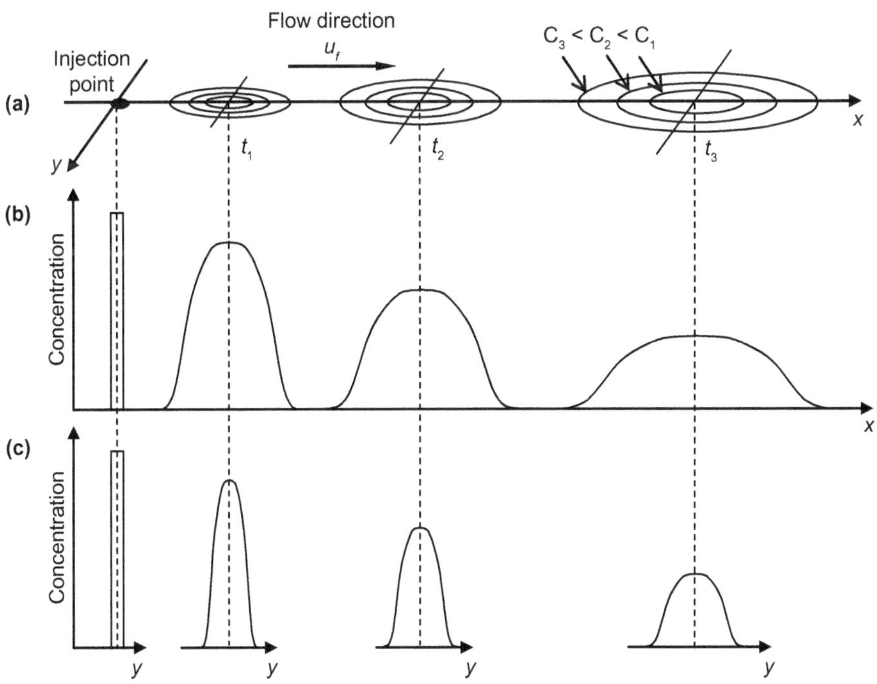

Figure 4.4. Schematic representation of the 2-dimensional spreading of a solute slug injected at a point: a) contours of iso-concentration; b) distribution of concentrations along the direction of flow; c) distribution of concentrations transverse to the flow direction. Plots are shown for 3 time moments t1 < t2 < t3. Modified from Bear (1972).

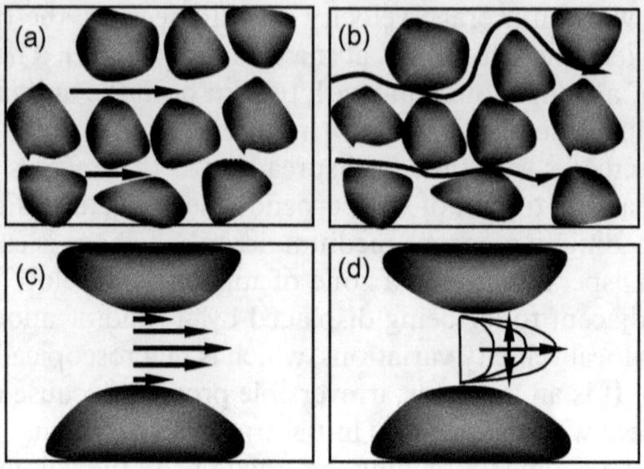

Figure 4.5. Factors causing solute dispersion at the microscale: a) pore size; b) flow path length; c) viscous friction in pores; and d) molecular diffusion. Modified from Bear (1972) and Fetter (1999).

occurs relative to the moving fluid. At the macroscopic level, the combined mechanism of solute spreading due to both mechanical dispersion and molecular diffusion is referred to as the *hydrodynamic dispersion* phenomenon.

The mechanical dispersion flux, $\vec{q}_{Dm}$, can be expressed as a Fickian-type law (Bear, 1972):

$$\vec{q}_{Dm} = -\theta_f \mathbf{D}_m \cdot \nabla C \tag{4.45}$$

where $\mathbf{D}_m$ is the mechanical dispersion coefficient.

In an isotropic porous medium the mechanical dispersion coefficient is a second rank tensor, components of which can be expressed in the form:

$$D_{ij} = \alpha_T u_f \delta_{ij} + (\alpha_L - \alpha_T)\frac{u_{fi} u_{fj}}{u_f} \tag{4.46}$$

where u_{fi} is the *i*-th component of the intrinsic velocity vector $\vec{u}_f$, $u_f = |\vec{u}_f|$, α_L and α_T are the longitudinal and transverse dispersivities, respectively.

Since the spreading of a tracer is higher in the direction of water flow compared to those in the transverse direction, the longitudinal dispersive flux is greater than the transverse flux. Field and laboratory experiment have shown that α_T is 5 to 20 times smaller than α_L (Anderson, 1979; and Domenico and Schwartz, 1998).

Various forms of the mechanical dispersion coefficient in anisotropic porous medium are discussed in detail by Bear and Cheng (2009).

For a fluid of variable density, the mechanical dispersion flux in porous medium is:

$$\vec{q}_{Dm} = -\theta_f \rho_f \mathbf{D}_m \cdot \nabla c_m \tag{4.47}$$

4.3.2 *Transport equations of conservative solutes*

Combining the three mechanisms of solute transport advection (4.39), diffusion (4.42), and dispersion (4.45) the total macroscopic solute flux has the form:

$$\vec{q}_{Dh} = \theta_f \left(\vec{u}_f C - \mathbf{D}_h \cdot \nabla C\right) \tag{4.48}$$

where $\mathbf{D}_h = \mathbf{D} + \mathbf{D}_m$ denotes the hydrodynamic dispersion tensor.

The mass balance equation for a conservative solute has the form:

$$\frac{\partial(\theta_f C)}{\partial t} + \vec{\nabla}\cdot(\vec{q}_{Dh}) = q_C \tag{4.49}$$

where q_C denotes the rate of extraction/injection of the solute per unit volume of porous medium.

Substituting the expression for the solute flux (4.48) into (4.49) results in the transport equation (Advection-Dispersion Equation, ADE) for a conservative solute in a porous medium:

$$\frac{\partial(\theta_f C)}{\partial t} = \vec{\nabla}\cdot\left[\theta_f\left(\mathbf{D}_h\cdot\nabla C - \vec{u}_f C\right)\right] + q_C \tag{4.50}$$

For a fluid of variable density, the ADE has the form:

$$\frac{\partial(\theta_f \rho_f c_m)}{\partial t} = \vec{\nabla}\cdot\left[\theta_f\left(\rho_f \mathbf{D}_h\cdot\nabla c_m - \vec{u}_f \rho_f c_m\right)\right] + q_C \tag{4.51}$$

4.4 HEAT TRANSPORT EQUATIONS

The transfer of thermal energy from one location to another (often called heat transport) occurs by some combination of the following three mechanisms: conduction, convection, and radiation. In a porous medium, heat transport by radiation is usually neglected; only the first two processes of heat transport are considered. Heating from viscous dissipation is also neglected. When fluids move through porous material, two different assumptions can be used to model heat transfer (Combarnous and Bories, 1975): (i) the fluid and solid phases are assumed in thermal equilibrium at each point of contact so that a single temperature can be assigned to all phases within the REV; (ii) the temperatures of the solid and fluid phases are assumed to differ, and the heat flux between phases is expressed by means of a heat transfer coefficient. The first approach may not be adequate at high flow rates (forced convection) through porous materials with relatively low thermal conductivity (Combarnous and Bories, 1973). Nevertheless, it is most frequently used in petroleum engineering and hydrology due to relatively low groundwater velocity. This assumption is used to derive the heat transport equation below.

4.4.1 *Conduction and convection*

4.4.1.1 *Heat conduction*

Heat conduction refers to the energy transfer from molecule to molecule through physical contact. When a temperature gradient is imposed in a porous medium with a stagnant saturating fluid phase, a conductive heat flux (thermal energy transfer per unit area and per unit time), $\vec{q}_T$, can be determined by Fourier's law:

$$\vec{q}_T = -\mathbf{k}_T\cdot\vec{\nabla}T \tag{4.52}$$

in which $\mathbf{k}_T$ is the effective thermal conductivity tensor of the porous medium.

The negative sign in (4.52) indicates that heat transfer occurs towards the lower temperature. For a homogeneous porous medium, the thermal conductivity is a scalar, k_T, which can be roughly estimated as the weighted geometric mean (Nield and Bejan, 2006):

$$k_T = (k_f)^{\varphi}(k_s)^{(1-\varphi)} \tag{4.53}$$

where k_f and k_s are the thermal conductivity of the fluid and solid phases, respectively.

In an actual porous medium, the isothermal surfaces are highly irregular on the pore scale, and to more accurately account the real structure of the porous medium, a number of geometrical and pure empirical models have been developed (Combarnous and Bories, 1975).

4.4.1.2 *Heat convection*

Heat convection refers to the thermal energy transported by a fluid in motion. Depending on the force that drives the fluid motion, two major mechanisms of convection are considered: free (or natural) convection when the circulation of fluids is due to buoyancy from the density changes induced by heating; forced convection when fluid flow is driven by a hydraulic gradient. In both cases, heat flux by convection, $\vec{Q}_f$, is determined as:

$$\vec{Q}_f = \rho_f \vec{u}_f h_f \tag{4.54}$$

where h_f denotes the specific enthalpy of the fluid.

Total heat flux in a porous medium, $\vec{Q}_f$, by heat conduction and convection mechanisms is the sum of (4.52) and (4.54):

$$\vec{Q}_T = -\mathbf{k}_T \cdot \vec{\nabla} T + \rho_f \vec{u}_f h_f \tag{4.55}$$

In a porous medium, in analogy with hydrodynamic dispersion for solute transport, the effective thermal conductivity is generalized by accounting for the thermal dispersion (Diersch, 2005, Šimůnek *et al.*, 2006). The components of thermal conductivity can then be expressed in the form:

$$k_{Tij} = \left(k_T + \beta_T u_f\right)\delta_{ij} + (\beta_L - \beta_T)\frac{u_{fi} u_{fj}}{u_f} \tag{4.56}$$

where β_L and β_T are the longitudinal and transverse thermal dispersivities, respectively; u_{fi} is the component of Darcy's velocity vector, and $u_f = \left|u_f\right|$.

4.4.2 *Heat transport in single fluid phase systems*

The equation for internal energy balance is derived from the principle of energy conservation similar to that for mass conservation (4.5):

$$\frac{\partial\left(\rho_f e_m\right)}{\partial t} + \vec{\nabla} \cdot (\rho \vec{Q}_T) = Q_H \tag{4.57}$$

where e_m is the internal energy (energy per unit mass), ρ_f is the fluid density, and Q_H is the volumetric rate of heat generation.

The internal energy for a fluid, e_{mf}, is:

$$e_{mf} = h_f - \frac{p_f}{\rho_f} = c_f T - \frac{p_f}{\rho_f} \tag{4.58}$$

where c_f denotes the specific heat capacity of the fluid at constant pressure, and where it is implicitly assumed that no phase changes occur. It is often assumed that the internal energy of the fluid is equal to its specific enthalpy, h_f, i.e., the last term in (4.59) can be neglected. Thus, the volumetric internal energy of a porous medium saturated with one fluid is:

$$\rho e_m = (1 - \varphi)\rho_s c_s T + \varphi \rho_f c_f T \tag{4.59}$$

where c_s denotes the specific heat capacity of the solid phase at constant pressure.

Substituting (4.56) and (4.60) into (4.58) yields the heat transport equation in a fluid saturated porous medium:

$$\frac{\partial\left[(1-\varphi)\rho_s c_s + \varphi\rho_f c_f\right]T}{\partial t} = \vec{\nabla}\cdot\left(\mathbf{k}_T\cdot\vec{\nabla}T\right) - \vec{\nabla}\cdot\left(\rho_f\vec{u}_f c_f T\right) + Q_H \tag{4.60}$$

4.4.3 *Heat transport in multiple fluid phases systems*

The energy conservation equation for a fluid phase i in a porous medium has the form:

$$\frac{\partial\left(S_i\rho_i e_{mi}\right)}{\partial t} + \vec{\nabla}\cdot\left(S_i\rho_i Q_{Ti}\right) = S_i Q_{Hi} \tag{4.61}$$

The energy conservation equation for the porous medium is obtained by summing (4.61) over all i phases, and adding the energy conservation for the solid matrix:

$$\frac{\partial\left[(1-\varphi)\rho_s c_s + \varphi\sum_i S_i\rho_i c_i\right]T}{\partial t} = \vec{\nabla}\cdot\left(\mathbf{k}_T\cdot\vec{\nabla}T\right) - \vec{\nabla}\cdot T\sum_i S_i\rho_i\vec{v}_i c_i + \sum_i S_i Q_{Hi} \tag{4.62}$$

When considering multiple fluid heat transport in a porous medium, the scalar part, k_T, of the thermal conductivity must be calculated by taking into account saturation of all phases. For example for a two-phase system, water-gas (or variable water saturation), and neglecting the thermal conductivity of the gas phase, the thermal conductivity as a function of water content, θ, is described by the simple equation (Chung and Horton, 1987):

$$k_T = b_1 + b_2\theta + b_2\theta^{0.5} \tag{4.63}$$

where b_1, b_2 and b_3 are empirical parameters.

The thermal dispersivities β_L and β_T must also be determined for each phase.

4.5 REACTIVE TRANSPORT

Simulations of reactive transport entail solving fluid flow and solute transport equations, discussed in the previous sections, together with chemical equations, presented in Chapter 3. The mathematical equations describing reactive transport are somewhat complex. A first question to ask is whether reactive transport simulations are really needed. This question is addressed in section 4.5.1, where simulations are shown mixing seawater and freshwater by diffusion across a 1-D calcite sample. This example will also be used to formulate reactive transport problems. The equations for reactive transport are presented in sections 4.5.2 and 4.5.3. The chapter concludes with the solution of a simple binary case, which illustrates some basic features of reactive transport.

4.5.1 *The need for reactive transport: calcite dissolution in the fresh-salt water mixing zone*

Complex and interesting geochemical processes occur in the freshwater-saltwater mixing zone of coastal carbonate aquifers. These lead to a variety of features such as large-scale dissolution and cave formation, replacement of aragonite by secondary calcite, and dolomitization (Runnels, 1969; Plummer, 1975; Hanshaw and Back, 1979; Tucker and Wright, 1990;

Raeisi and Mylroie, 1995). Many of these processes result from the fact that, when two solutions are mixed, concentrations in the mixture are volume-weighted (linear) averages of the two end-members, but the thermodynamic activities of the species controlling the water-mineral reactions are non-linear functions of the mixing ratio. Therefore, the mixing of two end-member solutions each in equilibrium with a solid phase (i.e., Gibbs free energy is at a minimum and the mass action law holds) can lead to mixtures that are either undersaturated or supersaturated with respect to the given solid. If the mixture additionally reacts to equilibrium, the solid phase dissolves or precipitates to compensate for the effect of mixing that tends to disrupt the equilibrium. Sanford and Konikow (1989) simulated reactive transport in the mixing zone of a generic coastal aquifer and concluded that appreciable carbonate dissolution can occur due to mixing of fresh water and seawater.

The results of Rezaei *et al.* (2005), who applied the solutions and geochemical reactions of Sanford and Konikow (1989) to a simple 1D diffusion problem, are shown below. The example sheds light on the processes controlling the dissolution rate, and on the need for reactive transport simulations.

Diffusion occurs across a 1D calcite porous medium, 1 cm long. The model represents a mixing line with prescribed concentrations at both ends: saline groundwater (that is, seawater entered into a carbonate aquifer) on the right-side and freshwater on the left-side, both equilibrated with calcite. The low salinity end-member solution is used as the initial condition for the entire domain, although final results are insensitive to initial conditions. The domain is divided in 40 linear elements of equal length and simulations run for 2 years with time steps of 1 day. Parameter values used are: initial porosity, 0.3; molecular diffusion coefficient (same for all species), 10^{-5} cm^2 s^{-1}. Temperature is 25°C in all the simulations.

To represent the calcite-water system 24 species were selected: H^+, OH^-, Ca^{2+}, CO_3^{2-}, HCO_3^-, CO_2(aq), $CaHCO_3^+$, $CaCO_3$(aq), Mg^{2+}, $MgHCO_3^+$, $MgCO_3$(aq), $MgOH^-$, K^+, Na^+, $NaHCO_3$(aq), $NaCO_3^-$, Cl^-, SO_4^{2-}, $CaSO_4$(aq), $MgSO_4$(aq), KSO_4^-, $NaSO_4^-$, CO_2(g) and $CaCO_3$(calcite). Calcite dissolution was assumed to quickly progress to saturation equilibrium, because typical dissolution rates are fast relative to residence times in coastal aquifers.

The RETRASO code (Saaltink *et al.*, 1998, 2004) was used, which solves simultaneously the equations for aqueous complexation, dissolution/precipitation, reactions and solute transport.

Figure 4.6 displays the distribution of calcite saturation states for varying proportions of a non-reactive mixture of saline groundwater ($P_{CO_2} = 10^{-1.96}$) and freshwater ($P_{CO_2} = 10^{-2.0}$), whose end-member waters are in equilibrium with calcite. This distribution is compared with the curve of cumulative change in calcite volumetric fraction (expressed as% porosity change), after 100 days of mixing in a carbonate reactive medium (Fig. 4.6b). Transport parameters such as diffusion coefficients are assumed not to be effected by this porosity change. Because of non-linear effects, mixed waters are undersaturated with respect to calcite throughout the domain, even though the end-members were in equilibrium with calcite. Maximum undersaturation occurs towards the freshwater side (mixing ratio around 0.15). However, when a calcite reactive medium is assumed (under local equilibrium), maximum dissolution is obtained at the freshwater end of the column (Fig. 4.6b). In other words, dissolution occurs in the mixing zone under much fresher conditions than predicted by nonreactive mixing. The fact that maximum dissolution rate is always found at the second node of the grid, regardless of grid size (recall that concentrations are prescribed to be in equilibrium at the first node) reflects that maximum dissolution actually occurs at the freshwater boundary. This observation implies that the interaction between transport processes and chemical reactions significantly affects dissolution patterns.

Figure 4.6 is somewhat paradoxical because one would expect the maximum dissolution of calcite to take place at the location of maximum undersaturation. Understanding the simulation results requires a detailed look at the distribution of the aqueous species involved (Fig. 4.7). It is apparent that the concentration of the major species, such as Ca^{2+} is barely affected by dissolution. Compared to the non reactive case, it increases only slightly because of calcite dissolution. CO_3^{2-} displays a similar trend, that is accentuated by changes

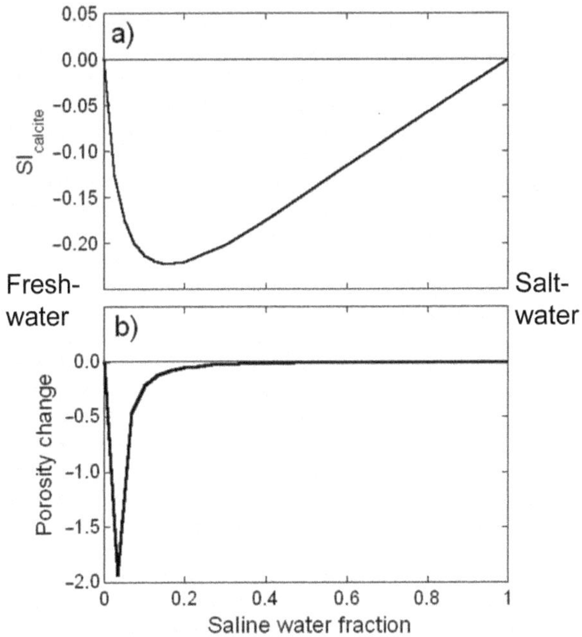

Figure 4.6. Linear mixing of fresh and saline groundwaters as they diffuse through a calcite porous medium (Rezaei *et al.*, 2005): (a) calcite saturation index for non-reactive mixing; and (b) cumulative change in calcite volumetric fraction (expressed as %porosity change) after 100 days of diffusion in equlibrium with calcite. Maximum undersaturation occurs for a mixing ratio around 20%; and yet most of the dissolution occurs closer to the freshwater end-member.

in speciation (the increase of pH causes an increase in CO_3^{2-}). However, the actual change is very small (units are μmol kg^{-1}_{water} for CO_3^{2-}). On the other hand, dissolution causes a major reduction of CO_2(aq) and H^+, thus dramatically increasing the concentration gradient (notice that the vertical scale is logarithmic) and, therefore, the diffusive transport of CO_2(aq) and H^+ at the freshwater side. Since undersaturation conditions occur immediately upon entrance, dissolution happens immediately, further enhancing the process. Dissolution also causes the gradients of CO_2(aq) and H^+ species to be reduced further down towards the saline end-member, thus reducing the dissolution rate progressively.

In summary, dissolution concentrates at the freshwater side of the column because of an imbalance in the transport of some species (Ca^{2+} and CO_3^{2-}) relative to that of others (CO_2(aq) and H^+). The dissolution rate is proportional to D, that is to the rate of diffusive mixing. Another, more intuitive, way of expressing this is that by the time solutes diffuse towards the middle of the column, the solution they form has been equilibrated with calcite. That is, the non-reactive mixing of Figure 4.6a does not occur in reality. From a mathematical perspective, dissolution occurs at the freshwater side because the rate of variation of activity (and hence solubility) with respect to salinity is largest there (section 4.5.4 will return to this issue).

Dissolution processes in the mixing zone are known to depend on a large number of factors (mainly the differences in species total concentration, P_{CO_2}, pH and ionic strength of end-members). Therefore, porosity development will be sensitive to the P_{CO_2} of the end-members as well as the chemistry of the entire system. This sensitivity was analyzed by reducing the P_{CO_2} in the freshwater end-member solution. This enhances calcite undersaturation (Fig. 4.8a). Therefore, one might expect dissolution rates to increase. Surprisingly, calcite dissolution displays the opposite behavior. That is, dissolution rates decrease with the decrease of P_{CO_2} of the freshwater end-member (Fig. 4.8b). In short, the behavior of the

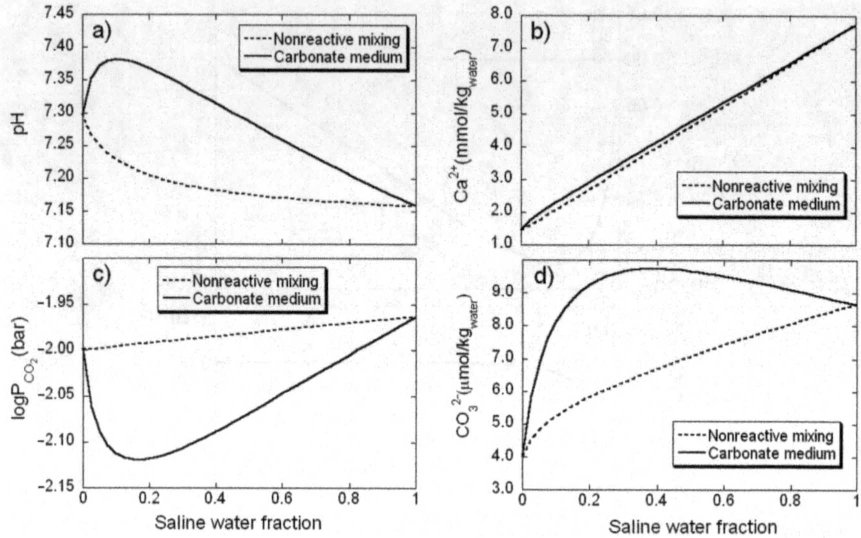

Figure 4.7. Distribution of *pH*, log P_{CO_2}, Ca^{2+} and CO_3^{2-} function of the saline water fraction for a non-reactive medium, and for a carbonate reactive medium as fresh- and sea-water mix by diffusion. Calcite dissolution causes an increase in *pH* and in the concentration of Ca^{2+} and CO_3^{2-} (notice the units), but a decrease in P_{CO_2}.

dissolution rate at equilibrium can not be deduced from the behavior of the saturation index without dissolution. This can also be attributed to the interplay between transport and reactions. As shown by Figure 4.7, dissolution enhances transport of H^+ and CO_2(aq), and thus enhances further dissolution. However, in the case of Figure 4.8, this effect is relatively minor when the P_{CO_2} of freshwater is $10^{-3.0}$ bar, because the freshwater boundary concentrations of CO_2 and H^+ are small. This causes the diffusive transport of acidity and, hence, the dissolution rate, to be small. On the other hand, the transport of acidity is relevant when the P_{CO_2} of freshwater is $10^{-2.0}$ bar. It is worth noting that the opposite occurs in the saline water half of the mixing, where dissolution rate decreases with the P_{CO_2} of freshwater (Fig. 4.8c). Here, transport distances are much larger and dissolution of calcite much smaller, so that the depletion in P_{CO_2} caused by dissolution does not cause a significant increase in H^+ and CO_2(aq) transport. Under these conditions, the dissolution rate is related to the extent of disequilibrium indicated by the saturation index (Fig. 4.8a), and the dissolution rate is larger in the case of the larger contrast of P_{CO_2} between fresh and saltwater as previouly stated by Wigley and Plummer (1976).

The most important conclusion from this 1D analysis is that one needs to couple chemistry and transport to understand dissolution dynamics. Chemistry (i.e., concentrations and saturation indices of the mixtures and end-members) controls dissolution potential at any particular location. Transport processes (mixing) control how much, when, and where dissolution occurs. In our example, the dissolution rate was primarily controlled by the mixing rate (D in our case). The interplay of transport and chemistry led to non-trivial conclusions or a-priori guesses. Enhanced CO_2 transport at the freshwater end caused the dissolution rate to be at a maximum there (far from where nonreactive mixing would have led to maximum undersaturation).

4.5.2 *Mass balance equations*

Reactive transport is the phenomenon resulting from the interaction and coupling of solute transport, explained in section 4.3, and chemical reactions, explained in Chapter 3.

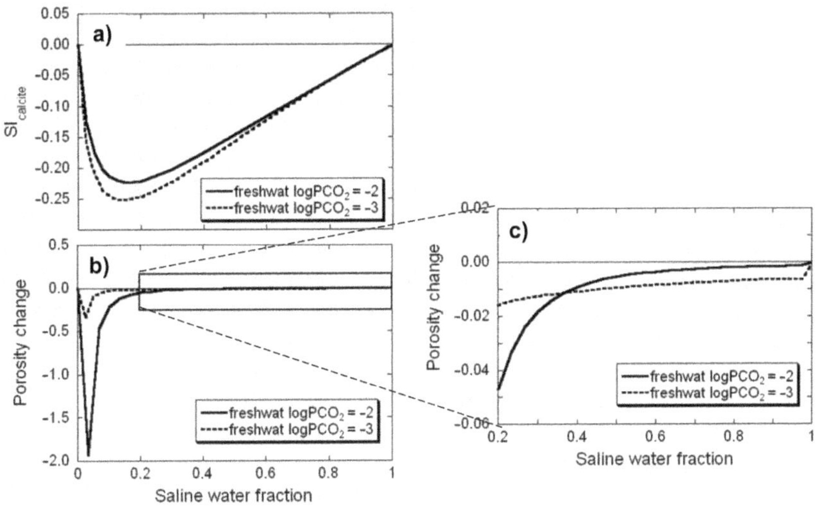

Figure 4.8. Sensitivity of mixing to the P_{CO_2} of the freshwater end-members: (a) calcite saturation index; (b) cumulative change in calcite volumetric fraction (expressed as % porosity change) after 100 days of mixture in a carbonate reactive medium; (c) blow-up of figure (b). Reducing P_{CO_2} makes the non reactive mixture slightly more undersaturated in calcite, but dramatically increases the dissolution rate.

Thus equations for reactive transport can be viewed starting from either the mass balance of reactive species in a closed system, without transport (eq. 3.12 in Chapter 3) or from the transport equation of a conservative species (eq. 4.50). The first approach requires adding terms for transport processes. The second one requires adding source/sink terms due to chemical reactions. Regardless of the starting point, it should be clear that reactive transport equations simply represent the mass balance of chemical species subject to both transport processes and chemical reactions. What is critical in these reactions is that they relate to elemental mass transfers between phases (mineral/aqueous/gas) that may have different transport potentials (or mobilities).

To write these mass balances, n_t chemical species are considered that can react with each other through n_e equilibrium reactions and n_k kinetic reactions. The mass balances of every species i can be written as:

$$\frac{\partial \theta_{f(i)} c_i}{\partial t} = L_{f(i)}(c_i) + \sum_{j=1}^{j=n_e} S_{e_{ji}} r_{e_j} + \sum_{j=1}^{j=n_k} S_{k_{ji}} r_{k_j} \quad i = 1, \cdots, n_s \tag{4.64}$$

where $L_{f(i)}(c_i)$ is a linear operator for the transport terms and non-chemical sources/sinks for phase $f_{(i)}$, that contains species i. Typically, $f = l$ for the liquid aqueous phase, g for gas, or s for solid, but other phases, such as subcritical and supercritical CO_2 or oil, may be present. In fact, the solid may consist of several phases. The transport operator for fluid phases can be written as:

$$L_{f(i)}(c_i) = \vec{\nabla} \cdot \left[\theta_{f(i)} \left(\mathbf{D}_{h_f} \cdot \nabla c_i - \vec{u}_f c_i \right) \right] + q_{c_i} \tag{4.65}$$

Note that, if one assumes the solid phases immobile, then $L_s = 0$.

The last two terms of equation (4.64) refer to sources/sinks due to equilibrium and kinetic reactions. Here equilibrium reactions are defined as reactions, whose Gibbs free energy are minimum or, in other words, which obey the mass action law (see Chapter 3, section 3.1.2). In closed systems without transport, the rate of such an equilibrium reaction (r_e) becomes zero.

However, it is important to understand that this is different for systems with transport. Transport processes tend to disrupt the equilibrium. Equilibrium reactions precisely occur to counteract transport and maintain equilibrium, leading to a nonzero equilibrium reaction rate (r_e).

It is more convenient to rewrite 4.64 using a matrix-vector notation, similar to section 3.1.1 in Chapter 3:

$$\frac{\partial}{\partial t}(\mathbf{m}) = \mathbf{L} + \mathbf{S}_e^T\mathbf{r}_e + \mathbf{S}_k^T\mathbf{r}_k \tag{4.66}$$

Since transport processes are phase dependent, it is convenient to subdivide vectors **m** and **L** according to the type of phase:

$$\mathbf{m} = \begin{pmatrix} \theta_l \mathbf{c}_l \\ \theta_g \mathbf{c}_g \\ \theta_s \mathbf{c}_s \end{pmatrix} \quad \mathbf{L} = \begin{pmatrix} L_l(\mathbf{c}_l) \\ L_g(\mathbf{c}_g) \\ \mathbf{0} \end{pmatrix} \tag{4.67}$$

where it is assumed that solid species are immobile ($L_s = 0$).

The kinetic reaction rates ($\mathbf{r}_k$) can be written explicitly as a function of concentration through the kinetic rate laws (see Chapter 3, section 3.5), which can be substituted into equation (4.66). Equilibrium reactions cannot. Therefore, equilibrium reaction rates $\mathbf{r}_e$ must be considered (initially) as unknowns, whose value will result from solving the whole problem (this issue will be addressed explicitly in section 4.5.3). Therefore, the basic formulation of reactive transport contains $n_s + n_e$ unknowns. These are matched by the n_s mass balances of all species, expressed by equation (4.66), plus the n_e mass action laws, that relate activities of reactants and products of the reaction (Chapter 3, section 3.1.2).

The large number of coupled non-linear equations hinders direct use of the formulation of equation (4.66). The number can be reduced by making linear combinations of these equations. Examples are the formulations of Friedly and Rubin (1992), Saaltink *et al.* (1998), Fang *et al.* (2003), Molins *et al.* (2004), Kräutle and Knabner (2005, 2007) and De Simoni *et al.* (2005, 2007). All these formulations share the concept of eliminating the equilibrium reaction rates by multiplying equation (4.66) by a component matrix. The concept was explained in Chapter 3 (section 3.1.4) for the case without transport. For the case with transport, multiplying equation (4.66) by a components matrix **U**, which is a $(n_t - n_r) \times n_t$ kernel matrix of $\mathbf{S}_e^T$ (that is, $\mathbf{U}\mathbf{S}_e^T = \mathbf{0}$) eliminates the equilibrium reactions rates ($\mathbf{r}_e$) and leads to:

$$\frac{\partial}{\partial t}(\mathbf{Um}) = \mathbf{UL} + \mathbf{US}_k^T\mathbf{r}_k \tag{4.68}$$

If one wishes to maintain a division between the various phases, we must split matrix **U** into submatrices for each fluid:

$$\mathbf{U} = (\mathbf{U}_a \mid \mathbf{U}_g \mid \mathbf{U}_s) \tag{4.69}$$

which is used to define the total concentration of components in every fluid:

$$\mathbf{u}_l = \mathbf{U}_l\mathbf{c}_l \quad \mathbf{u}_g = \mathbf{U}_g\mathbf{c}_g \quad \mathbf{u}_s = \mathbf{U}_s\mathbf{c}_s \tag{4.70}$$

so that:

$$\begin{aligned} \mathbf{Um} &= \theta_l\mathbf{U}_l\mathbf{c}_l + \theta_g\mathbf{U}_g\mathbf{c}_g + \theta_s\mathbf{U}_s\mathbf{c}_s \\ &= \theta_l\mathbf{u}_l + \theta_g\mathbf{u}_g + \theta_s\mathbf{u}_s \end{aligned} \tag{4.71}$$

With these definitions one can rewrite equation (4.68):

$$\frac{\partial \theta_l \mathbf{u}_l}{\partial t} + \frac{\partial \theta_g \mathbf{u}_g}{\partial t} + \frac{\partial \theta_s \mathbf{u}_s}{\partial t} = L(\mathbf{u}_l) + L(\mathbf{u}_g) + \mathbf{U}\mathbf{S}_k^T \mathbf{r}_k \tag{4.72}$$

Note, that the number of transport equations have been reduced from n_t in equation (4.66) to $(n_t - n_e)$ in equation (4.68) or (4.72), as matrix $\mathbf{U}$ has $(n_t - n_e)$ rows. Moreover, there are n_e mass action laws. So the total number of equation is n_t, which equals the number of variables (the concentrations of n_t species). As explained in Chapter 3 (section 3.1.3), the mass action laws can be written in such a way that the concentrations of n_e secondary species are a function of the concentrations of $(n_t - n_e)$ primary species. So, these mass action laws can be substituted into equation (4.72), leading to $(n_t - n_e)$ equations and the same number of variables (the concentrations of $(n_t - n_e)$ primary species). However, this substitution has to be done with care. Mass action laws can be written in a form such that the *activities* of secondary species are an explicit function of the *activities* of the primary species (eq. 3.8), but this is not strictly true when mass action laws are written in terms of *concentrations* (eq. 3.18), because the activity coefficients ($\boldsymbol{\gamma}$) normally depend on all concentrations, including those of primary species.

A special case of (4.72) occurs when all species of the chemical system pertain to the same fluid, and all reactions are in equilibrium. For instance, the example presented in Chapter 3 (section 3.1) only considers species of the liquid fluid. In that case (4.72) reduces to:

$$\frac{\partial \theta_l \mathbf{u}_l}{\partial t} = L(\mathbf{u}_l) \tag{4.73}$$

This equation, written in terms of concentrations of components in liquid ($\mathbf{u}_l$), has the same appearance as the transport equation for conservative solutes (eq. 4.50), that is, chemical reactions do not affect $\mathbf{u}_l$ and one could calculate $\mathbf{u}_l$ by using algorithms and models for conservative transport. Moreover, for the special case of (4.73) transport of one component does not affect transport of the others. The concentrations of the individual species ($\mathbf{c}_l$) could be calculated from $\mathbf{u}_l$ by means of the equations explained in Chapter 3 (section 3.2). Section 4.5.4 will return to this when explaining the conceptual meaning of reactive transport equations.

Unfortunately, most reactive transport problems must consider heterogeneous chemical reactions (defined in Chapter 3, section 3.3). This makes the transport equations (4.72) of the components dependent on each other. Moreover, they are highly non-linear due to the combination of linear partial differential equations (i.e., transport equations) and non-linear algebraic equations (i.e., mass action laws).

To illustrate the components and component matrices with more than one fluid, the chemical system presented in Chapter 3 (section 3.1) is extended here with a gaseous and solid species and corresponding heterogeneous reactions:

(R1) $H_2O \rightleftarrows H^+ + OH^-$

(R2) $HCO_3^- \rightleftarrows H^+ + CO_3^{2-}$

(R3) $CO_2(aq) + H_2O \rightleftarrows H^+ + HCO_3^-$

(R4) $CO_2(g) \rightleftarrows CO_2(aq)$

(R5) $CaCO_3(s) \rightleftarrows Ca^{2+} + CO_3^{2-}$

This system has the following stoichiometric matrix:

$$\mathbf{S} = \begin{pmatrix} & H_2O & Cl^- & CO_2(aq) & H^+ & Ca^{2+} & OH^- & HCO_3^- & CO_3^{2-} & CO_2(g) & CaCO_3(s) \\ R1 & -1 & 0 & 0 & 1 & 0 & 1 & 0 & 0 & 0 & 0 \\ R2 & 0 & 0 & 0 & 1 & 0 & 0 & -1 & 1 & 0 & 0 \\ R3 & -1 & 0 & -1 & 1 & 0 & 0 & 1 & 0 & 0 & 0 \\ R4 & 0 & 0 & 1 & 0 & 0 & 0 & 0 & 0 & -1 & 0 \\ R5 & 0 & 0 & 0 & 0 & 1 & 0 & 0 & 1 & 0 & -1 \end{pmatrix}$$

By applying Gauss-Jordan elimination (eq. 3.15) and using H_2O, Cl^-, $CO_2(aq)$, H^+, and Ca^{2+} as primary species, one obtains the following component matrix:

$$\mathbf{U} = \left(\begin{array}{cccccc|ccccc} & H_2O & Cl^- & CO_2(aq) & H^+ & Ca^{2+} & OH^- & HCO_3^- & CO_3^{2-} & CO_2(g) & CaCO_3(s) \\ \sum H_2O & 1 & 0 & 0 & 0 & 0 & 1 & 1 & 1 & 0 & 1 \\ \sum Cl & 0 & 1 & 0 & 0 & 0 & 0 & 0 & 0 & 0 & 0 \\ \sum C & 0 & 0 & 1 & 0 & 0 & 0 & 1 & 1 & 1 & 1 \\ \sum H^+ & 0 & 0 & 0 & 1 & 0 & -1 & -1 & -2 & 0 & -1 \\ \sum Ca & 0 & 0 & 0 & 0 & 1 & 0 & 0 & 0 & 0 & 1 \end{array}\right)$$

This gives the following total concentration of components in liquid, gas and solid:

$$\mathbf{u}_l = \begin{pmatrix} c_{H_2O} + c_{OH^-} + c_{HCO_3^-} + c_{CO_3^{2-}} \\ c_{Cl} \\ c_{CO_2(aq)} + c_{HCO_3^-} + c_{CO_3^{2-}} \\ c_{H^+} - c_{OH^-} - c_{HCO_3^-} - 2c_{CO_3^{2-}} \\ c_{Ca^{2+}} \end{pmatrix} \quad \mathbf{u}_g = \begin{pmatrix} 0 \\ 0 \\ c_{CO_2(g)} \\ 0 \\ 0 \end{pmatrix} \quad \mathbf{u}_s = \begin{pmatrix} c_{CaCO_3(s)} \\ 0 \\ c_{CaCO_3(s)} \\ -c_{CaCO_3(s)} \\ c_{CaCO_3(s)} \end{pmatrix}$$

In this example, water was included as a species and as a component because it is involved in several of the chemical reactions. Usually, as water is a major constituent of the porous medium, the effect of reactions on the mass balance of water can often be neglected. If so, it is more convenient to substitute this equation by the flow equation, that is, the mass balance equation of the liquid fluid. Of course, if effects of reactions are neglected, this equation can be solved independently from the mass balance equations of the other components.

4.5.3 *Constant activity species*

Besides mass action laws, activities are also controlled by constraints for each phase. This constraint depends on the type of phase. A very dilute aqueous liquid can be seen as (almost) pure water and, therefore, the activity of water can be assumed unity. For moderately saline liquids, the activity of water equals its molar fraction. The activity of water in high salinity solutions needs to be calculated from the osmotic coefficient used in the model of Pitzer (1973). For a gas phase the sum of the partial pressures of all gaseous species must equal the total gas pressure. If the phase is pure (it only consists of one species), the partial pressure equals the total pressure. Major minerals are usually considered to precipitate as pure phases (see also Chapter 3, section 3.3.3). Therefore, their activities are usually assumed unity, although exceptions can occur (Glynn, 2000).

If these constraints fix the activities (e.g., unit activity of water and/or minerals), the number of coupled transport equations can be reduced. Several methods have been proposed (Saaltink *et al.*, 1998; Molins *et al.*, 2004; De Simoni *et al.*, 2007). The most simple and illustrative consists of calculating the component matrix, by applying Gauss-Jordan elimination as in the previous section, but using the species of constant activity as primary species. The component matrix of equation (3.15) is divided into a part for the constant activity species (indicated by subscript 1) and a part for the rest (indicated by subscript 2):

$$\mathbf{U} = \left(\mathbf{I} \mid \left(\mathbf{S}_1^*\right)^T\right) = \begin{pmatrix} \mathbf{U}_1 \\ \mathbf{U}_2 \end{pmatrix} = \begin{pmatrix} \mathbf{I} & \mathbf{0} & \left(\mathbf{S}_{1.1}^*\right)^T \\ \mathbf{0} & \mathbf{I} & \left(\mathbf{S}_{1.2}^*\right)^T \end{pmatrix} \quad (4.74)$$

Likewise, the vectors of concentrations are divided:

$$\mathbf{m} = \begin{pmatrix} \mathbf{m}_1 \\ \mathbf{m}_2 \end{pmatrix} = \begin{pmatrix} \mathbf{m}_{1.1} \\ \mathbf{m}_{1.2} \\ \mathbf{m}_2 \end{pmatrix} \quad \mathbf{c} = \begin{pmatrix} \mathbf{c}_1 \\ \mathbf{c}_2 \end{pmatrix} = \begin{pmatrix} \mathbf{c}_{1.1} \\ \mathbf{c}_{1.2} \\ \mathbf{c}_2 \end{pmatrix} \quad (4.75)$$

With these definitions one can rewrite equation (4.68) as:

$$\frac{\partial}{\partial t}\left(\mathbf{m}_{1.1} + \left(\mathbf{S}_{1.1}^*\right)^T \mathbf{m}_2\right) = \mathbf{L}\left(\mathbf{c}_{1.1} + \left(\mathbf{S}_{1.1}^*\right)^T \mathbf{c}_2\right) + \mathbf{U}\mathbf{S}_k^T \mathbf{r}_k \quad (4.76)$$

$$\frac{\partial}{\partial t}\left(\mathbf{m}_{1.2} + \left(\mathbf{S}_{1.1}^*\right)^T \mathbf{m}_2\right) = \mathbf{L}\left(\mathbf{c}_{1.2} + \left(\mathbf{S}_{1.2}^*\right)^T \mathbf{c}_2\right) + \mathbf{U}\mathbf{S}_k^T \mathbf{r}_k \quad (4.77)$$

Assuming that rate laws do not depend on constant activity species, the transport equations of components having constant activity species as primary species (4.76) do not affect the other transport equations (4.77). The fact that the solution of equation (4.77) is independent of $\mathbf{c}_{1.1}$, the solution of equation (4.76), results from two properties. First, being primary, these species do not form part of the other components. Second, they do not affect any secondary species through the mass action law because their activities are fixed, independent of their concentration. Therefore, first equation (4.77) can be solved, in which the mass action laws have been substituted. Then, (4.76) can be solved.

This is illustrated by means of the example of the previous section, where it is assumed that H_2O and $CaCO_3$(s) have fixed activities equal to unity. For primary species these two fixed activity species are chosen plus Cl^-, CO_2(aq) and H^+. As explained in Chapter 3 (section 3.1.3) the mass action law can be written in a form such that the activities of secondary species are an explicit function of primary species:

$$\begin{pmatrix} \log a_{Ca^{2+}} \\ \log a_{OH^-} \\ \log a_{HCO_3^-} \\ \log a_{CO_3^{2-}} \\ \log a_{CO_2(g)} \end{pmatrix} = \begin{pmatrix} -1 & 1 & 0 & -1 & 2 \\ 1 & 0 & 0 & 0 & 1 \\ 1 & 0 & 0 & 1 & -1 \\ 1 & 0 & 0 & 1 & -2 \\ 0 & 0 & 0 & 1 & 0 \end{pmatrix} \begin{pmatrix} \log a_{H_2O} \\ \log a_{CaCO_3(s)} \\ \log a_{Cl^-} \\ \log a_{CO_2(aq)} \\ \log a_{H^+} \end{pmatrix} + \log \mathbf{k}^*$$

where $\mathbf{k}^*$ is the vector of equilibrium constants.

As the activities of H_2O and $CaCO_3$(s) equal unity, one can also write it removing these species:

$$\begin{pmatrix} \log a_{Ca^{2+}} \\ \log a_{OH^-} \\ \log a_{HCO_3^-} \\ \log a_{CO_3^{2-}} \\ \log a_{CO_2(g)} \end{pmatrix} = \begin{pmatrix} 0 & -1 & 2 \\ 0 & 0 & 1 \\ 0 & 1 & -1 \\ 0 & 1 & -2 \\ 0 & 1 & 0 \end{pmatrix} \begin{pmatrix} \log a_{Cl^-} \\ \log a_{CO_2(aq)} \\ \log a_{H^+} \end{pmatrix} + \log \mathbf{k}^*$$

It gives the following component matrix:

$$\mathbf{U} = \left(\begin{array}{c|cc|ccc|ccccc} & H_2O & CaCO_3(s) & Cl^- & CO_2(aq) & H^+ & Ca^{2+} & OH^- & HCO_3^- & CO_3^{2-} & CO_2(g) \\ \sum H_2O & 1 & 0 & 0 & 0 & 0 & -1 & 1 & 1 & 1 & 0 \\ \sum CaCO_3 & 0 & 1 & 0 & 0 & 0 & 1 & 0 & 0 & 0 & 0 \\ \hline \sum Cl & 0 & 0 & 1 & 0 & 0 & 0 & 0 & 0 & 0 & 0 \\ \sum C-Ca & 0 & 0 & 0 & 1 & 0 & -1 & 0 & 1 & 1 & 1 \\ \sum H^+ & 0 & 0 & 0 & 0 & 1 & 2 & 1 & -1 & -2 & 0 \end{array} \right)$$

Due to the structure of the component matrix (the left-hand part is the identity matrix), the last three components (labeled ΣCl, ΣC–Ca, and ΣH^+) do not contain the constant activity species H_2O and $CaCO_3$(s). This, together with the fact that the mass action laws can be written independent of the constant activity species, makes the transport equations of these last three components independent of the first two. So one can first calculate the concentrations of Cl^-, CO_2(aq), H^+ and all secondary species from the transport equations of the last three components and the mass action laws. Then, one can calculate the concentrations of H_2O and $CaCO_3$(s) from the transport equations of the first two components.

A situation can occur where some phases are present in only part of the domain. This is especially true for minerals, which may be present only in some small fraction in part of the domain. They may disappear upon later reaction. This means that the component matrix and/or the set of primary and secondary species may depend on space and time. Although, still possible to solve, it complicates the numerical solution. Also it affects some methods of numerical solution more than others, as will be explained in Chapter 5 (section 5.3). Still, the simplicity of the above formulation will be used below to illustrate some basic features of reactive transport.

4.5.4 *Analytical solution for a binary system: equilibrium reaction rates*

4.5.4.1 *Problem statement*

A pure dissolution/precipitation reaction at equilibrium is considered, where an immobile solid mineral S_{3s} dissolves reversibly to yield ions B_1 and B_2 in a saturated porous medium ($\theta_l = \varphi$). This system was considered by de Simoni *et al.* (2005) to obtain the analytical solution to a reactive transport problem, which is outlined in this section. The basic reaction is:

$$B_1 + B_2 \;=\; S_{3s} \tag{4.78}$$

It is further assumed that the mineral, S_{3s}, is a pure phase, so that its activity equals 1. As discussed in the previous section, vector $\mathbf{m}_a$ contains the mass per unit volume of medium, $m_1 = \varphi c_1$ and $m_2 = \varphi c_2$, of the aqueous species B_1 and B_2 respectively, while $\mathbf{m}_c$ contains the mass per unit volume of medium, m_3, of the solid mineral, S_{3s}. The stoichiometric matrix of the system described by (4.78) is:

$$\mathbf{S}_e = \begin{pmatrix} -1 & -1 & 1 \end{pmatrix} \tag{4.79}$$

For this system, $\mathbf{S}_e$ can be split as:

$$\mathbf{S}_{ec} = (1), \tag{4.80a}$$

$$\mathbf{S}_{ea} = \left(\mathbf{S}'_{ea} \mid \mathbf{S}''_{ea}\right) = \left(-1 \mid -1\right) \tag{4.80b}$$

Matrices $\mathbf{S}_{ec}$ and $\mathbf{S}_{ea}$ contain the stoichiometric coefficients related to the constant activity species and to the remaining aqueous species, respectively. Equation (4.80b) implicitly identifies B_1 as primary species and B_2 as secondary.

Thus, the mass action law for the considered system is expressed as:

$$\log c_1 + \log c_2 = \log K' \tag{4.81}$$

The apparent equilibrium constant, K', is strictly related to solubility of the solid phase, S_{3s}, and usually depends on temperature and pressure. However, since activity coefficients are not included in equation (4.81), K' will also depend on the chemical composition of the solution. For very diluted solutions the apparent equilibrium constant, K', equals the normal equilibrium constant, K.

Recalling equation (4.72), the mass balance equations for the three species are:

$$\frac{\partial(\varphi c_1)}{\partial t} - L_t(c_1) = -r, \tag{4.82a}$$

$$\frac{\partial(\varphi c_2)}{\partial t} - L_t(c_2) = -r, \tag{4.82b}$$

$$\frac{\partial(m_3)}{\partial t} = r \tag{4.82c}$$

where r is the rate of the reaction 4.78. The reaction rate expresses the moles of B_1 (and B_2) that precipitate in order to maintain equilibrium conditions at all points in the moving fluid (as seen by eq. 4.82a and 4.82b) and coincides with the rate of change of m_3 (as seen by 4.82c).

The non-linear problem described by the mass balances of the aqueous species (eq. 4.82a and 4.82b) and the local equilibrium condition (eq. 4.81) needs to be solved to obtain the space-time distribution of the concentrations of the two aqueous species and the reaction rate. The rate of change in the solid mineral mass is then provided by equation (4.84c).

4.5.4.2 *Methodology of solution*

The general procedure to solve a multi-species transport process was outlined in section 4.5.3. Its application to the binary system described above is shown here, along with a closed-form analytical solution. The solution procedure develops according to the following steps:

i. *Definition of mobile conservative components of the system.* As explained in section 4.5.3, the component matrix is:

$$\mathbf{U} = \begin{pmatrix} 1 & -1 \end{pmatrix} \tag{4.83}$$

This implies that one needs to solve the transport problem for only one conservative component $u = Uc_a = c_1 - c_2$. Notice that simply subtracting equation (4.82b) from (4.82a) leads to the equation which governs the transport of the conservative component $u = (c_1 - c_2)$. In other words, dissolution or precipitation of the mineral S_{3s} equally affects c_1 and c_2 so that the difference $(c_1 - c_2)$ is not altered.

ii. *Transport of conservative components.* Only one component is necessary in this case and the conservative transport equation is solved for u.

iii. *Speciation.* Here one needs to compute the concentrations of $(N_s - N_c)$ mobile species from the concentrations of the components. In our binary system, this implies solving

$$c_1 - c_2 = u, \tag{4.84a}$$

$$\log c_1 + \log c_2 = \log K' \tag{4.84b}$$

In case of very diluted solutions (K' equals K and is independent of c_1 and c_2), the solution of (4.84a) and (4.84b) is:

$$c_1 = \frac{u + \sqrt{u^2 + 4K'}}{2}, \tag{4.85a}$$

$$c_2 = \frac{-u + \sqrt{u^2 + 4K'}}{2} \tag{4.85b}$$

iv. *Evaluation of the reaction rate.* Substitution of the concentration of the secondary species, B_2, into its transport equation (4.82b) leads to:

$$\frac{r}{\varphi} = \frac{\partial^2 c_2}{\partial u^2}(\nabla u)^T \cdot \mathbf{D} \cdot (\nabla u) + 2\frac{\partial^2 c_2}{\partial u \partial K'}(\nabla u)^T \cdot \mathbf{D} \cdot (\nabla K') + \frac{\partial^2 c_2}{\partial K'^2}(\nabla K')^T \cdot \mathbf{D} \cdot (\nabla K') \tag{4.86}$$

When K' is a function of conservative quantities such as salinity, s, equation (4.86) can be rewritten as:

$$\frac{r}{\varphi} = \frac{\partial^2 c_2}{\partial s^2}(\nabla s)^T \cdot \mathbf{D} \cdot (\nabla s) + \frac{\partial^2 c_2}{\partial u^2}(\nabla u)^T \cdot \mathbf{D} \cdot (\nabla u) + 2\frac{\partial^2 c_2}{\partial u \partial s}(\nabla u)^T \cdot \mathbf{D} \cdot (\nabla s) \tag{4.87}$$

In general, when many components and equilibrium reactions occur simultaneously, the vector of reaction rates is (de Simoni *et al.*, 2005):

$$\frac{\mathbf{r}}{\varphi} = \mathbf{H}(\nabla \mathbf{u})^T \cdot \mathbf{D} \cdot (\nabla \mathbf{u}) \tag{4.88}$$

where **r** is the vector of (equilibrium) reaction rates, **u** is the vector of components (possibly extended to include temperature and/or pressure in the case of non-isothermal and/or non-isobaric systems) and H is the Hessian matrix (or vector) of secondary species with respect to components. That is, $H_{ij}^k = \partial^2 c_{2k} / \partial u_i \partial u_j$, where c_{2k} is the concentration of the secondary species associated to the kth equilibrium reaction. In the case of the binary system considered here, the Hessian reduces to:

$$\frac{\partial^2 c_2}{\partial u^2} = \frac{2K'}{\left(u^2 + 4K'\right)^{3/2}} \tag{4.89}$$

In general, reaction rates should be used to compute the mass change of solid mineral. In turn, this would cause a modification in the medium properties (such as, as porosity and permeability). Here, however, these changes are neglected.

The results encapsulated in equation (4.86), (4.87) and (4.88) deserve discussion. First of all, under chemical equilibrium conditions, they provide a way to compute directly the rate of dissolution/precipitation as a function of quantities such as the concentrations of components, the equilibrium constants, and the dispersion coefficients, without the need to actually evaluate the concentrations of the dissolved species.

Furthermore, the reaction rate is always positive (i.e., precipitation occurs) in systems where K is constant. This is consistent with the comments of Rubin (1983), who points out that, in the case of the reaction 4.78 under very dilluted conditions ($K' = K$), reactive transport processes cannot result in dissolution of the solid mineral. This is also evident from Figure 4.9. Figure 4.9 displays another interesting feature. The equilibrium point can be obtained by drawing a line from the conservative mixing point towards the equilibrium line. The slope of this line is equal to the ratio of the stoichiometric coefficients and is equal to 1 in our example.

Equations (4.86) and 4.88 clearly demonstrate that the reaction rate depends on chemistry, which controls $\partial^2 c_2/\partial u^2$, but also on transport processes, controlling the gradient of u. Figure 4.10 qualitatively shows a synthetic example, which allows evaluation of the relative importance of the two factors (chemistry and transport) in the reaction rate. It is difficult to make general statements about which one is more important, which is consistent with the sensitivity analysis of Tebes-Stevens *et al.* (2001). One of the most paradoxical features of Figure 4.10 is that reaction does not necessarily take place where concentrations attain their maximum values. In fact, the reaction rate equals zero when u is maximum or minimum (i.e., when c_1 or c_2 reach their maximum value, respectively). In general (see eq. 4.88), the reaction rate depends on both the gradients of u and K'. A non-zero gradient of K', for example,

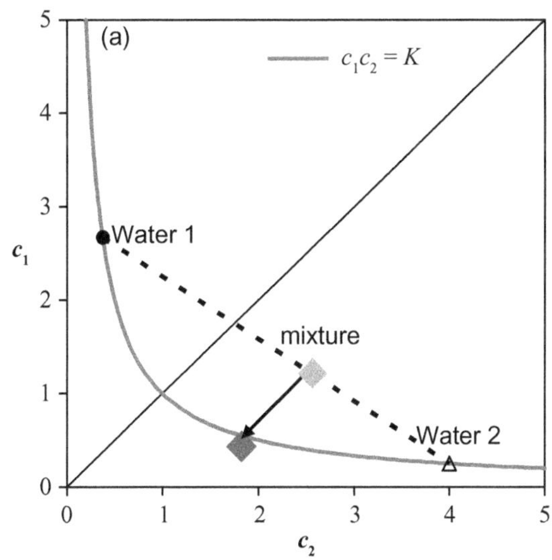

Figure 4.9. Mixing of two waters. The concentrations of the dissolved species, c_1 and c_2, in the mixing waters satisfy the equilibrium condition. From the conservative mixing point, concentrations weighted averages of end-members, the mixture evolves toward the equilibrium curve following a straight line (whose slope is given by the ratio of the stoichiometric coefficients and is equal to 1 in the simple binary case presented in the text).

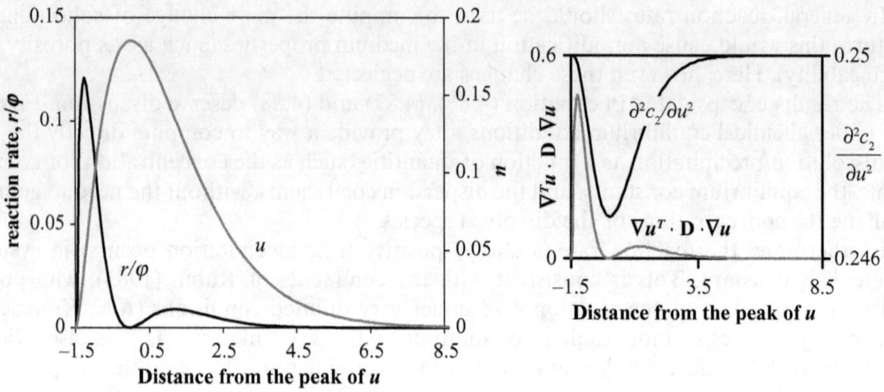

Figure 4.10. An example of the evaluation of (4.90a). Function u (arbitrarily chosen, for illustrative purposes) and the two factors in (4.90a) are displayed versus the distance from the location at which u reaches its maximum value. Note the different scales.

can occur when the mixing of different waters induces spatial variability in temperature or salinity (Berkowitz *et al.*, 2003; Rezaei *et al.*, 2005), as studied in section 4.5.1.

It is interesting that, in the absence of the first contribution on the right hand side of equation (4.86), all terms are proportional to the dispersion tensor, **D**, thus strengthening the relevance of mixing processes to the development of such reactions. In particular, the term $(\nabla u)^T \cdot \mathbf{D} \cdot (\nabla u)$ can be used as a measure of the mixing rate, which is consistent with the concept of dilution index, as defined by Kitanidis (1994) on the basis of entropy arguments. This result also suggests that evaluating mixing rates may help to properly identify not only the sources of water (Carrera *et al.*, 2004), but also the geochemical processes occurring in the system.

4.5.4.3 *An analytical solution: pulse injection in a binary system*

The general methodology is applied to derive a closed-form solution for a mixing-driven precipitation reaction. A three-dimensional homogeneous porous formation is considered, with constant porosity, ϕ, under uniform and saturated flow conditions. The system is affected by an instantaneous point-like injection of water containing the same constituents than the initial resident water. This problem can be used as a kernel for other injection functions. The case of constant K is considered. It is assumed that the velocity is aligned with the x-coordinate, $u_f = v_f / \varphi$, and that the dispersion tensor is diagonal, D_L and D_T respectively being its longitudinal and transverse components. It is further assumed that the time of the reaction is small as compared to the typical time of transport (i.e., the local equilibrium assumption applies).

The reactive transport system is governed by (4.81) and (4.82). The boundary conditions are:

$$c_{i\infty} = c_i(\mathbf{x} \to \infty, t) = c_{i0}, \quad i = 1;2 \tag{4.90}$$

Initially, resident water is displaced by injecting a volume V_e of solution with concentration $c_{iext} = c_{i0} + c_{ie}$. In order to find an analytical solution for small V_e, it is mathematically convenient to write the initial concentration condition, after equilibrium is reached at the injection point, as:

$$c_i(\mathbf{x}, t = 0) = c_{ie} V_e \delta(\mathbf{x}) / \varphi + c_{i0}, \quad i = 1;2 \tag{4.91}$$

where equilibrium must be satisfied at all points in the aquifer. This implies that (a) $c_{10}c_{20} = K$, and (b) $(c_{10} + c_{1e})(c_{20} + c_{2e}) = K$.

With this in mind, the steps followed are detailed in section 4.5.4.2. The (conservative) component $u = c_1 - c_2$ satisfies equation (4.74) with boundary and initial conditions:

$$u_\infty = u(\mathbf{x} \to \infty, t) = c_1(\mathbf{x} \to \infty, t) - c_2(\mathbf{x} \to \infty, t) = u_0, \tag{4.92a}$$

$$u(\mathbf{x}, t = 0) = u_e V_e \delta(\mathbf{x}) / \varphi + u_0 \tag{4.92b}$$

With this definition, u_e is then the excess of the injected component u that remains in the aquifer immediately after injection. The solution for conservative transport is (Domenico and Schwartz, 1997, page 380)

$$u(\mathbf{x}, t) = u_0 + \frac{18}{(2\pi)^{1/2}} \frac{u_e}{\bar{V}_d} \exp\left[-\frac{1}{2}\rho^2\right] \tag{4.93}$$

where ρ is the normalized radial distance from the centre of the plume, defined as:

$$\rho = \sqrt{\frac{(x - u_f t)^2}{2tD_L} + \frac{y^2 + z^2}{2tD_T}} \tag{4.94}$$

The dimensionless quantity $\bar{V}_d$ is the ratio between the volume containing about 99.7% of the excess of injected mass (Domenico and Schwartz, 1997) and the injection volume:

$$\bar{V}_d = \frac{72\pi n\sqrt{2}\, t^{3/2} \sqrt{D_L D_T}}{V_e} \tag{4.95}$$

and is a measure of the temporal evolution of the dispersive effect.

Substituting equation (4.93) into (4.85a) and (4.85b), one obtains the concentrations, c_1 and c_2, of the dissolved species. Finally the expression of the local mineral mass precipitation rate, r, per unit volume of medium is derived from (4.89) as:

$$r(\mathbf{x}, t) = \frac{nK}{t} \frac{\left(\frac{18}{(2\pi)^{1/2}} \frac{u_e}{\bar{V}_d}\right)^2 \rho^2 \exp\left[-\rho^2\right]}{\left\{\left(u_0 + \frac{18}{(2\pi)^{1/2}} \frac{u_e}{\bar{V}_d} exp\left[-\frac{1}{2}\rho^2\right]\right)^2 + 4K\right\}^{3/2}} \tag{4.96}$$

The rate, r, vanishes in the trivial case of $u_e = 0$, since the concentrations of the two injected species are the same as those in the background water and there is no excess of injected mass in the solution right after the injection. The rate vanishes also when $\bar{V}_d \to \infty$. The latter situation describes a scenario where gradients of the component u are negligible because of large dispersion or because a long time has elapsed since injection.

To illustrate the features of this solution the following problem is considered: one starts with a background water characterized by $c_{10}/\sqrt{K} = 0.25$ and $c_{20}/\sqrt{K} = 4.0$ ($u_0/\sqrt{K} = -3.75$); then water is injected from an external source, characterized by $c_{1ext}/\sqrt{K} = 0.184$, $c_{2ext}/\sqrt{K} = 5.434$ ($u_{ext}/\sqrt{K} = -5.25$). The equilibrium condition is satisfied when $c_{1e}/\sqrt{K} = -0.066$, $c_{2e}/\sqrt{K} = 1.434$ ($u_e/\sqrt{K} = -1.5$). Figure 4.11a depicts the dependence of the dimensionless concentrations $\tilde{c}_1 = c_1/\sqrt{K}$ and $\tilde{c}_2 = c_2/\sqrt{K}$ on the normalized distance from the center of the (moving) plume, $(x - Vt)/\sqrt{2tD_L}$ (while $z = y = 0$), and $\bar{V}_d = 3.5$. Dimensionless concentrations ($\tilde{c}_{1NR}$, $\tilde{c}_{2NR}$) for the corresponding non-reactive system are also shown for comparison. While both concentrations

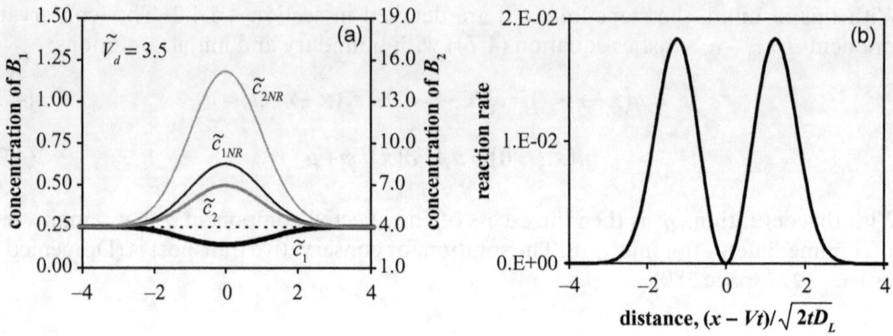

Figure 4.11. Dimensionless concentrations (a) and reaction rate (b) in response to a pulse injection in a binary system. Notice that precipitation causes a more significant decrease in concentrations in the reactive case than in the non reactive (NR) case. In fact c_1 falls below initial and input values. Notice also that precipitation occur on the plume edges, but not in the center.

are higher than and proportional to the initial concentrations in the non-reactive case, in the reactive case $\tilde{c}_1$ decreases while $\tilde{c}_2$ increases (in agreement with the fact that $c_{1e} < 0$ and $c_{2e} > 0$).

The spatial distribution of the (local) dimensionless reaction rate, $\tilde{r} = rt/(\varphi\sqrt{K})$, for the same conditions of Figure 4.11a, is depicted in Figure 4.11b. A comparison of these figures clearly shows that the system is chemically active (i.e., the chemical reaction rate is significant) at locations where concentration gradients of both species are relevant. This implies, in turn, that strong gradients of the component u give rise to significant reaction rates, in agreement with equation (4.89). In general, no reactions occur within the system when concentration gradients vanish. As a consequence, no reaction occurs at the (moving) centre of the plume, that is, at the points of highest (or lowest) concentration values.

According to equation (4.96), a change in D_T has stronger effects rather than a modification in D_L. This is because D_T is not squared in the definition of $\bar{V}_d$, and because dispersion is enhanced along two directions by an increase in D_T, while an increase in D_L affects only one spatial direction. This topic will be discussed further below.

The sensitivity of the reaction rate to u_e for a given u_0 is presented in Figure 4.12, with reference to a resident water with $u_0/\sqrt{K} = -20$ ($c_{10}/\sqrt{K} = 0.05$; $c_{20}/\sqrt{K} = 20.05$). The case with a negative u_e ($u_e/\sqrt{K} = -30$) is considered first. The dimensionless mixing volume is set as $\bar{V}_d = 3.5$. The shape of the reaction rate function is displayed in Figure 4.12a at the plane $z = 0$. From the plot it is clear that, at any given time, precipitation concentrates in a (three-dimensional) aureole around the moving center of the plume. The actual location and shape of this aureole is governed by equation (4.96), and depends on D_L, D_T and u_e. One should note that Figure 4.12a displays an artificial symmetry, as coordinate x is normalized by $\sqrt{2tD_L}$, while the y direction is normalized by $\sqrt{2tD_T}$. Figure 4.12b depicts radial profiles of the reaction rate, r, for different u_e. Due to the symmetry of the solution with respect to the normalized coordinates, only radial profiles are displayed starting from the plume center (Fig. 4.12b) and organized in such a way that while the curves resulting from positive values of u_e are displayed on the larger scale, those arising by negative values of u_e are displayed within the insert. Figure 4.12b reveals that the reaction rate is larger when $u_e/\sqrt{K}$ and $u_0/\sqrt{K}$ have opposite signs.

All information related to dispersion is concentrated in $\tilde{V}_d$. The sensitivity of r_Ω, total cumulative precipitate, on dispersion can be analyzed from the following expression (de Simoni *et al.*, 2005):

$$\frac{\partial r_\Omega}{\partial D_L} = \frac{\partial r_\Omega}{\partial \tilde{V}_d}\frac{\partial \tilde{V}_d}{\partial D_L} = \frac{\partial r_\Omega}{\partial \tilde{V}_d}\frac{\tilde{V}_d}{2D_L} = \frac{D_T}{2D_L}\frac{\partial r_\Omega}{\partial D_T} \tag{4.97}$$

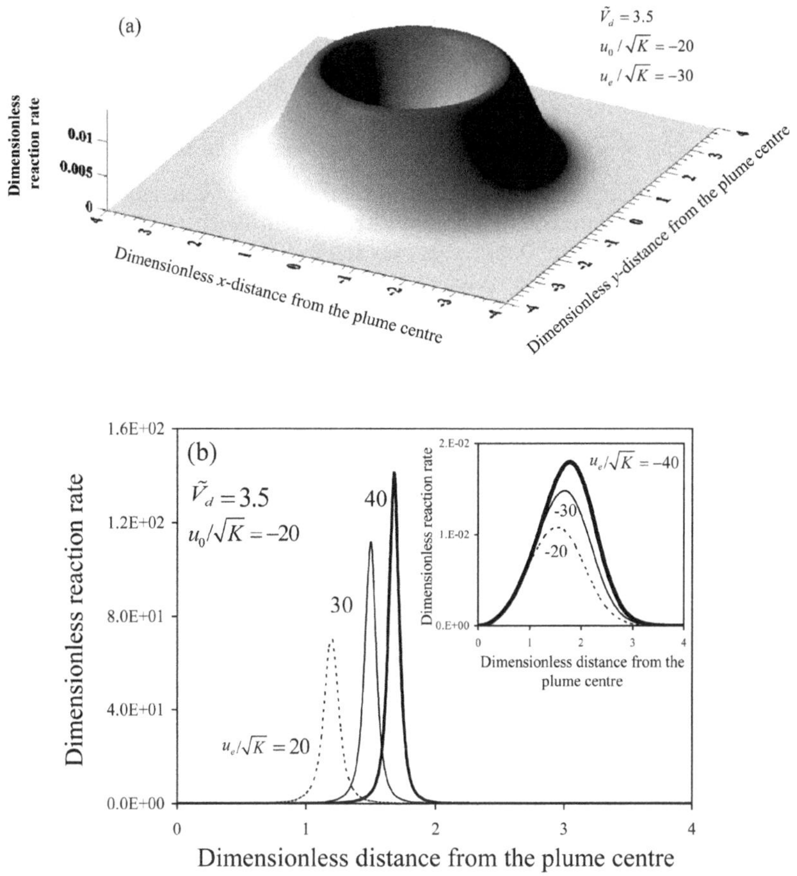

Figure 4.12. Spatial distribution of precipitation rate (a), and cross section of precipitation rate for several values of the input concentration.

Equation (4.97) shows that the overall reaction rate, r_{Ω}, is much more sensitive to D_T than to D_L. Considering that longitudinal dispersion is often taken to be about five to ten times greater than transverse dispersion, equation (4.97) implies that the overall reaction rate would be ten to twenty times more sensitive to D_T than to D_L. Furthermore, the input of solutes is frequently continuous in time and reactants do not enter directly into the flow domain, as they are laterally driven into it. Both factors would tend to enhance the role of transverse dispersion in reactive transport problems. This is particularly relevant in view of the uncertainties surrounding the actual values of transverse dispersion. While some argue that transverse dispersion tends to zero in three-dimensional domains for large travel times (Dagan, 1989), others (Neuman and Zhang, 1990; Zhang and Neuman, 1990; Attinger *et al.*, 2004) show that this is not the case, thus corroborating the idea that the actual transverse dispersion is linked to the interplay between spatial heterogeneity and time fluctuations of velocity (Cirpka and Attinger, 2003; Dentz and Carrera, 2003).

4.6 THE EFFECT OF HETEROGENEITY AND NON-LOCAL FORMULATIONS

The formulations presented in previous sections for flow (section 4.2), solute transport (section 4.3), heat transport (section 4.4) and reactive transport (section 4.5) display severe limitations when applied to geological media. This section is devoted to analyzing the limitations of those formulations (hereinafter termed "traditional" formulations) and to outlining alternative formulations.

4.6.1 *The limitations of traditional formulations and the need for upscaling*

The need for upscaling can be best illustrated with an example. Watching a movie of boats in the ocean, alert spectators immediately and intuitively identify whether the boats were filmed in a pool or in the actual ocean. Pool waves may look as large as ocean ones, but they never look as foamy. The reason is that surface tension, which is an important force for small water bodies, becomes negligible when compared the inertial forces of true ocean waves. While it is evident that dominant processes may change with scale, the conventional continuum mechanics approach does not acknowledge this. The approach follows several steps. First, processes are observed at the laboratory scale (e.g., column experiments such as Darcy's experiment or the one in Figure 4.3). Second, observations are generalized into constitutive laws (e.g., Darcy's or Fick's laws). The third step is to invoke some conservation law to derive a partial differential equation (e.g., fluid mass conservation, equation (4.15), or energy conservation, equation (4.61)). This equation is then used to predict the behavior of the system at different scales, under different conditions than that of the laboratory observations.

The continuum mechanics approach is fine as long as the main processes remain unaltered. However, this is rarely the case in natural phenomena. In the ocean waves example, viscous and surface tension forces are important when considering small (laboratory) waves. They hinder the formation of eddies and small bubbles. However, inertial forces become increasingly important as the size of the waves increases, to the point that dissipating some energy into formation of bubbles or eddies becomes unimportant. The end result is not only the foamy and turbulent appearance of large waves, but also a decrease in density and viscosity of the water bubbles, which causes their process dynamics to be essentially different.

In geological media, spatial variability (heterogeneity) of medium properties is the main driving force behind scale effects. Variability is important not only because of the uncertainty it brings (it is impossible to know medium properties at every point), but also because large-scale behavior of a spatially variable phenomenon may be significantly different from the small-scale behavior. That is, changes of spatial (or time) scale may (i) lead to changes in effective parameters, (ii) cause new processes to emerge, and therefore (iii) require that the governing equations be changed.

The term "scale effect" refers to any of these changes. Upscaling is the procedure to derive these changes (in parameters, relevant processes, or governing equations) assuming that parameters, processes, and governing equations are known at a smaller scale. Spatial and temporal variability are (i) very important, and (ii) impossible to describe accurately. Therefore, scale effects should be generally expected. Scale effects are described below for flow and for conservative and reactive transport.

Single phase flow is governed by the flow equation (4.15), which is based on conservation of fluid mass and momentum. The latter is expressed by Darcy's law, whose only parameter is hydraulic conductivity. While the flow equation is broadly accepted, it is also well known that hydraulic conductivity displays significant scale effects, that is, its representative value grows with the size of the problem (Fig. 4.13). This implies that the hydraulic conductivity measured at one scale need not be equal to that described at a different, larger, scale. Measurements are often made at small scales (e.g., ms), while flow may be need to be described at much larger (e.g., kms) scales. Procedures for upscaling flow are described by Sanchez-Vila *et al.* (2006).

The situation is much worse for solute transport. Traditionally transport through homogeneous media has been represented by the ADE (advection-dispersion equation, 4.50). The solution of this equation for constant dispersion and velocity in response to a unit input was shown in Figure 4.5. The main point in this figure was that the solute input was displaced and dispersed.

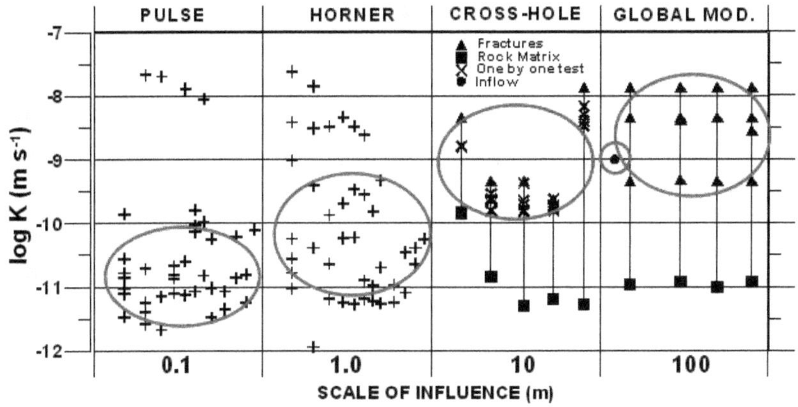

Figure 4.13. Hydraulic conductivity measured at different scales in a fractured granite formation (Martinez-Landa and Carrera, 2005). Notice that the representative hydraulic conductivity increases by two orders of magnitude when going from the 0.1 to the 100 m scale.

These phenomena (displacement and dispersion) do occur. However, also implicit in Figure 4.5, is the assumption of equivalence between spreading and mixing. That is, the ADE implies that all the domain between the leading and tailing edges of the plume is occupied by fluid with (some) solute. Actually, some portions of the pore volume may have not been reached by the solute. The initial fluid will likely lag behind the leading solute front in zones of small permeability (Fig. 4.14).

Given these limitations, it should not come as a surprise that observed chemical transport in heterogeneous media does not behave as implied by Figure 4.5 (e.g., Adams and Gelhar, 1992; Neuman and Zhang, 1990; Carrera, 1993; Berkowitz and Scher, 1997; Steefel *et al.*, 2005; Meile and Tuncay, 2006). For one thing both apparent dispersivity (Fig. 4.15) and porosity (Fig. 4.16) display marked scale effects. Dispersivity grows (almost) linearly with the scale of the problem, whereas porosity in fractured media grows with residence time (almost linearly with the square root of time).

Differences are not restricted to transport parameters. The shape of breakthrough curves observed in fields experiments observations differs from ADE predictions (Fig. 4.17). It is therefore clear that the ADE is not adequate for simulating transport. A number of alternatives have been proposed in recent years. They will be discussed in section 4.5.2. However, the driving force in most recent developments has been reactive transport. When dealing with reactive transport, it is important to distinguish between fast reactions, which tend to proceed to equilibrium, and slow reactions, whose rate depends somewhat on the extent of disequilibrium (such as expressed by saturation indices), but also on a variety of factors affecting reaction mechanistics, pathways and rates (e.g., sorption, catalysis, inactivation).

Rezaei *et al.* (2005) showed that the actual rate of equilibrium reactions in porous media is affected by mixing. The interplay between transport and chemistry is non trivial. Specifically, performing geochemical calculations is needed, but by itself it will not predict where reactions will take place, what will be their rate, or under which conditions the rate of reaction processes will be at a maximum. These outcomes are affected by mixing, which can be affected in its turn by spreading and water flow. In fact, shortly thereafter, de Simoni *et al.* (2005) found a relatively simple expression to quantify mixing as a result of Fickian dispersion and diffusion. A result from these and many other works is that proper representation of reactive transport requires a proper representation of mixing.

One might argue that slow reactions, whose rate is controlled by local chemistry through kinetic rate laws, should not suffer from scale effects. As it turns out, kinetic reaction rates observed in the field are much slower than what might be expected from laboratory

Figure 4.14. Mixing refers to the rate at which different waters blend together. As such, it is controlled by local gradients and tends to destroy local irregularities in concentration. Spreading, on the other hand, tends to increase the extent of the plume or the overall width of an advancing front in both longitudinal and lateral direction. They are linked, because irregular spreading tends to generate gradients perpendicular to the flow direction, but they are different concepts. Yet, both are equated in the ADE.

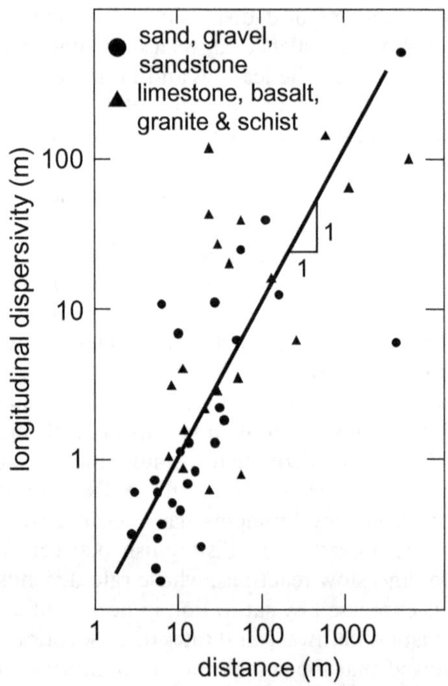

Figure 4.15. Dispersivity values collected from tracer tests and pollution plumes worldwide (Lallemand-Barres and Peaudecerd, 1978). Dispersivity grows linearly with the scale of the test.

measurements (Fig. 4.18). This implies that local concentrations are not well represented in the ADE either. In fact, most of the alternatives to the ADE discussed adopt non local formalisms. That is, transport processes at one point in space and time depend not only on the concentration field at that point, but also on its spatial variability, possibly far away, and on its time history. These alternative transport representations contribute to explaining scale effects on reaction rates, which lends further support to these non local representations of transport.

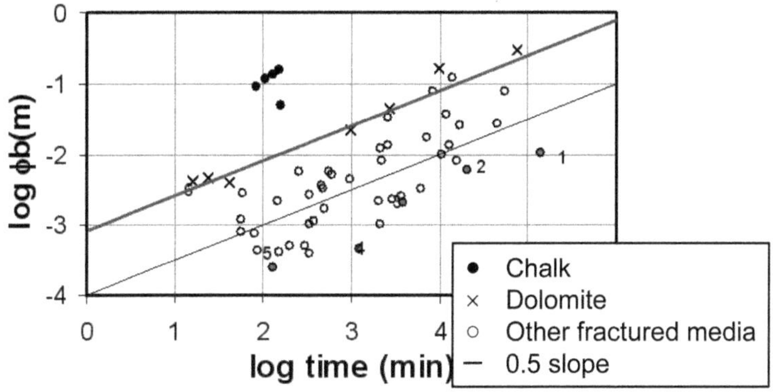

Figure 4.16. Effective kinematic porosities (water flux divided by tracer velocity, as measured from peak arrival times) derived from tracer tests performed in fractured rock environments worldwide (Guimera and Carrera, 2000). The effective porosity increases with residence time as $t_{1/2}$. Particular data points at the bottom of the trend are expressed in grey circles: 1. Birgersson *et al.* (1993); 2. García-Gutiérrez *et al.* (1997); 3. Chalk River dipole 6–11 after Raven *et al.* (1988); 4. Chalk River convergent after Raven *et al.* (1988); 5. Chalk River after Novakowski *et al.* (1985); and 6. Shapiro and Nicholas (1989), three results in a single fracture in dolomite.

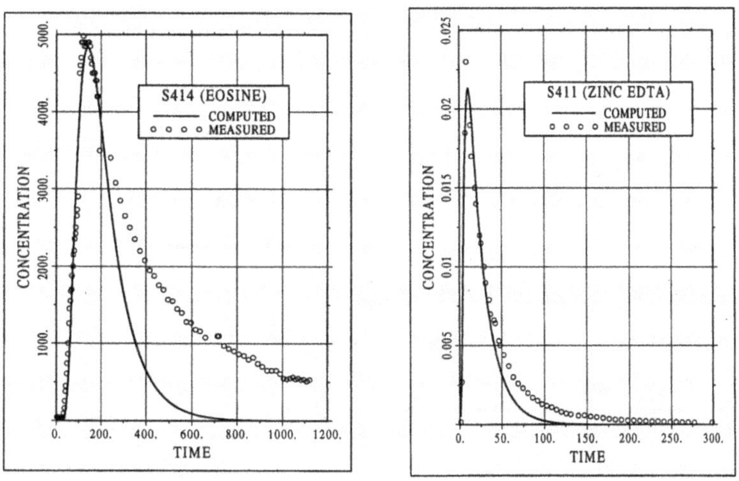

Figure 4.17. Breakthrough curves measured (dots) and computed with the ADE (lines) for a tracer test performed in metamorphic rocks at El Cabril (Spain) (Carrera, 1993). The ADE model, which had been calibrated with early time data (up to shortly after the peak) clearly fails to reproduce tailing at later times. This behavior is quite frequent.

4.6.2 *Solution of reactive transport in MRMT formulations*

The Multi Rate Mass Transfer (MRMT) system can be viewed as depicted in Figure 4.19. That is, water flows through a mobile region but solutes can diffuse into (a set of) immobile regions. That is:

$$\varphi_m \frac{\partial c_m}{\partial t} = -\vec{\nabla} \cdot \vec{v}_f c_m + \vec{\nabla} \cdot (\mathbf{D}_h \cdot \nabla c_m) - \sum_i F_i \tag{4.98a}$$

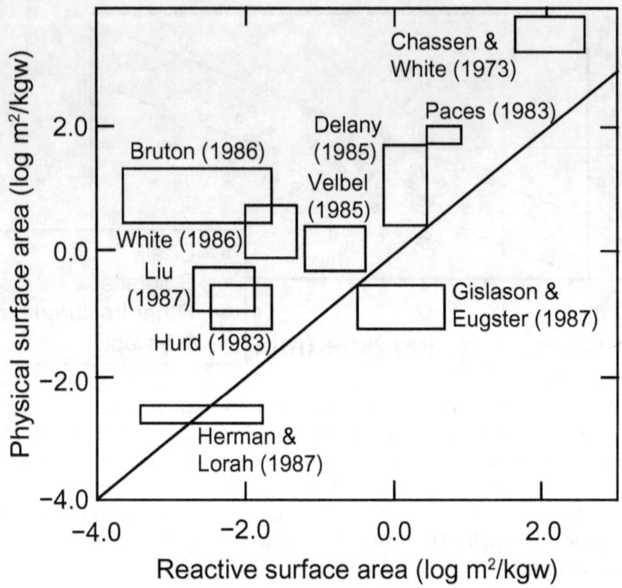

Figure 4.18. Reaction rates expected from laboratory kinetic rates and physical measurements of surface areas (vertical axis) versus those measured in the field (as quantified by the effective reactive surface areas). The scales are logarithmic, so that field reaction rates are usually 1 and 4 orders of magnitude faster than their laboratory counterparts. (From White and Peterson (1990)).

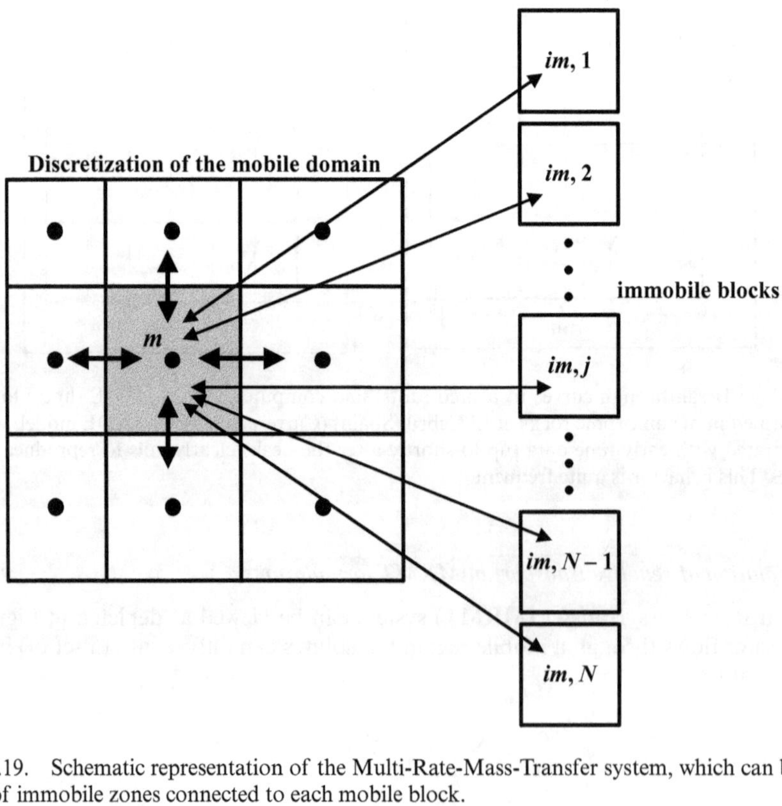

Figure 4.19. Schematic representation of the Multi-Rate-Mass-Transfer system, which can be viewed as a set of immobile zones connected to each mobile block.

$$F_i = \alpha_i(c_m - c_{imi}) \tag{4.98b}$$

$$\beta_i \frac{\partial c_{imi}}{\partial t} = F_i \tag{4.98c}$$

where c_m is the concentration in the mobile region and c_{imi} are the concentrations in the immobile regions. β_i is the immobile porosity of the *i-th* immobile region and α_i is a mass transfer coefficient.

The solution of these equations and the distributions of α's and β's are discussed extensively by Silva *et al.* (2009).

These equations can be easily extended to the reactive case by simply adding the reaction rate terms to equations (4.98a) and (4.98b). A solution algorithm for this case is discussed by Donado *et al.* (2009).

A particularly interesting application is presented by Willmann *et al.* (2008, 2010). These authors first simulated conservative transport through a heterogeneous medium. They then computed the breakthrough curve (BTC) and adjusted the α's and β's in equation (4.98) to fit this BTC with a homogeneous model. Equations (4.98) fitted well the BTC, which suggests that they can be used to represent the effect of heterogeneity, but may be considered a fitting exercise. Still, the authors then simulated reactive transport both with the heterogeneous medium and with the MRMT formulation. Local reaction rates displayed large differences but overall rates (the spatial distributions of cumulative precipitates) agreed. As chemical parameters (i.e., equilibrium constants) were not used in fitting the BTC, the mechanisms were well captured by equation (4.98).

APPENDIX

Notation

Symbol	Meaning	Units (SI)
a_i	Activity of species i	
c_i	Concentration of species i (mol per volume of fluid)	mol m^{-3}
	Concentration (mass of solute for ...... mass of solid)	mol m^{-3}
c_f	Specific heat capacity of the fluid	J kg^{-1} K^{-1}
c_m	Concentration (mass of solute for ...... mass of fluid)	mol m^{-3}
c_s	Specific capacity of the solid	J kg^{-1} K^{-1}
$\mathbf{c}$	Vector (n_t) of concentrations of all species (mol per vol. of fluid)	mol m^{-3}
$\mathbf{c}_1$	Vector (n_r) of concentrations of all primary species (mol per volume of fluid)	mol m^{-3}
$\mathbf{c}_{1.1}$	Vector (n_c) of concentrations of primary species having constant activity (mol per volume of fluid)	mol m^{-3}
$\mathbf{c}_{1.2}$	Vector (n_r) of concentrations of primary species not having constant activity (mol per volume of fluid)	mol m^{-3}
$\mathbf{c}_2$	Vector (n_r) of concentrations of secondary species (mol per volume of fluid)	mol m^{-3}
$\mathbf{c}_g$	Vector (n_g) of concentrations of species in the gas phase (mol per volume of fluid)	mol m^{-3}
$\mathbf{c}_l$	Vector (n_l) of concentrations of species in the liquid phase (mol per volume of fluid)	mol m^{-3}
$\mathbf{c}_s$	Vector (n_s) of concentrations of species in the solid phase (mol per volume of fluid)	mol m^{-3}
D	Molecular diffusion coefficient	m^2 s^{-1}
D_h	Hydrodynamic dispersion coefficient	m^2 s^{-1}

D_m	Mechanical dispersion coefficient	$m^2\ s^{-1}$
g	Gravity acceleration	$m\ s^{-2}$
h_f	Enthalpy of the fluid	$J\ kg^{-1}$
$\mathbf{K}$	Permeability	m^2
K_H	Hydraulic conductivity	$m\ s^{-1}$
k_T	Effective thermal conductivity	$J\ m\ K^{-1}\ s^{-1}$
m_f	Mass of fluid	kg
$\mathbf{m}$	Vector (n_t) of concentrations of all species (mol per volume of porous media)	$mol\ m^{-3}$
$\mathbf{m}_1$	Vector (n_t) of concentrations of all primary species (mol per volume of porous media)	$mol\ m^{-3}$
$\mathbf{m}_{1.1}$	Vector (n_c) of concentrations of primary species having constant activity (mol per volume of porous media)	$mol\ m^{-3}$
$\mathbf{m}_{1.2}$	Vector (n_t) of concentrations of primary species not having constant activity (mol per volume of porous media)	$mol\ m^{-3}$
$\mathbf{m}_2$	Vector (n_t) of concentrations of secondary species (mol per volume of porous media)	$mol\ m^{-3}$
n_c	Number of species with constant activity	
n_e	Number of equilibrium reactions	
n_g	Number of gas species	
n_k	Number of kinetic reactions	
n_l	Number of liquid species	
n_s	Number of solid species	
n_t	Number of species ($n_t = n_l + n_g + n_s$)	
p	Pressure	Pa
q_{adv}	Advection flux	$mol\ m^{-2}\ s^{-1}$
q_c	rate of extraction/injection of solute per unit volume of porous medium	$mol\ m^{-3}\ s^{-1}$
q_D	Diffusion flux	$mol\ m^{-2}\ s^{-1}$
q_{DM}	Mechanical dispersion flux	$mol\ m^{-2}\ s^{-1}$
Q_T	Heat flux	$J\ m^2\ s^{-1}$
$\mathbf{r}_e$	Vector (n_e) of reaction rates of all equilibrium reaction per volume of porous media	$mol\ m^{-3}\ s^{-1}$
$\mathbf{r}_k$	Vector (n_k) of reaction rates of all kinetic reaction per volume of porous media	$mol\ m^{-3}\ s^{-1}$
$S_{e_{ji}}$	Stoichiometric coefficient of species *i* in equilibrium reaction *j*	
S_f	Fluid saturation of the media	
$S_{k_{ji}}$	Stoichiometric coefficient of species *i* in kinetic reaction *j*	
$\mathbf{S}_e$	Matrix ($n_e \times n_t$) of stoichiometric coefficients of equilibrium reactions	
$\mathbf{S}_k$	Matrix ($n_k \times n_t$) of stoichiometric coefficients of kinetic reactions	
T	Temperature	K
u_f	Velocity of the fluid	$m\ s^{-1}$
$\mathbf{U}$	Component matrix for all species (dimensions $(n_t - n_e) \times n_t$)	
$\mathbf{U}_g$	Component matrix for gas species (dimensions $(n_t - n_e) \times n_g$)	
$\mathbf{U}_l$	Component matrix for liquid species (dimensions $(n_t - n_e) \times n_l$)	
$\mathbf{U}_s$	Component matrix for solid species (dimensions $(n_t - n_e) \times n_g$)	
v_f	Specific fluid discharge or Darcian velocity	$m^3\ m^{-2}\ s^{-1}$
α_L, α_T	Longitudinal and transverse mechanical dispersivities	m
β_L, β_T	Longitudinal and transverse thermal dispersivities	m
θ_f	Volumetric fluid fraction	
μ	Fluid kinematic viscosity	Pa s
ρ_f	Density of fluid	$kg\ m^{-3}$
τ_f	Tortuosity of the porous medium	
φ	Porosity	

REFERENCES

Adams, E.E. & Gelhar, L.W.: Field-study of dispersion in a heterogeneous aquifer, 2. Spatial moments analysis. *Water Resour. Res* 28:12 (1992), pp. 3293–3307.

Anderson, D.E.: Diffusion in electrolyte mixtures. *Rev. Mineral.* 8 (1981), pp. 211–260.

Anderson, D.E. & Graf, D.L.: Multicomponent electrolyte diffusion. *Ann. Rev. Environ. Earth Planet Sci.* 4 (1976), pp. 95–121.

Anderson, M.P.: Using models to simulate the movement of contaminants through groundwater flow systems. *Crit. Rev. Environ. Control* 9 (1979), pp. 97–156.

Attinger, S., Dentz, M. & Kinzelbach, W.: Exact transverse macro dispersion coefficient for transport in heterogeneous porous media. *Stoch. Environ. Res. Risk Assess.* 18 (2004) pp. 9–15.

Bear, J.: *Dynamics of fluids in porous media*. American Elsevier, New York, NY, 1972.

Bear, J.: *Hydraulics of groundwater*. McGraw-Hill, London, UK, 1979.

Bear, J. & Bachmat, Y.: *Introduction to modeling of transport phenomena in porous media*. Kluwer Academic Publishers, Dordrecht, The Netherlands, 1990.

Bear, J. & Cheng, A.H.-D.: *Modeling groundwater flow and contaminant transport*. Springer, Heidelberg, Germany, 2009.

Berkowitz, B. & Scher, H.: Anomalous transport in random fracture networks. *Phys. Rev. Lett.* 79 (1997), pp. 4038–4041.

Berkowitz, B., Singurindy, O. & Lowell, R.P.: Mixing-driven diagenesis and mineral deposition: $CaCO_3$ precipitation in salt water—fresh water zones. *Geophys. Res. Lett.* 30:5 (2003), 1016.

Birgersson, L., Moreno, L., Neretnieks, I., Widén, H. & Ägrén, T.: A tracer migration experiment in a small fracture zone in granite. *Water Resour. Res.* 29:12 (1993), pp. 3867–3878.

Brinkman, H.C.: On applied scientific research section a-mechanics heat chemical. *Eng. Math. Methods* 1:2 (1948). pp. 81–86.

Bundschuh, J. & Suárez Arriaga, M.-C.: *Introduction to the numerical modeling of groundwater and geothermal systems: Fundamentals of mass, energy and solute transport in poroelastic rocks.* CRC Press, Boca Raton, FL, 2010.

Carrera, J.: An overview of uncertainties in modeling groundwater solute transport. *J. Contam. Hydrol.* 13:1–4 (1993), pp. 23–48.

Carrera, J., Vazquez-Sune, E., Castillo, O. & Sanchez-Vila, X.: A methodology to compute mixing ratios with uncertain end-members. *Water Resour. Res.* 40:12 (2004), W12101.

Chadam, J., Ortoleva, P. & Sen, A.: Reactive-infiltration instabilities. *IMA J. Appl. Math.* 36 (1986), pp. 207–220.

Chen, J.-S. & Liu, C.-W.: Numerical simulation of the evolution of aquifer porosity and species concentrations during reactive transport. *Comput. Geosci.* 28 (2002), pp. 485–499.

Chung S.-O. & Horton, R.: Soil heat and water flow with a partial surface mulch. *Water Resour. Res.* 23:12 (1987), pp. 2175–2186.

Cirpka, O.A. & Attinger, S.: Effective dispersion in heterogeneous media under random transient flow conditions. *Water Resour. Res.* 39:9 (2003), 1257.

Combarnous, M.A. & Bories, S.A.: Thermal convection in saturated porous media. Groupe d'etude sur les milieux poreux, Rapport G.E.11, Institute Francais Du Petrol, Toulouse, France, 1973.

Combarnous, M.A. & Bories, S.A.: Hydrothermal convection in saturated porous media. In: Ven Te Chow (ed): *Advances in hydroscience* 10, 1975, pp. 231–305.

Dagan, G.: *Flow and transport in porous formations*. Springer-Verlag, New York, NY, 1989.

De Simoni, M., Carrera, J., Sánchez-Vila, X. & Guadagnini: A procedure for the solution of multicomponent reactive transport problems. *Water Resour. Res.* 41 (2005), W11410.

De Simoni, M., Sanchez-Vila, X. Carrera, J. & Saaltink, M.W.: A mixing ratios-based formulation for multicomponent reactive transport. *Water Resour. Res.* 43 (2007), W07419.

Dentz, M. & Carrera, J.: Effective dispersion in temporally fluctuating flow through heterogeneous porous medium. *Phys. Rev. E* (2003), 036310.

Diersch, H.-J.G.: FEFLOW Reference manual. Wasy—Institute for Water Resources Planning and System Research Ltd., Berlin, Germany, 2005.

Diersch, H.-J.G. & Kolditz, O.: Coupled groundwater flow and transport: 2. Thermohaline and 3D convection systems. *Adv. Water Resour.* 21 (1998), pp. 401–425.

Diersch, H.-J.G. & Kolditz, O.: Variable-density flow and transport in porous media: approaches and challenges. *Adv. Water Resour.* 25 (2002), pp. 899–944.

Domenico, P.A. & Schwartz, F.W.: *Physical and chemical hydrogeology*. 2nd ed., John Wiley and Sons, Inc., New Yorkm NY, 1998.

Donado, L., Sanchez-Vila, X., Dentz, M., Carrera, J. & Bolster, D.: Multicomponent reactive transport in multicontinuum media. *Water Resour. Res.* 45 (2009), W11402.

Durlofsky, L. & Brady, J.F.: Analysis of the Brinkman equation as a model for flow in porous media. *Phys. Fluids* 30 (1987), pp. 3329–3341.

Fang, Y., Yeh, G.T. & Burgos, W.D.: A general paradigm to model reaction-based biochemical processes in batch systems. *Water Resour. Res.* 39 (2003), 1083.

Fetter, C.W.: *Contaminant hydrogeology*. 2nd ed. Prentice Hall, NJ, 1999.

Forchheimer, P.: Water movement through the ground. *Zeitschrift des Vereines Deutscher Ingenieure* 45 (1901), pp. 1781–1788.

Friedly, J.C. & Rubin, J.: Solute transport with multiple equilibrium-controlled or kinetically controlled chemical reactions. *Water Resour. Res*. 28:6 (1992), pp. 1935–1953.

García-Gutiérrez, M., Guimerà, J., Yllera de Llano, A., Hernández-Benítez, A., Humm, J. & Saaltink, M.: Tracer test at El Berrocal site. *J. Contam. Hydrol.* 26 (1997), pp. 179–188.

Glynn, P.: Solid-solution solubilities and thermodynamics: Sulfates, carbonates and halides. *Reviews in Mineralogy and Geochemistry* 40 (2000), pp. 481–511.

Guimera, J. & Carrera, J.: A comparison of hydraulic and transport parameters measured in low-permeability fractured media. *J. Contam. Hydrology* 41 (2000), pp. 261–281.

Hanshaw, B.B. & Back, W.: The major geochemical processes in the evolution of carbonate-aquifer system. *J. Hydrology* 43:20 (1979), pp. 287–312.

Hassanizadeh, S.M. & Gray, W.G.: General conservation equations for multi-phase systems: 1. Averaging procedure. *Adv. Water Resour.* 2:3 (1979a), pp. 131–144.

Hassanizadeh S.M. & Gray, W.G.: General conservation equations for multi-phase systems: 2. Mass, momenta, energy, and entropy equations. *Adv. Water Resour*. 2:4 (1979b), pp. 191–203.

Hassanizadeh, S.M. & Gray, W.G.: General conservation equations for multi-phase systems: 3. Constitutive theory for porous media flow. *Adv. Water Resour.* 3:1 (1980), pp. 25–40.

Josef, D.D., Nield, D.A. & Papanicolaou, G.: Nonlinear equation governing flow in a saturated porous medium. *Water Resour. Res.* 18:4 (1982), pp. 1049–1052.

Katz, G.E., Berkowitz, B., Guadagnini, A. & Saaltink, M.W.: Experimental and modeling investigation of multicomponent reactive transport in porous media. *J. Contam. Hydrol.* 120–121 (2011), pp. 27–44.

Kaviani, M.: *Principles of heat transfer in porous media.* Springer-Verlag New-York, Inc., 1995.

Kitanidis, P.K.: The concept of the dilution index. *Water Resour. Res.* 30:7 (1994), pp. 2011–2026.

Kräutle, S. & Knabner, P.: A new numerical reduction scheme for fully coupled multicomponent transport-reaction problems in porous media. *Water Resour. Res.* 41 (2005), W09414.

Kräutle, S. & Knabner, P.: A reduction scheme for coupled multicomponent transport-reaction problems in porous media: Generalization to problems with heterogeneous equilibrium reactions. *Water Resour. Res.* 43, (2007), W03429.

Li, Y.S., Wang, Y.G., Pennell, K.D. & Abriola, L.M.: Investigation of the transport and deposition of fullerene (C60) nanoparticles in quartz sands under varying flow conditions. *Environ. Sci. Technol.* 42 (2008), pp. 7174–7180.

Martínez-Landa, L. & Carrera, J.: An analysis of hydraulic conductivity scale effects in granite (Full-scale Engineered Barrier Experiment (FEBEX), Grimsel, Switzerland). *Water Resour. Res.* 41 (2005), W03006.

Meile, C. & Tuncay, K.: Scale dependence of reaction rates in porous media. *Adv. Water Resour.* 29 (2006), pp. 62–71.

Millington, R.J.: Gas diffusion in porous media. *Science* 130 (1959), pp. 100–102.

Molins, S., Carrera, J. Ayora, C. & Saaltink, M.W.: A formulation for decoupling components in reactive transport problems. *Water Resour. Res.* 40 (2004), W10301.

Neuman, S.P. & Zhang, Y.-K.: A quasilinear theory of non-Fickian and Fickian subsurface dispersion: 1. Theoretical analysis with application to isotropic media. *Water Resour. Res.* 26:5 (1990), pp. 887–902.

Nield, D.A. & Bejan, A.: *Convection in porous media.* 3rd ed., Springer-Verlag + Buisness Media, New York, NY, 2006.

Novakowski, K.S., Evans, G.V., Lever, D.A. & Raven, K.G.: A field example of measuring hydrodynamic dispersion in a single fracture. *Water Resour. Res.* 21:8 (1985), pp. 1165–1174.

Oldenburg, C.M. & Pruess, K.: Dispersive transport dynamics in a strongly coupled groundwater–brine flow system. *Water Resour. Res.* 31 (1995), pp. 289–302.

Oldenburg, C.M. & Pruess, K.: Layered thermohaline convection in hypersaline geothermal systems. *Transp. Porous Media* 33:1–2 (1998) pp. 29–63.

Ortoleva, P., Chadam, J., Merino, E. & Sen, A.: Selforganization in water–rock interaction systems II: the reactive-infiltration instability. *Amer. J. Sci.* 287 (1987), pp. 1008–1040.

Pachepsky, Y.A., Rawls, W.J. & Lin, H.S.: Hydropedology and pedotransfer functions. *Geoderma* 131 (2006), pp. 308–316.

Pitzer, S.: Thermodynamics of electrolytes. I. Theoretical basis and general equations. *J. Phys. Chem.* 77:2 (1973), pp. 268–277.

Poling, B.E., Prausnitz, J.M. & O'Connell, J.P.: Properties of gases & liquids. 5th ed., McGraw-Hill, New York, NY, 2000.

Plummer, L.N.: Mixing of seawater with calcium carbonate groundwater. In: E.H.T. Whitten (ed): *Quantitative studies in the geological sciences*. Geological Society of America Memoir 142 (1975), pp. 219–236.

Raeisi, E. & Mylroie, J.E.: Hydrodynamic behavior of caves formed in the fresh-water lens of carbonate island. *Carbonates and Evaporites* 10:2 (1995), pp. 207–214.

Raven, K.G., Novakowski, K.S. & Lapcevic, P.S.: Interpretation of field tracer tests of a single fracture using a transient solute storage model. *Water Resour. Res*. 24:12 (1988), pp. 2019–2032.

Rezaei, M., Sanz, E., Raeisi, E., Vázquez-Suñé, E., Ayora, C. & Carrera, J.: Reactive transport modeling of calcite dissolution in the salt water mixing zone. *J. Hydrol.* 311 (2005), pp. 282–298.

Richards, L.A.: Capillary conduction of liquids through porous medium. *Physics* 1 (1931), pp. 18–33.

Rubin, H.: Onset of thermohaline convection in a cavernous aquifer. *Water Resour. Res*. 12 (1976), pp. 141–147.

Rubin, J.: Solute transport with multisegment, equilibrium-controlled classical reactions: problem solvability and feed forward method's applicability for complex segments of at most binary participants. *Water Resour. Res*. 28:6 (1992), pp. 1681–1702.

Rubin, H. & Roth, C.: Thermohaline convection in flowing groundwater. *Adv. Water Resour*. 6 (1983), pp. 146–56.

Rubinstein, J.: Effective equations for flow in random porous media with large number of scales. *J. Fluid Mech.* 170 (1986), pp. 379–383.

Runnels, D.D.: Diagenesis, chemical sediments, and the mixing of natural waters. *J. Sedim. Petrol.* 39 (1969), pp. 1188–1201.

Saaltink, M.W., Ayora, C. & Carrera J.: A mathematical formulation for reactive transport that eliminates mineral concentrations. *Water Resour. Res.* 34:7 (1998), pp. 1649–1656.

Saaltink, M.W., Batlle, F., Ayora, C., Carrera, J. & Olivella, S.: RETRASO, a code for modeling ractive transport in saturated and unsaturated porous media. *Geologica Acta* 2:3 (2004), pp. 235–251.

Sánchez-Vila, X., Guadagnini, A. & Carrera, J.: Representative hydraulic conductivities in saturated groundwater flow. *Rev. Geophysics* 44:3 (2006), Art. RG3002.

Sanford, W.E. & Konikow, L.F.: Simulation of calcite dissolution and porosity changes in saltwater mixing zones in coastal aquifers. *Water Resour. Res.* 25 (1989), pp. 655–667.

Shapiro, A.M. & Nicholas, J.R.: Assessing the validity of the channel model of fracture aperture under field conditions. *Water Resour. Res.* 25:5 (1989), pp. 817–828.

Silva, O., Carrera, J., Kumar, S., Dentz, M., Alcolea, A. & Willmann, M.: A general real-time formulation for multi-rate mass transfer problems. *Hydrol. Earth System Sci.* 13:8 (2009), pp. 1399–1411.

Sorek, S., Levi-Hevroni, D., Levy A. & Ben-Dor, G. Extensions to the macroscopic Navier-Stokes equation. *Transp. Porous Media* 61 (2005), pp. 215–233.

Sorek, S., Ronen, D. & Gitis, V.: Scale-dependent macroscopic balance equations governing transport through porous media: Theory and observations. *Transp. Porous Media* 81:1 (2010), pp. 61–72.

Steefel, C.I., DePaolo, D.J. & Lichtner, P.C.: Reactive transport modeling: An essential tool and a new research approach for the earth sciences. *Earth Planet. Sci. Lett.* 240 (2005), pp. 539–558.

Tebes-Stevens, C.L., Espinoza, F. & Valocchi, A.J.: Evaluating the sensitivity of a subsurface multicomponent reactive transport model with respect to transport and reaction parameters. *J. Contam. Hydrol.* 52:1–4 (2001), pp. 3–27.

Tucker, M.E. & Wright, V.P.: *Carbonate sedimentology*. Blackwell Scientific Publication, Oxford, UK, 1990.

Vafai, K. & Tien, C.L.: Boundary and inertia effects on flow and heat-transfer in porous-media. *Int. J. Heat Mass Transfer* 24:2 (1981), pp. 195–203.

van Genuchten, M.Th.: A closed-form equation for predicting the hydraulic conductivity of unsaturated soils. *Soil Sci. Soc. Am. J.* 44 (1980), pp. 892–898.

Voss, C.I. & Provost, A.M.: SUTRA: A Model for Saturated-Unsaturated, Variable-Density Ground-Water Flow with Solute or Energy Transport. Water-Resources Investigations Report 02–4231, Reston, VA, 2008.

Whitaker, S.: The Forchheimer equation: A theoretical development. *Transp. Porous Media* 25:1 (1996), pp. 27–61.

Whitaker, S.: *The method of volume averaging*. Kluwer Publisher, Dordrecht, the Netherlands, 1999.

White, A.F. & Peterson, M.-L.: Role of reactive-surface-area characterization in geochemical kinetic-models. *ACS Symposium Series* 416, 1990, pp. 461–475.

Wigley, T.M.L. & Plummer, L.N.: Mixing of carbonate waters. *Geochim. Cosmochim. Acta.* 40:9 (1976), pp. 989–995.

Willmann, M., Carrera, J. & Sanchez-Vila, X.: Transport upscaling in heterogeneous aquifers: What physical parameters control memory functions? *Water Resour. Res.* 44:12 (2008), W12437.

Willmann, M., Carrera, J., Sanchez-Vila, X., Silva, O. & Dentz, M.: Coupling of mass transfer and reactive transport for nonlinear reactions in heterogeneous media. *Water Resour. Res.* 46 (2010), W07512.

Worsten, J.H.M., Pachepsky, Ya.A. & Rawls, W.J.: Pedotransfer functions: bridging the gap between available soil data and missing soil hydraulic characteristics. *J. Hydrology* 251 (2001), pp. 123–150.

Xu, T.F. & Pruess, K.: Modeling multiphase non-isothermal fluid flow and reactive geochemical transport in variably saturated fractured rocks: 1. Methodology. *Amer. J. Sci.* 301:1 (2006), pp. 16–33.

Zhang, Y.-K. & Neuman, S.P.: A quasilinear theory of non-Fickian and Fickian subsurface dispersion: 2. Application to anisotropic media and the Borden site. *Water Resour. Res.* 26:5 (1990), pp. 903–913.

CHAPTER 5

Numerical solutions of reactive transport equations

Maarten W. Saaltink, Jesus Carrera & Carlos Ayora

"Life is difficult because it is non-linear"

Anonymous

5.1 INTRODUCTION

This chapter discusses numerical solution methods for the reactive transport equations presented in the previous chapter. Methods for discretization of the spatial and temporal domain are explained in section 5.2. As there already are excellent textbooks on the subject (e.g., Pinder and Gray, 1977; Huyakorn and Pinder, 1983; Zheng and Bennett, 1995; Bundschuh and Suárez-Arriaga, 2010), only a brief description of the two most important methods, finite differences and finite elements, is provided here. Following this description, methods used to solve the non-linear equations after the discretization are discussed in section 5.3.

5.2 METHODS FOR DISCRETIZING SPACE AND TIME

5.2.1 *Finite differences*

5.2.1.1 *Fundamentals*

The method of finite differences discretizes space into rectangular cells or blocks. Nodes are placed either at the center of the cells (cell centered) or at the edges of the cells (grid centered), as shown in Figure 5.1. State variables, such as pressure or concentration, are calculated at the nodes. Properties of the medium, such as permeability or porosity, are usually given for the cells, but can also be given at the cell edges. Derivatives of the state variables can be also approximated at the nodes. It is convenient to formulate and analyze these approximated derivatives using the Taylor series, which yields information on the error of the approximation. The Taylor series approximates the value of any function (f) at certain point ($x + \Delta x$) from the values and derivatives at another, implicitly assumed close, point (x):

$$f_{x+\Delta x} = f_x + \Delta x \frac{\partial f}{\partial x}\bigg|_x + \frac{\Delta x^2}{2!}\frac{\partial^2 f}{\partial x^2}\bigg|_x + \frac{\Delta x^3}{3!}\frac{\partial^3 f}{\partial x^3}\bigg|_x + \cdots \tag{5.1}$$

An expression can also be used for the point $x - \Delta x$:

$$f_{x-\Delta x} = f_x - \Delta x \frac{\partial f}{\partial x}\bigg|_x + \frac{\Delta x^2}{2!}\frac{\partial^2 f}{\partial x^2}\bigg|_x - \frac{\Delta x^3}{3!}\frac{\partial^3 f}{\partial x^3}\bigg|_x + \cdots \tag{5.2}$$

The first derivatives can be approximated by truncating after the second term in equations (5.1) and (5.2), and solving for $\partial f/\partial x$:

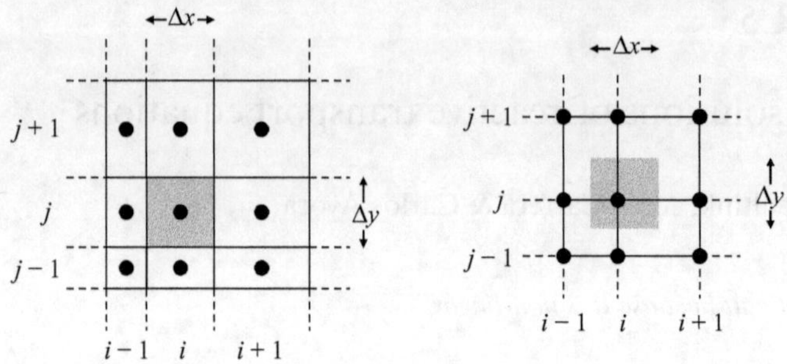

Figure 5.1. Two-dimensional finite difference grids, one cell centered (left) and grid centered (right).

$$\left.\frac{\partial f}{\partial x}\right|_x = \frac{f_{x+\Delta x} - f_x}{\Delta x} + O(\Delta x) \tag{5.3}$$

$$\left.\frac{\partial f}{\partial x}\right|_x = \frac{f_x - f_{x-\Delta x}}{\Delta x} + O(\Delta x) \tag{5.4}$$

where $O(\Delta x)$ denotes the error introduced due to the truncation. As truncation occurred after the second term, the error depends on Δx, and, therefore, the truncation error is of the first order. The first derivative can also be estimated by truncating after the third term in equations (5.1) and (5.2) and subtracting the equations, which yields:

$$\left.\frac{\partial f}{\partial x}\right|_x = \frac{f_{x+\Delta x} - f_{x-\Delta x}}{2\Delta x} + O(\Delta x^2) \tag{5.5}$$

The truncation error then depends on Δx^2 and equation (5.5) therefore represents an approximation of the second order. As is illustrated in Figure 5.2, a second order error is to be preferred to a first order error.

The second derivative can be approximated by summing equations (5.1) and (5.2), yielding:

$$\left.\frac{\partial^2 f}{\partial x^2}\right|_x = \frac{f_{x+\Delta x} - 2f_x + f_{x-\Delta x}}{\Delta x^2} + O(\Delta x^2) \tag{5.6}$$

which also has a second order truncation error.

The derivatives at each node of a finite difference grid can now be calculated. For simplicity, only one-dimensional grids with equal grid size, Δx, are presented here. The first derivatives at node i can be approximated by applying equations (5.3), (5.4) or (5.5):

$$\left.\frac{\partial f}{\partial x}\right|_i = \frac{f_{i+1} - f_i}{\Delta x} \tag{5.7}$$

$$\left.\frac{\partial f}{\partial x}\right|_i = \frac{f_i - f_{i-1}}{\Delta x} \tag{5.8}$$

$$\left.\frac{\partial f}{\partial x}\right|_i = \frac{f_{i+1} - f_{i-1}}{2\Delta x} \tag{5.9}$$

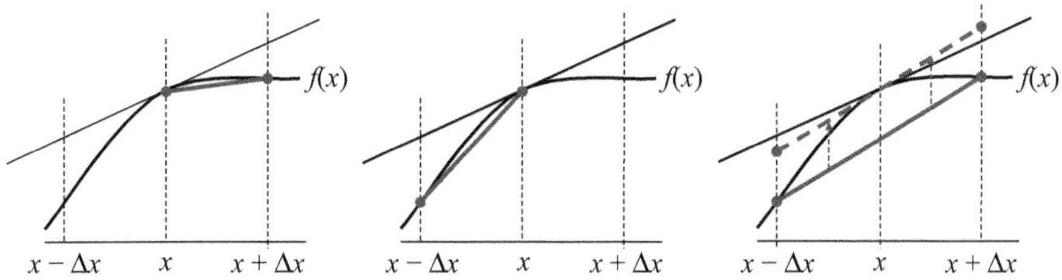

Figure 5.2. Approximation of the derivatives according to forward (left), backward (middle) and centered (right) finite differences. The centered finite difference approximates better the true derivative (indicated by the tangent of function f).

Equations (5.7), (5.8) and (5.9) represent the forward, backward and centered finite difference approximations, respectively. Second derivatives can be approximated by equation (5.6):

$$\left.\frac{\partial^2 f}{\partial x^2}\right|_i = \frac{f_{i+1} - 2f_i + f_{i-1}}{\Delta x^2} \tag{5.10}$$

5.2.1.2 *Application to conservative transport*

The finite difference approximations are now applied to the transport equation of a conservative species as given in Chapter 4 (section 4.3.2; eq. 4.51). For simplicity, a homogeneous one-dimensional system is assumed. This allows writing the conservative transport equation as:

$$\theta_f \frac{\partial c}{\partial t} = -\theta_f u_f \frac{\partial c}{\partial x} + \theta_f D_h \frac{\partial^2 c}{\partial x^2} + q_c \tag{5.11}$$

Applying equation (5.9) for the spatial derivatives ($\partial c/\partial x$), that is the centered finite difference approximation, and equation (5.10) for the second spatial derivatives ($\partial^2 c/\partial x^2$), results in:

$$\theta_f \frac{\partial c_i}{\partial t} = -\theta_f u_f \frac{c_{i+1} - c_{i-1}}{2\Delta x} + \theta_f D_h \frac{c_{i+1} - 2c_i + c_{i-1}}{\Delta x^2} + q_c \quad i = 1,\ldots,n \tag{5.12}$$

or:

$$\theta_f \frac{\partial c_i}{\partial t} = \left(\frac{\theta_f u_f}{2\Delta x} + \frac{\theta_f D_h}{\Delta x^2}\right) c_{i-1} - \frac{2\theta_f D_h}{\Delta x^2} c_i + \left(-\frac{\theta_f u_f}{2\Delta x} + \frac{\theta_f D_h}{\Delta x^2}\right) c_{i+1} + q_C \quad i = 1,\ldots,n \tag{5.13}$$

As it uses the centered finite difference approximation, equation (5.13) is called centered weighted. For reasons explained later, it may also be convenient to use either backward or forward approximations depending on the direction of the fluid flow:

$$\frac{\partial c}{\partial x} = \begin{cases} \dfrac{c_i - c_{i-1}}{\Delta x} & \text{if } u_f \geq 0 \\ \dfrac{c_{i+1} - c_i}{\Delta x} & \text{if } u_f < 0 \end{cases} \tag{5.14}$$

which leads to:

$$\begin{cases} \theta_f \dfrac{\partial c_i}{\partial t} = -\theta_f u_f \dfrac{c_i - c_{i-1}}{\Delta x} + \theta_f D_h \dfrac{c_{i+1} - 2c_i + c_{i-1}}{2\Delta x} + q_c & \text{if } u_f \geq 0 \quad i = 1,\ldots,n \\ \theta_f \dfrac{\partial c_i}{\partial t} = -\theta_f u_f \dfrac{c_{i+1} - c_i}{\Delta x} + \theta_f D_h \dfrac{c_{i+1} - 2c_i + c_{i-1}}{2\Delta x} + q_c & \text{if } u_f < 0 \quad i = 1,\ldots,n \end{cases} \tag{5.15}$$

or:

$$\begin{cases} \theta_f \dfrac{\partial c_i}{\partial t} = \left(\dfrac{\theta_f u_f}{\Delta x} + \dfrac{\theta_f D_h}{\Delta x^2} \right) c_{i-1} + \left(-\dfrac{\theta_f u_f}{\Delta x} - \dfrac{2\theta_f D_h}{\Delta x^2} \right) c_i + \dfrac{\theta_f D_h}{\Delta x^2} c_{i+1} + q_C & u_f \geq 0 \quad i = 1,\ldots,n \\ \theta_f \dfrac{\partial c_i}{\partial t} = \dfrac{\theta_f D_h}{\Delta x^2} c_{i-1} + \left(\dfrac{\theta_f u_f}{\Delta x} - \dfrac{2\theta_f D_h}{\Delta x^2} \right) c_i + \left(-\dfrac{\theta_f u_f}{\Delta x} + \dfrac{\theta_f D_h}{\Delta x^2} \right) c_{i+1} + q_C & u_f < 0 \quad i = 1,\ldots,n \end{cases} \tag{5.16}$$

Equation (5.16) is said to be upstream weighted, because it places more weight on the nodes located upstream, than on the nodes located downstream.

Both equations (5.13) and (5.16) represent a set of mathematical equations that can be represented in a vector-matrix notation as:

$$\mathbf{F} \frac{\partial \mathbf{c}}{\partial t} = \mathbf{E}\mathbf{c} + \mathbf{g} \tag{5.17}$$

where **F** is a diagonal matrix of storage terms containing the volumetric fluid fraction (θ_f) of each cell, **E** a (non-diagonal) matrix containing advection and dispersion terms, and **g** is a vector containing the external source-sink terms. Simple inspection of equations (5.13) or (5.16) identifies the components of these matrices as:

$$\begin{aligned} F_{i,i} &= \theta_f \\ E_{i,i-1} &= \frac{\theta_f u_f}{2\Delta x} + \frac{\theta_f D_h}{\Delta x^2} \\ E_{i,i} &= \frac{2\theta_f D_h}{\Delta x^2} \\ E_{i,i-1} &= -\frac{\theta_f u_f}{2\Delta x} + \frac{\theta_f D_h}{\Delta x^2} \\ b_i &= q_c \end{aligned} \tag{5.18}$$

All the other components in the i-th equation are zero.

Equation (5.17) stills needs an approximation for the first derivative of time. Time is discretized in the same way as space was. Time is divided in time steps, Δt, and a forward finite difference is used:

$$\frac{\partial \mathbf{c}}{\partial t} = \frac{\mathbf{c}^{k+1} - \mathbf{c}^k}{\Delta t} \tag{5.19}$$

where superscript k refers to the time step number. It might be tempting to approximate the time derivative using a centered second order scheme in time, similar to equation (5.9). This would be disastrous, however, because the resulting numerical scheme would be unconditionally unstable (see section 5.2.3).

The concentration in the right-hand-side of (5.17) has to be evaluated at some time between time step k and $k + 1$. The following general equation is used to do this:

$$\mathbf{c} = \vartheta \mathbf{c}^{k+1} + (1 - \vartheta)\mathbf{c}^{k} \tag{5.20}$$

where ϑ is a time weighting factor ranging between 0 and 1. A value of 0 for ϑ evaluates the concentration at time k and a value of 1 at time $k + 1$. Substitution of (5.19) and (5.20) into (5.17) yields:

$$\mathbf{F}\frac{\mathbf{c}^{k+1} - \mathbf{c}^{k}}{\Delta t} = \mathbf{E}\left(\vartheta \mathbf{c}^{k+1} + (1 - \vartheta)\mathbf{c}^{k}\right) + \mathbf{g} \tag{5.21}$$

which can be rewritten as a linear set of equations that has be solved in order to obtain the concentrations of all cells at some time step given those of the previous time step:

$$\mathbf{A}\mathbf{c}^{k+1} = \mathbf{b} \tag{5.22}$$

with

$$\mathbf{A} = \left(\frac{\mathbf{F}}{\Delta t} - \vartheta \mathbf{E}\right) \tag{5.23}$$

$$\mathbf{b} = \left((1 - \vartheta)\mathbf{E} + \frac{\mathbf{F}}{\Delta t}\right)\mathbf{c}^{k} + \mathbf{g} \tag{5.24}$$

Note that if ϑ equals 0 the system matrix (**A**) is diagonal, because in that case it only depends on the diagonal matrix **F**. This means that the concentration can be calculated directly without the need to solve a linear system of equations. If, on the other hand, ϑ is larger than 0, the system matrix (**A**) is not diagonal because it depends on the non-diagonal matrix **E**: a linear equation solver is then required. For this reason, a solution with ϑ equal to 0 is called an explicit solution, whereas a solution with ϑ equal to 1 is called an implicit solution. A solution with ϑ is between 0 and 1 is referred to as a Crank-Nicholson solution. An explicit solution requires less computational time but it may have numerical disadvantages as will be explained in section 5.2.3.

5.2.2 *Finite elements*

The finite element method discretizes a spatial domain into nodes and elements. State variables, such as pressure or concentration, are calculated at the nodes. Properties of the medium, such as permeability or porosity, must be given at the elements. The elements connect the nodes and can have various shapes such as one-dimensional segments, two-dimensional triangles and quadrilaterals and three-dimensional tetrahedrons and prisms. A finite element mesh may also contain a mixture of finite element types (see Figure 5.3).

The finite element method consists of (i) interpolating a state variable through a shape function and (ii) minimizing a residual function. These two features are explained by applying them to the transport equation of a conservative species, as was done above for the finite difference method.

The concentrations in the whole domain are approximated by interpolating from the values at the nodes:

$$\hat{c}(x) = \sum_{i=1}^{n} N_i(x)c_i \tag{5.25}$$

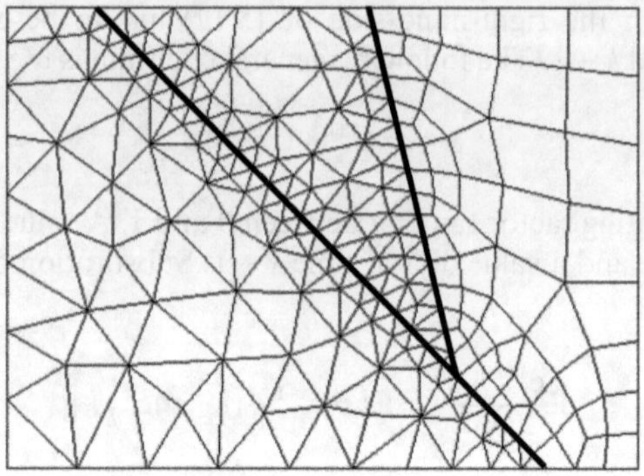

Figure 5.3. Example of a finite element mesh, consisting of triangles (on the left), quadrilaterals (on the right) and one-dimensional segments (indicated by thick lines).

where the hat (^) indicates the estimated value and N_i is the shape function of node *i*. Shape function are given for the nodes pertaining to each element and must be summed for all the elements. Each type of element has specific shape (also called basis, interpolation or weighting) functions. For instance, the two nodes of a one-dimensional segment (the simplest type of element) have the following shape functions (Fig. 5.4):

$$N_1 = \frac{x_2 - x}{x_2 - x_1} \quad N_2 = \frac{x - x_1}{x_2 - x_1} \tag{5.26}$$

It can be noticed that N_1 and N_2 have the following properties, which are shared by all shape functions of all types of elements: (i) N_i equals 1 at node *i* but zero at all other nodes of the element; (ii) The sum of all shape functions equals 1 at every point of the element; (iii) The shape functions vary smoothly (they are linear in eq. 5.25) within the element. Applying equation (5.26) to (5.25) simply gives a linear interpolation between the concentrations of the nodes of the element.

If the estimated concentration ($\hat{c}$) is substituted into the transport equation of a conservative species, there will be an error expressed by the residual function (ε):

$$\varepsilon = \theta_f \frac{\partial \hat{c}}{\partial t} + \theta_f \vec{u}_f \cdot \nabla \hat{c} - \vec{\nabla} \cdot \left(\theta_f \mathbf{D}_h \nabla \hat{c}\right) - q_c + q_f \hat{c} \tag{5.27}$$

ε equals 0 if $\hat{c}$ is exact. The finite element method minimizes the residuals by imposing that the weighted average of the residuals equal zero:

$$\int_V W_i \varepsilon(\hat{c}) dV = 0 \quad i = 1,\ldots,n \tag{5.28}$$

where *V* is the domain and *W* is called the weighting function. Various weighting functions can be used. The Galerkin weighting function is the one most commonly used and explained here. It simply equals the weighting function to the shape function ($W_i = N_i$).

Substitution of (5.25) and (5.27) into (5.28) yields:

$$\begin{aligned}\sum_j \frac{\partial c_j}{\partial t} \int_V N_i \theta_f N_j dV + \sum_j c_j \int_V N_i \theta_f \vec{u}_f \cdot \nabla N_j \theta_f dV - \sum_j c_j \int_V N_i \nabla \cdot \left(\theta_f \mathbf{D}_h \nabla N_j\right) dV \\ - \int_V N_i q_c dV + \sum_j c_j \int_V N_i q_f N_j dV = 0 \quad i = 1,\ldots,n\end{aligned} \tag{5.29}$$

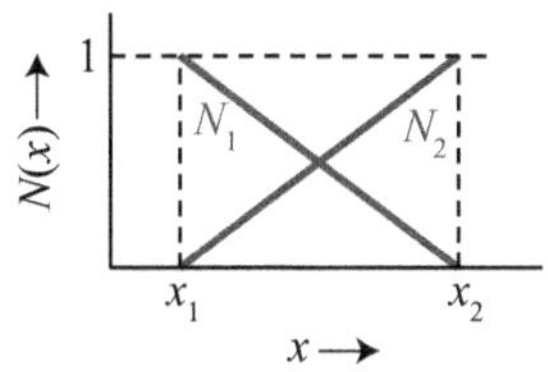

Figure 5.4. Shape functions of a one-dimensional segment.

Normally Green's theorem is applied to the dispersive term, such that:

$$\int_{\Gamma} N_i \theta_f \mathbf{D}_h \nabla N_j \cdot \vec{n} d\Gamma = \int_{V} N_i \nabla \cdot \left(\theta_f \mathbf{D}_h \nabla N_j\right) dV + \int_{V} \theta_f \nabla N_i \cdot \mathbf{D}_h \nabla N_j dV \tag{5.30}$$

where $\vec{n}$ is the vector normal to a boundary Γ. The boundary integral can be used to define boundary conditions:

$$\int_{\Gamma} N_i \theta_f \mathbf{D}_h \nabla N_j \cdot \vec{n} d\Gamma = \int_{\Gamma} N_i \left(q_c^b - q_f^b \hat{c}\right) d\Gamma \tag{5.31}$$

where the superscript b refers to fluxes at the boundary.

This leads to:

$$\begin{aligned} &\sum_j \frac{\partial c_j}{\partial t} \int_V N_i \theta_f N_j dV + \sum_j c_j \int_V N_i \theta_f \vec{u}_f \cdot \nabla N_j \theta_f dV + \sum_j c_j \int_V \theta_f \nabla N_i \cdot \mathbf{D}_h \nabla N_j dV \\ &+ \sum_j c_j \int_V N_i q_f N_j dV + \sum_j c_j \int_\Gamma N_i q_f^b N_j d\Gamma = \int_V N_i q_c dV + \int_\Gamma N_i q_c^b d\Gamma \quad i = 1,\ldots,n \end{aligned} \tag{5.32}$$

As every element has an explicitly given shape function (e.g., equation (5.26) for one-dimensional segment), integrals of equation (5.32) can be calculated for each element, and are expressed by a matrix whose rows and columns refer to the node numbers of the element. For example, the integral of the third term for a one-dimensional segment would lead to a 2×2 matrix (because the element has 2 nodes), $\mathbf{E}^e$:

$$\mathbf{E}^e = \begin{pmatrix} \int_{x_1}^{x_2} \theta_f \nabla N_1 D_h \nabla N_1 dx & \int_{x_1}^{x_2} \theta_f \nabla N_1 D_h \nabla N_2 dx \\ \int_{x_1}^{x_2} \theta_f \nabla N_2 D_h \nabla N_1 dx & \int_{x_1}^{x_2} \theta_f \nabla N_2 D_h \nabla N_2 dx \end{pmatrix} = \frac{\theta_f D_h}{\left(x_2 - x_1\right)} \begin{pmatrix} 1 & -1 \\ -1 & 1 \end{pmatrix} \tag{5.33}$$

The matrices of all elements have to be summed into a global matrix for the whole mesh with the number of columns and rows equal to the number of nodes, as is shown in Figure 5.5 for a mesh of two one-dimensional segments. Equation (5.32) can be written by means of these global matrices:

$$\mathbf{F} \frac{\partial \mathbf{c}}{\partial t} + \mathbf{E}\mathbf{c} = \mathbf{g} \tag{5.34}$$

This is the same as equation (5.17) that was derived by the method of finite differences, although the matrices **E**, **F** and **g** were calculated differently. Usually a finite difference approximation is used to approximate the time derivatives. This is explained in the previous section.

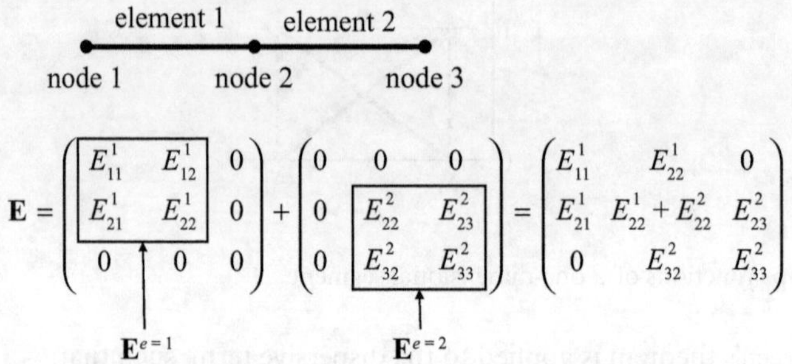

Figure 5.5. Summing the matrices of each element into a matrix for the whole mesh in case of a finite element mesh of two one-dimensional elements and three nodes.

5.2.3 *Instability and numerical dispersion*

Both finite elements and finite difference methods are Eulerian methods that suffer from instability and numerical dispersion. To illustrate instability, consider a simple case: a one-dimensional domain without dispersion or diffusion and a uniform velocity, u_f. When an explicit centered finite difference approximation is used, the concentration at a node i and time step $k+1$ is calculated as follows:

$$\theta_f \frac{c_i^{k+1} - c_i^k}{\Delta t} = -\theta_f u_f \frac{c_{i+1}^k - c_{i-1}^k}{2\Delta x} \Rightarrow c_i^{k+1} = -\frac{u_f \Delta t}{2\Delta x}\left(c_{i+1}^k - c_{i-1}^k\right) + c_i^k \tag{5.35}$$

Imagine a concentration profile with a sharp front between nodes i and $i+1$ (Fig. 5.6). As there is no dispersion, after a time step, $\Delta t = \Delta x / u_f$, the exact solution is simply a shift of the front, by one cell downstream. However, this is not what is calculated by equation (5.35). If c_{up} refers to the concentration upstream of the front ($c_{i-1}^k = c_i^k = c_{up}$) and c_{down} the concentration downstream ($c_{i+1}^k = c_{down}$), then:

$$c_i^{k+1} = \frac{u_f \Delta t}{2\Delta x}\left(c_{up} - c_{down}\right) + c_{up} \tag{5.36}$$

It can be easily seen that if the upstream concentration is higher than the downstream concentration, the concentration at time $k+1$ will always be higher than the upstream concentration as shown in Figure 5.6. This results in spurious oscillations. For centrally weighted approximation, oscillations are eliminated if the following condition is met (Huyakorn and Pinder; 1983):

$$P_e = \frac{u_f \Delta x}{D_h} \leq 2 \tag{5.37}$$

where P_e is called the Peclet number. Note, that for the case without dispersion ($D_h = 0$) this condition can never be met. Incidentally, this is the reason why a centrally finite difference approximation cannot be used for the time derivative.

On the other hand, if an upstream weighting is used, the concentration at a node i and time step $k+1$ is calculated according to:

$$c_i^{k+1} = -\frac{u_f \Delta t}{\Delta x}\left(c_i^k - c_{i-1}^k\right) + c_i^k = c_{up} \tag{5.38}$$

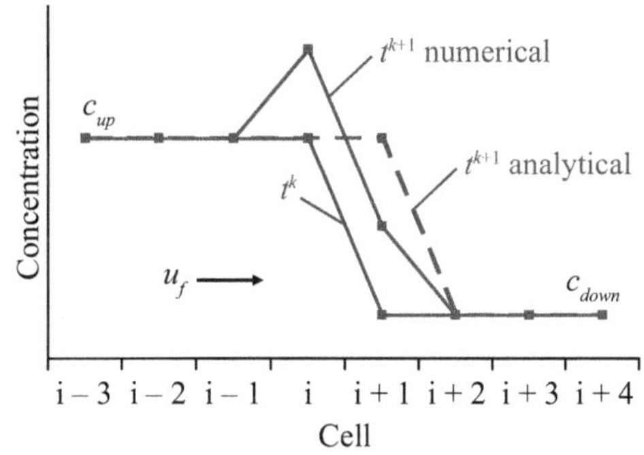

Figure 5.6. Illustration of oscillation due to numerical errors.

This is the correct solution and it does not give oscillations, at least, if $\Delta t = \Delta x/u_f$.

However, upstream weighting can suffer from numerical errors, specifically numerical dispersion. To illustrate this, consider the approximation of the advection term by upstream weighting as is done in equation (5.14). However, the Taylor's series is truncated after the third term instead of after the second:

$$-\theta_f u_f \frac{\partial c}{\partial x} = \begin{cases} -\theta_f u_f \dfrac{c_i - c_{i-1}}{\Delta x} - \dfrac{\theta_f u_f \Delta x}{2} \dfrac{\partial^2 c}{\partial x^2} & \text{if } u_f \geq 0 \\ -\theta_f u_f \dfrac{c_{i+1} - c_i}{\Delta x} + \dfrac{\theta_f u_f \Delta x}{2} \dfrac{\partial^2 c}{\partial x^2} & \text{if } u_f < 0 \end{cases} \tag{5.39}$$

Comparing this equation to the original conservative transport equation (5.11) demonstrates that an artificial or numerical dispersion term, with $D_{num} = |u_f|\,\Delta x/2$, was added due to the truncation error. In fact, the solution of equations (5.15) or (5.16) is identical to that of equations (5.12) or (5.13), if D_h in the latter is substituted by $D_h u_f \Delta x/2$ (the proof consists of adding and subtracting $u_f/(2\Delta x)$ to equation (5.12), and rearranging terms to get (5.15), but with the greater dispersion). Because the truncation error of the central weighting approximation is of second order, it does not produce numerical dispersion.

A similar analysis can be made for truncation errors in the time derivative. This gives a numerical dispersion for upstream weighting depending on the time weighting factor, ϑ (e.g., Zheng and Bennett, 1995):

$$D_{num} = (\vartheta - 0.5)\,\theta_f u_f^2 \Delta t + 0.5\theta_f u_f \Delta x \tag{5.40}$$

The first term of the right-hand-side is caused by the truncation error in the time derivative, and the second term is due to the error in the spatial derivatives. For central weighting, truncation error only occurs in the time derivative:

$$D_{num} = (\vartheta - 0.5)\,\theta_f u_f^2 \Delta t \tag{5.41}$$

5.3 METHODS FOR SOLVING REACTIVE TRANSPORT EQUATIONS

Finite differences or finite elements (or some other method) are first applied to the reactive transport equations formulated in Chapter 4 (section 4.5). The number of equations given for each node or cell equals the number of components. These equations are as a rule highly nonlinear. As previously explained in Chapter 3 (section 3.2.1), methods to solve nonlinear equations can be divided into two families: Picard and Newton-Raphson. When applied to reactive transport, the Picard approach is also called the *sequential iteration approach* (SIA), *operator splitting* or *two-step approach*. The *Newton-Raphson approach* is called the *direct substitution approach* (DSA) or *global implicit approach*. The following two sections explain the application of these techniques to reactive transport modeling. Both methods are then compared with each other.

5.3.1 *Sequential Iteration Approach (SIA)*

The *sequential iteration approach* consists of two steps, repeated iteratively until some convergence criterion is met. In the first step, the transport equation is solved for the total aqueous concentrations of every chemical component $\mathbf{u}_l$ (e.g., Xu *et al.*, 1999). The concentrations of other phases and kinetic rate terms are computed at the previous iteration. Using the matrices **E**, **F** and vector **g** defined in section 5.2, the reactive transport equations after discretization can be written as follows:

$$\mathbf{F}_l \frac{\mathbf{u}_{l,j}^{k+1,i+1} - \mathbf{u}_{l,j}^{k}}{\Delta t} + \mathbf{F}_g \frac{\mathbf{u}_{g,j}^{k+1,i} - \mathbf{u}_{g,j}^{k}}{\Delta t} + \mathbf{F}_s \frac{\mathbf{u}_{s,j}^{k+1,i} - \mathbf{u}_{s,j}^{k}}{\Delta t}$$
$$= \mathbf{g}_j + \mathbf{E}_l \mathbf{u}_{l,j}^{k+\vartheta,i+1} + \mathbf{E}_g \mathbf{u}_{g,j}^{k+\vartheta,i} + \mathbf{U}\mathbf{S}_k^T \mathbf{r}_k^{k+\vartheta,i} \quad j = 1,\ldots,n_c \tag{5.42}$$

where matrices **E** and **F** have a subscript indicating the phase to which they pertain (*l* liquid, *g* gas and *s* solid). Subscript *i* refers to the iteration and script *j* to the component. Therefore, $\mathbf{u}_{l,j}$, $\mathbf{u}_{g,j}$ and $\mathbf{u}_{s,j}$ are vectors of concentrations of component *j* of all nodes or cells of the mesh (note they have a slightly different meaning than in Chapter 3 and 4) and n_c is the number of components ($n_c = n_s - n_e$, n_s being the number of species and n_e the number of equilibrium reactions). This can be rewritten as n_c sets of linear equations:

$$\mathbf{A}\mathbf{u}_{l,j}^{k+1,i+1} = \mathbf{b}_j^i = \mathbf{b}_{cons,j} + \mathbf{b}_{reac,j}^i \quad j = 1,\ldots,n_c \tag{5.43}$$

where matrix **A** is the same as for conservative transport (both in case of finite differences and finite elements):

$$\mathbf{A} = \left(\frac{\mathbf{F}_l}{\Delta t} - \vartheta \mathbf{E}_l\right) \tag{5.44}$$

Vector **b** is a sum of the vector of the conservative linear system ($\mathbf{b}_{cons,j}$) and a vector that represents chemical reactions ($\mathbf{b}^i_{reac,j}$):

$$\mathbf{b}_{cons,j} = \mathbf{g}_j + \left(\mathbf{F}_l + (1-\vartheta)\mathbf{E}_l\right)\mathbf{u}_{l,j}^k \quad j = 1,\ldots,n_c \tag{5.45}$$

$$\mathbf{b}_{reac,j}^i = \mathbf{F}_s \frac{\mathbf{u}_{s,j}^{k+1,i} - \mathbf{u}_{s,j}^{k}}{\Delta t} + \mathbf{F}_g \frac{\mathbf{u}_{g,j}^{k+1,i} - \mathbf{u}_{g,j}^{k}}{\Delta t}$$
$$+ \mathbf{E}_g \mathbf{u}_{g,j}^{k+\vartheta,i} + \mathbf{U}_j \mathbf{S}_k^T \mathbf{r}_k^{k+\vartheta,i} \quad j = 1,\ldots,n_c \tag{5.46}$$

Comparing equation (5.43) to the general Picard's approach ($\mathbf{x}^{i+1} = \mathbf{g}(\mathbf{x}^i)$, see Chapter 3, section 3.2.1) demonstrates that the SIA is a special form of the Picard approach, function **g** being $\mathbf{A}^{-1}\mathbf{b}_j$. Note that contrary to vector $\mathbf{b}_{cons,j}$, matrix **A** is the same for each component. Moreover, if the definition of the components (i.e., the component matrix) does not vary in space, the set of equations (5.43) of each component is independent of the other components, that is, it can be solved for each component separately. This is an important characteristic for the computational performance as is discussed in section 5.3.3. However, the definition of components may vary in space. Component matrices that eliminate constant activity species such as minerals (see sections 3.1.4 and 4.5.3 in Chapers 3 and 4, respectively) are used, and these species exist only in part of the domain. In that case the total concentration for one component would have to be written as a function of that according for another component, making matrix **A** depending on the unknowns ($\mathbf{u}_l$), and making the components depend on each other. Therefore, the elimination of constant activity species is not particularly useful for the SIA.

In the second step, the chemical reaction term ($\mathbf{b}^i_{reac,j}$) is updated for a new iteration $i + 1$. To do so, the concentrations of all species are first calculated according to:

$$\mathbf{u}_{l,j}^{k+1,i+1} + \mathbf{u}_{g,j}^{k+1,i} + \mathbf{u}_{s,j}^{k+1,i} = \mathbf{U}\mathbf{c}_j^{k+1,i+1} \quad j = 1,\ldots,n_c \tag{5.47}$$

which has to be solved together with the mass action equations. The resulting non-linear system is usually solved by means of the Newton-Raphson approach. Actually, it is a speciation calculation, which has been explained in Chapter 3 (section 3.2.1) and will not be repeated here. Notice that the set of equations of each node or cell is independent to those of other nodes or cells. That is, each node or cell can be solved for separately. Once, the concentrations $\mathbf{c}_j^{i+1}$ have been obtained, the chemical reaction term ($\mathbf{b}^{i+1}_{reac,j}$) can be calculated by iterating equation (5.46) until convergence . The two steps are also iterated until convergence is reached.

A variant of the SIA uses the total *analytical* concentrations (**u**) instead of the total *aqueous* concentrations of every chemical component ($\mathbf{u}_l$) as unknowns (e.g., Yeh and Tripathi, 1991). For simplicity, all concentrations are assumed to be expressed using the same units, e.g., moles per unit volume of porous media or moles per unit mass of water. Then the storage matrices of all phases become identical ($\mathbf{F}_l = \mathbf{F}_g = \mathbf{F}_s$), and the transport step can be written as:

$$\mathbf{F}_l \frac{\mathbf{u}_j^{k+1,i+1} - \mathbf{u}_j^k}{\Delta t} = \mathbf{g}_j + \mathbf{E}_l \mathbf{u}_j^{k+\vartheta,i+1} + \left(\mathbf{E}_g - \mathbf{E}_l\right)\mathbf{u}_{g,j}^{k+\vartheta,i} - \mathbf{E}_l \mathbf{u}_j^{k+\vartheta,i} + \mathbf{U}_j \mathbf{S}_k^T \mathbf{r}_k^{k+\vartheta,i} \quad j = 1,\ldots,n_c \tag{5.48}$$

which can be rewritten as:

$$\mathbf{A}\mathbf{u}_j^{k+1,i+1} = \mathbf{b}_j^i = \mathbf{b}_{cons,j} + \mathbf{b}^i_{reac,j} \quad j = 1,\ldots,n_c \tag{5.49}$$

with:

$$\mathbf{A} = \left(\frac{\mathbf{F}_l}{\Delta t} - \vartheta \mathbf{E}_l\right) \tag{5.50}$$

$$\mathbf{b}_{cons,j} = \mathbf{g}_j + \left(\mathbf{F}_l + (1-\vartheta)\mathbf{E}_l\right)\mathbf{u}_j^k \quad j = 1,\ldots,n_c \tag{5.51}$$

$$\mathbf{b}^i_{reac,j} = \left(\mathbf{E}_g - \mathbf{E}_l\right)\mathbf{u}_{g,j}^{k+\vartheta,i} - \mathbf{E}_l \mathbf{u}_j^{k+\vartheta,i} + \mathbf{U}_j \mathbf{S}_k^T \mathbf{r}_k^{k+\vartheta,i} \quad j = 1,\ldots,n_c \tag{5.52}$$

Species not pertaining to the liquid phase are transported as if they were in the aqueous phase. This is corrected by the reactive term ($\mathbf{b}_{cons,j}$). In the chemistry step, the concentrations of all species from the total analytical concentrations are:

$$\mathbf{u}_j^{k+1,i+1} = \mathbf{U}\mathbf{c}_j^{k+1,i+1} \quad j = 1,\ldots,n_c \tag{5.53}$$

which as for equation (5.47) must be solved together with equations representing the mass action laws, following which the reactive term can be updated directly by equation (5.52). An advantage of using total analytical concentrations instead of total aqueous concentrations may be that equation (5.53) has total analytical concentrations evaluated at iteration $i + 1$, whereas equation (5.47) has only the aqueous part evaluated at that iteration.

The algorithm for the SIA proceeds as follows:

1. Set time step $k = 0$ and set or read all variables ($\mathbf{u}^{k=0}$, $\mathbf{u}_l^{k=0}$, $\mathbf{c}^{k=0}$) to their initial values.
2. Increment the time step $k = k + 1$.
3. Calculate the conservative transport matrix **A** and the vector $\mathbf{b}_{cons,j}$.
4. Reactive transport calculations:
   - 4.1. Initialize Picard's approach: set $i = 0$ and the chemical reaction terms ($\mathbf{b}^{i=0}_{reac,j}$). The latter can be given a value equal to those at the previous time step or, if at the first time step, a value of zero.
   - 4.2. Increment the iteration step: $i = i + 1$.
   - 4.3. Transport step: For each component solve equation (5.43) or (5.49).
   - 4.4. Speciation: For each node or cell solve equation (5.47) or (5.53) as explained in Chapter 3 (section 3.2.1). This is an iterative process and requires an initialization of the concentrations of the primary species. They can be assigned to the concentrations of the previous iteration or, if at the first iteration, to those of the previous time step.
   - 4.5. Update the chemical reaction terms ($\mathbf{b}^{i+0}_{reac,j}$) by means of equation (5.46) or (5.52).
   - 4.6. Check for convergence: If $\mathbf{u}_l^{i+1}$ is close enough to $\mathbf{u}_l^i$ (e.g., $|(\mathbf{u}_l^{i+1}-\mathbf{u}_l^i)/\ \mathbf{u}_l^{i+1}| < \varepsilon$) for every node and component, where ε is a convergence criterion), then convergence has been reached. If so, go to 5, the end of the reactive transport calculations; if not, go to 4.2, the next iteration.
5. Go to 2, the next time step.

The *sequential non-iterative approach* or SNIA is an alternative to the algorithm described above. As its name suggests, it performs only one transport and chemistry step without iterating between the two. This introduces an error, which can only be neglected if the size of the time step is sufficiently small. Hence, the use of the SNIA demands a careful choice of the time step size.

5.3.2 *Direct Substitution Approach (DSA)*

The *direct substitution approach* consists essentially of applying the method of Newton-Raphson (explained in section 3.2.1 of Chapter 3) to the discretized reactive transport equations of the components of each node or cell. Instead of writing these equations for each component as was done for the SIA, for the DSA it is more convenient to write them for each node j:

$$\mathbf{f}_j = \sum_m F_{l,jm}\frac{\mathbf{u}_{l,m}^{k+1}-\mathbf{u}_{l,m}^{k}}{\Delta t} + \sum_m F_{g,jm}\frac{\mathbf{u}_{g,m}^{k+1}-\mathbf{u}_{g,m}^{k}}{\Delta t} + \sum_m F_{s,jm}\frac{\mathbf{u}_{s,m}^{k+1}-\mathbf{u}_{s,m}^{k}}{\Delta t}$$
$$-\mathbf{g}_j - \sum_m E_{l,jm}\mathbf{u}_{l,m}^{k+\vartheta,i+1} - \sum_m E_{g,jm}\mathbf{u}_{g,m}^{k+\vartheta,i+1} - \mathbf{U}_j\mathbf{S}_k^T\mathbf{r}_k^{k+\vartheta,i} = \mathbf{0} \quad j = 1,\ldots,n \qquad (5.54)$$

where $\mathbf{u}_{l,j}$, $\mathbf{u}_{g,j}$ and $\mathbf{u}_{s,j}$ are vectors of concentrations of the components of node j. Subscripts j and m refer to the rows and columns of matrices **E** and **F**. These matrices have a subscript indicating the phase to which they pertain (l liquid, g gas and s solid). The equations written as a function of some unknown, most commonly the concentrations of primary species ($\mathbf{c}_1$, recall section 3.1.3), because the calculation of the various geochemical properties ($\mathbf{u}_l$, $\mathbf{u}_g$, $\mathbf{u}_s$, $\mathbf{r}_k$) is easier when written as a function of this unknown. Applying the method of

Newton-Raphson to equation (5.54) leads to the following set of equations to be solved at each iteration i:

$$\begin{pmatrix} \frac{\partial \mathbf{f}^i_{j=1}}{\partial \mathbf{c}_{1,j=1}} & \frac{\partial \mathbf{f}^i_{j=1}}{\partial \mathbf{c}_{1,j=2}} & \cdots & \frac{\partial \mathbf{f}^i_{j=1}}{\partial \mathbf{c}_{1,j=n}} \\ \frac{\partial \mathbf{f}^i_{j=2}}{\partial \mathbf{c}_{1,j=1}} & \frac{\partial \mathbf{f}^i_{j=2}}{\partial \mathbf{c}_{1,j=2}} & \cdots & \frac{\partial \mathbf{f}^i_{j=2}}{\partial \mathbf{c}_{1,j=n}} \\ \vdots & \vdots & & \vdots \\ \frac{\partial \mathbf{f}^i_{j=n}}{\partial \mathbf{c}_{1,j=1}} & \frac{\partial \mathbf{f}^i_{j=n}}{\partial \mathbf{c}_{1,j=2}} & \cdots & \frac{\partial \mathbf{f}^i_{j=n}}{\partial \mathbf{c}_{1,j=n}} \end{pmatrix} \begin{pmatrix} \mathbf{c}^{k+1,i+1}_{1,j=1} - \mathbf{c}^{k+1,i}_{1,j=1} \\ \mathbf{c}^{k+1,i+1}_{1,j=2} - \mathbf{c}^{k+1,i}_{1,j=2} \\ \vdots \\ \mathbf{c}^{k+1,i+1}_{1,j=n} - \mathbf{c}^{k+1,i}_{1,j=n} \end{pmatrix} = - \begin{pmatrix} \mathbf{f}^i_{j=1} \\ \mathbf{f}^i_{j=2} \\ \vdots \\ \mathbf{f}^i_{j=n} \end{pmatrix} \tag{5.55}$$

where the matrix at the left-hand side is the Jacobian matrix consisting of the derivatives of the transport equations at all nodes with respect to the unknowns ($\mathbf{c}_1$) at all nodes. Likewise, the vector on the right-hand side, called the residual, consists of the transport equations for all nodes. The system to be solved is much larger than that for the SIA.

Before calculating the residual, the various geochemical properties ($\mathbf{u}_l$, $\mathbf{u}_g$, $\mathbf{u}_s$, $\mathbf{r}_k$) need to be calculated from the concentrations of the primary species through mass action and kinetic rate equations. As explained in section 3.2.1 (Chapter 3), the dependence of activity coefficients on individual species concentrations is usually weak. Therefore, the concentrations of secondary species can be written as (almost) explicit functions of the concentrations of primary species. Hence, a simple Picard approach is often sufficient. When all concentrations are calculated, the kinetic reaction rates ($\mathbf{r}_k$) and the total component concentrations ($\mathbf{u}_l$, $\mathbf{u}_g$, $\mathbf{u}_s$) can be directly calculated.

Computing the Jacobian requires the derivatives of the geochemical properties with respect to the concentrations of the primary species ($\mathbf{c}_1$). In turn, these derivatives require the derivatives of the concentrations of secondary species ($\mathbf{c}_2$). These can be obtained from the mass action equations (see eq. 3.23). As these are algebraic equations, there are only derivatives of $\mathbf{c}_2$ with respects to $\mathbf{c}_1$ at the same node, that is, the derivative of $\mathbf{c}_2$ of one node with respect to $\mathbf{c}_1$ of another node equal zero. Inclusion of the mass action equations gives the following for each node:

$$\begin{aligned} \frac{d\log \mathbf{c}_2}{d\log \mathbf{c}_1} &= \mathbf{S}_1^* + \mathbf{S}_1^* \frac{\partial \log \gamma_1}{\partial \log \mathbf{c}_1} + \mathbf{S}_1^* \frac{\partial \log \gamma_1}{\partial \log \mathbf{c}_2} \frac{d\log \mathbf{c}_2}{d\log \mathbf{c}_1} - \frac{\partial \log \gamma_2}{\partial \log \mathbf{c}_1} - \frac{\partial \log \gamma_2}{\partial \log \mathbf{c}_2} \frac{d\log \mathbf{c}_2}{d\log \mathbf{c}_1} \\ \Rightarrow \left(\mathbf{I} - \mathbf{S}_1^* \frac{\partial \log \gamma_1}{\partial \log \mathbf{c}_2} + \frac{\partial \log \gamma_2}{\partial \log \mathbf{c}_2} \right) \frac{d\log \mathbf{c}_2}{d\log \mathbf{c}_1} &= \mathbf{S}_1^* + \mathbf{S}_1^* \frac{\partial \log \gamma_1}{\partial \log \mathbf{c}_1} - \frac{\partial \log \gamma_2}{\partial \log \mathbf{c}_1} \end{aligned} \tag{5.56}$$

The equation is in logarithmic form, but can easily be transformed to non-logarithmic form (through $dc_2/dc_1 = (c_2 d\log c_2)/(c_1 d\log c_1)$). Formally, a linear system has to be solved. However, as the activity coefficients do not usually change much with concentration, their derivatives can be neglected (i.e., $\partial \log\gamma/\partial \log c \approx 0$), or updated in an iterative manner, which would lead to an explicit function without the need to solve a linear system. Once the derivatives are calculated, the derivatives of geochemical properties ($\mathbf{u}_l$, $\mathbf{u}_g$, $\mathbf{u}_s$, $\mathbf{r}_k$) can also be calculated without difficulty.

As explained in sections 3.1.4 and 4.5.3, it may be convenient to use component matrices that eliminate constant activity species, such as minerals. As these species may exist in only portions of the domain, the definition of components can vary spatially. In contrast to the SIA, this is not a problem for the DSA. The vectors of component concentrations are calculated differently for the equation of each node. In equation (5.54), the component concentrations of node *m* for the equation of node *j* must be calculated as:

$$\mathbf{u}_{l,m} = \mathbf{U}_{l,j}\mathbf{c}_{l,m} \quad \mathbf{u}_{g,m} = \mathbf{U}_{g,j}\mathbf{c}_{g,m} \quad \mathbf{u}_{s,m} = \mathbf{U}_{s,j}\mathbf{c}_{s,m} \tag{5.57}$$

that is, using the component matrix of node *j* and the concentrations of node *m*. Similarly, the derivatives can be calculated.

The algorithm for DSA proceeds as follows:

1. Set the time step $k = 0$, and set or read the unknowns ($\mathbf{c}_1^{k=0}$) to their initial values.
2. Increment the time step $k = k + 1$.
3. Calculate conservative transport matrices **E**, **F** vector **g**.
4. Initialize the Newton-Raphson process: Set $i = 0$ and the unknowns ($\mathbf{c}_1^{k+1,i=0}$). The latter can be given a value equal to the previous time step ($\mathbf{c}_1^{k+1,i=0} = \mathbf{c}_1^{k}$).
5. Increment the iteration step: $i = i + 1$.
6. Calculate the components concentrations and kinetic rates ($\mathbf{u}_l$, $\mathbf{u}_g$, $\mathbf{u}_s$, $\mathbf{r}_k$) and their derivatives. This can be done for each node independently.
7. Assemble the Jacobian matrix and residual.
8. Solve the linear system of equations (5.55).
9. Update the unknown values.
10. Check for convergence: If $\mathbf{c}_1^{i+1}$ is close enough to $\mathbf{c}_1^{i}$ and/or the residual is sufficiently small (e.g., $|(\mathbf{c}_1^{i+1} - \mathbf{c}_1^{i})/\mathbf{c}_1^{i+1}| < \varepsilon_1$) and/or $\mathbf{f}^{i+1} < \varepsilon_2$ for every node and component, where ε_1 and ε_2 are convergence criteria), then convergence has been reached. If so, go to 11, the end of the reactive transport calculations; if not, go to 5, the next iteration.
11. Go to 2, the next time step.

5.3.3 *Comparison between SIA and DSA*

The SIA and DSA approaches differ with respect to coding and CPU memory and time requirements, depending on the type of problem that has to be solved. In this section, the advantages and disadvantages of each approach are discussed.

The first advantage of the SIA or SNIA is the separation of the geochemical from the transport calculations. It is relatively easy to develop a reactive transport code by simply coupling a geochemical code to a conservative transport code. For instance, the reactive transport code PHAST (Parkhurst *et al.*, 2010) uses the SNIA to couple the geochemical code PHREEQC (Parkhurst and Appelo, 1999) with the transport code HST3D (Kipp, 1997). In contrast, implementing the DSA is more difficult because the calculation of the Jacobian matrix requires considerable additional coding.

A second advantage of the SIA is that the size of the linear system to be solved (eq. 5.43 or 5.49) equals the number of nodes. The DSA, on the other hand, requires the solution of a system equal to the number of nodes times the number of components. As a result, the DSA demands larger CPU memory and more CPU time for each reactive transport iteration than the SIA. This led Yeh and Tripathi (1989) to conclude that the DSA is not feasible for 2D and 3D cases with large numbers of nodes.

However, Yeh and Tripathi (1989) did not take into account that the SIA generally converges more slowly than the DSA (that is, it requires more iterations to reach convergence), and it is stiffer (that is, it may only reach convergence for very small time steps). So, the DSA may outperform the SIA due to the smaller amount of total iterations. Indeed, Reeves and Kirkner (1988) and Steefel and MacQuarrie (1996) reported more numerical problems for the SIA and/or SNIA than for the DSA in the case of small 1D grids. Saaltink *et al.* (2001) compared the SIA and the DSA for larger 2D cases and concluded that the DSA runs faster than the SIA in chemically difficult cases (i.e., highly non-linear and/or very retarded), but that the SIA may become faster for chemically simple problems with large 2D and 3D grids. Moreover, in a comprehensive comparison of various codes using SIA/SNIA or DSA, Carrayrou *et al.* (2010) showed that the DSA was very competitive compared to the SIA/SNIA for a number of benchmark cases.

The DSA can be further improved by use of more efficient linear solvers. For instance, Hammond *et al.* (2005) applied the Jacobian-free Newton-Krylov solver with physics-based preconditioning to an iterative solver (GMRES), which avoids the explicit computation and storage of the Jacobian matrix. They found good performance if the concentration fronts were not too retarded. Moreover, Fahs *et al.* (2008) found that the DSA becomes more efficient than the SIA, even for large number of unknowns, when a sparse direct solver is applied.

Also DSA performance can be improved by using formulations that are based on independent components (i.e., components that do not affect each other), such as those proposed by Molins *et al.* (2004) and Kräutle and Knabner (2005, 2007). This divides the large system of equations to be solved (eq. 5.55) into smaller ones, leading to less CPU memory and CPU time for each iteration. Initial results indicate that this method seems to work well (Carrayrou *et al.*, 2010; Hoffmann *et al.*, 2010).

REFERENCES

Bundschuh, J. & Suárez Arriaga, M.-C.: *Introduction to the numerical modeling of groundwater and geothermal Systems: Fundamentals of mass, energy and solute transport in poroelastic rocks.* CRC Press, Boca Raton, FL, 2010.

Carrayrou, J., Hoffmann, J., Knabner, P., Kräutle, S., de Dieuleveult, C., Erhel, J., Van der Lee, J., Lagneau, V., Mayer, K.U. & MacQuarrie, K.T.B.: Comparison of numerical methods for simulating strongly nonlinear and heterogeneous reactive transport problems—the MoMaS benchmark case. *Comput. Geosci.* 14 (2010), pp. 483–502.

Fahs, M., Carrayrou, J., Younes, A. & Ackerer, P.: On the efficiency of the direct substitution approach for reactive transport problems in porous media. *Water Air Soil Pollut.* 193 (2008), pp. 299–308.

Hammond, G.E., Valocchi, A.J. & Lichtner, P.C.: Application of Jacobian-free Newton–Krylov with physics-based preconditioning to biogeochemical transport. *Adv. in Water Resour.* 28:4 (2005), pp. 359–376.

Hoffmann, J., Kräutle, S. & Knabner, P.: A parallel global-implicit 2-D solver for reactive transport problems in porous media based on a reduction scheme and its application to the MoMaS benchmark problem. *Comput. Geosci.* 14 (2010), pp. 421–433.

Huyakorn, P.S. & Pinder, G.F.: *Computational methods in subsurface flow*. Academic Press, New York, NY, 1983.

Kräutle, S. & Knabner, P.: A new numerical reduction scheme for fully coupled multicomponent transport-reaction problems in porous media. *Water Resour. Res.* 41 (2005), W09414.

Kräutle, S. & Knabner, P.: A reduction scheme for coupled multicomponent transport-reaction problems in porous media: Generalization to problems with heterogeneous equilibrium reactions. *Water Resour. Res.* 43 (2007), W03429.

Kipp, K.L.: Guide to the revised heat and solute transport simulator HST3d—Version 2. Tech. Rep. Water-Resources Investigations Report 97-4157, U.S. Geological Survey, 1997.

Molins, S., Carrera, J., Ayora, C. & Saaltink, M.W.: A formulation for decoupling components in reactive transport problems. *Water Resour. Res.* 40 (2004), W10301.

Parkhurst, D.L. & Appelo, C.A.J.: Users guide to PHREEQC (Version 2)—a computer program for speciation, batch-reaction, one-dimensional transport, and inverse geochemical calculations. Tech. Rep. Water-Resources Investigations Report 99-4259, U.S. Geological Survey, 1999.

Parkhurst, D.L., Kipp, K.L. & Charlton, S.R.: PHAST Version 2—A program for simulating groundwater flow, solute transport, and multicomponent geochemical reactions: U.S. Geological Survey Techniques and Methods 6–A35, 2010.

Pinder, G.F. & Gray W.G.: *Finite element simulation in surface and subsurface hydrology*. Interscience, New York, NY, 1977.

Reeves, H. & Kirkner, D.J.: Multicomponent mass transport with homogeneous and heterogeneous chemical reactions: effect of the chemistry on the choice of numerical algorithm: 2. Numerical results. *Water Resour. Res.* 24:10 (1988), pp. 1730–1739.

Saaltink, M.W., Carrera, J. & Ayora, C.: On the behavior of approaches to simulate reactive transport. *J. Contam. Hydrol.* 48:3–4 (2001), pp. 213–235.

Steefel, C.I. & MacQuarrie, K.T.B.: Approaches to modeling reactive transport. In: P.C. Lichtner, C.I. Steefel, C.I. & E.H. Oelkers (eds): Reactive transport in porous media. *Reviews in Mineralogy* 34, Mineralogical Society of America, Washington, DC, 1996, pp. 83–129.

Xu, T., Samper, J., Ayora, C., Manzano, M. & Custodio, E.: Modeling of non-isothermal multi-component reactive transport in field scale porous media flow systems. *J. Hydrol.* 214:1–4 (1999), pp. 144–164.

Yeh, G.T. & Tripathi, V.S.: A critical evaluation of recent developments in hydrogeochemical transport models of reactive multichemical components. *Water Resour. Res.* 25:1 (1989), pp. 93–108.

Yeh, G.T. & Tripathi, V.S.: A model for simulating transport of reactive multispecies components: Model development and demonstration, *Water Resour. Res.* 27:12 (1991), pp. 3075–3094.

Zheng, C. & Bennett, G.D.: *Applied contaminant transport modeling, theory and practice*. Van Nostrand Reinhold, New York, NY, 1995.

CHAPTER 6

Elaboration of a geochemical model

Michael Zilberbrand

"Make everything as simple as possible, but not simpler."

Albert Einstein

6.1 INTRODUCTION

Elaboration and running of geochemical models can expand our knowledge of both natural and artificial systems and of processes occurring in these systems. Modeling presents each of such complex systems in a schematized form described by theoretical equations, and solves these equations using specialized software. As a result, additional information about the system state and development is received. Sometimes this information can be verified by existing measurements, but sometimes such measurements are absent or cannot be obtained.

Following the Draft Guidance on the Development, Evaluation, and Application of Regulatory Environmental Models prepared by EPA's Council for Regulatory Environmental Modeling (EPA CREM, 2008), model development can be viewed as a process that is achieved by following four main steps: (i) identify the environmental issue (or set of issues) the model is intended to address; (ii) develop the conceptual model; (iii) construct the model framework (develop the mathematical model), and (iv) parameterize the model to develop the application tool. Identifying the environmental issue or issues is site-specific. Such issues could be contamination and water resource problems. Development of the conceptual model is the most important and creative procedure. It requires the complete knowledge and estimation of the studied system. There are a lot of mathematical models exteriorized in tens of available computer programs. The choice of the most appropriate software is usually not difficult. Model parametrization includes setting reliable parameters of the system (temperature, pressures, redox potential, *pH* and concentrations of different components in liquid, solid and gas) and a choice of a suitable thermodynamic database.

6.2 MODEL TYPES AND THE MOST POPULAR EXISTING SOFTWARE PACKAGES

There are four different types of hydrogeochemical models: (i) chemical speciation and solubility models; (ii) reaction-path models (forward modeling); (iii) inverse (mass balance) models and (iv) reactive transport models. Usually, several types of models need to be coupled. Comparison of abilities of the most popular hydro-geochemical models is shown in Table 6.1.

6.2.1 *Speciation-solubility models*

Speciation-solubility models deal with closed, static, batch or beaker-type systems (Zhu and Anderson, 2002). These models evaluate concentrations/activities of chemical species (free ions and complexes) existing either in a water sample or in a solution at mutual (homogeneous) equilibrium with solids and/or gases of known composition at certain

Table 6.1. Main features of the most popular hydro-geochemical codes.

Software ability	PHREEQC	NETPATHXL, NETPATH-WIN	Visual MINTEQ	MINEQL+*	EQ3/6, EQ6*	GWB*	OLI analyzers*	SOLMINEQ.88 PC/SHELL*	MIN3P	PHSAT	PHT3D	TEQUIL	WINHUMIC V	SCALE 2000*	CHEAQS	MIN3P	GMIN	CHILLER, SOLVEQ*	SHEMAT*	FREZCHEM	TOUGHREACT*	FRACHEM*	HP1	RETRASO
Temperature range °C	0–300	0–50	0–50	0–50	0–300	0–300	0–300	0–350	0–50	0–300	0–300	0–250	0–25	25–250	0–50	0–50	0–300	0–350	0–300	–113–+25	0–300	0–300	0–300	0–300
Treating fresh and brackish water	+	+	+	+	+	+		+	+	+	+		+		+	+		+	+		+		+	+
Treating brines	+				+	+	+	+		+	+	+		+			+		+	+	+	+	+	
Phase transition: boiling (b) or freezing (f)					b		b	b										b		f	b	b		
Reaction progress	+	+			+	+	+	+	+	+	+			+		+	+	+	+		+	+	+	+
Kinetics consideration	+				+	+	+	+	+	+	+			+		+			+		+	+	+	+
Ion exchange consideration	+	+	+	+	+	+	+	+	+	+	+		+			+					+		+	+
Surface complexation	+		+	+	+	+	+	+	+	+	+		+		+	+	+				+		+	+
Complexation on natural organic matter	+									+	+		+		+								+	+
Solid solutions consideration	+		+	+	+	+	+	+		+	+						+						+	+
Isotopes fractionation	+	+			+	+				+	+										+		+	
Transport in the saturated zone	+ 1D				+ 1D	+ 1D-2D			+ 2D-3D	+ 1D-3D	+ 2D-3D			+		+ 2D-3D			+ 2D-3D		+ 2D-3D	+ 2D-3D	+ 1D	+ 2D
Transport in the unsaturated zone									+							+					+ 2D-3D	+ 2D-3D	+ 1D	+ 2D
Multi-component diffusion in clays	+																							
Inverse modeling	+	+																						
Consideration of uncertainties	+																						+	
User-friendly interface	+	+	+	+		+	+	+		+	+	+	+	+	+				+	+				+

*commercial software.

thermodynamic conditions. In addition, these models estimate the saturation of the solution with respect to different minerals and gases at these conditions.

When considering ion exchange or surface complexation, speciation-solubility models can predict the composition of stable species on ion-exchangers and adsorbing surfaces. Results of equilibrium calculations can be useful for evaluating kinetic rates as a function of deviation from equilibrium (Zhu and Anderson, 2005).

6.2.2 *Reaction-path models*

Reaction-path models deal with stepwise-equilibrium or kinetic calculations of an irreversible reaction progress (forward modeling). Definite masses of reactants (minerals, aquatic species or gases) could be both added to the system and removed from it at steady or changing temperature and pressure. Reaction progress may be complicated by mixing solutions, cation exchange and surface complexation.

6.2.3 *Inverse (mass-balance) models*

Inverse (mass-balance) models calculate mole transfers and, sometimes, isotope fractionation between solutions and both minerals and gases along a stable flow path from one point with known chemical compositions to another. Thermodynamics in such a calculation are neglected. For use of these models, dispersion and diffusion influence on solution chemistry are required to be small, and chemical composition of solutions in both initial and final points—constant over a long time.

6.2.4 *Reactive transport models*

Reactive transport models calculate a change in chemical composition of moving water caused by advection, diffusion and reactions with the solid phase. Coupling both thermodynamic and transport calculations, these models are the most complex and require a good knowledge of both hydrogeology and geochemistry.

6.3 DATA REQUIRED FOR GEOCHEMICAL MODELING

Data required for geochemical modeling depend on its goal, a type of a model that will be used and a decision how detailed this model will be. Actually, the goal of the modeling defines a type of a model.

6.3.1 *Data for speciation-solubility models*

For speciation-solubility models field measurements of the main parameters of state (temperature and, sometimes, pressure), redox potential and *pH* of groundwater are needed. Reliable measurements of these parameters are especially important for deep pressurized groundwater. Pressure drop during the sampling can cause degassing of groundwater and the corresponding strong change in concentrations of Ca^{2+}, Mg^{2+}, Sr^{2+}, HCO_3^-, CO_3^{2-} and other ions. At such conditions, special methods of sampling and measurements *in-situ* should be applied (Meng *et al.*, 1997; Horie *et al.*, 1997; Ding and Seyfried, 2007). It should be taken into consideration that the degassing takes place also during vacuum sampling of pore water from the unsaturated zone. The natural water composition in such cases could be restored by "putting back" the released gases and solids into the water composition detected in the lab, by using the reaction-path geochemical modeling. This operation should be performed at the measured in field temperature and pressure. The only problem is receiving reliable information about gases released and solids precipitated during the sampling.

An example of distortion of the natural composition of deep pressurized groundwater at its sampling is shown in Table 6.2. It represents results of modeling (by using PHREEQC software with llnl.dat database) of changes in the carbonate system during the sampling of the Mediterranean Sea water with sulfate reduced by 1 mole of CH_2O, entrapped at the depth of 3 km at temperature of 95°C in a carbonate aquifer. A rise of the artesian water to the land surface by a self-pressure was considered. During this rise, the water cooled by 10°C, brought in equilibration with atmospheric air and sampled to a closed vessel that was further opened in the lab and its water equilibrated with atmospheric air at 25°C during the analysis. The modeling shows that speciation of carbonate system in the sampled water was changed drastically. About 16 liters of CO_2 released, and about 230 g of calcite precipitated from each kg of water. As a result, concentrations of HCO_3^- and $H_2CO_3^0$ drop by two and five orders of magnitude, respectively. The ratio of concentrations of free ion Ca^{2+} and total Ca changes from 0.79 *in-situ* to about 0.9 in the sample. The similar would happen if groundwater, sampled at the depth into a special hermetic vessel, releases for analysis in the lab, but in this case, a possibility of analyzing all the released gases and precipitated solids exists. That is, theoretically, original composition of groundwater could be restored, and plausible speciation could be received.

For the speciation modeling all important measured components of alkalinity (HCO_3^-, CO_3^{2-}, sometimes, NH_3 and others) should be considered. Concentration of major ions should be measured precisely in order to provide reaction error less than 3–5%. Minor and trace components are included depending on their possible significance and the goal of the study. If redox-sensitive components are involved, both *Eh* and dissolved oxygen measured in field can be helpful. Among other dissolved gases, often CO_2 and H_2S concentrations are important. Neglecting some of the above chemical data can cause severe distortion in speciation results (Appelo and Postma, 2005; Merkel and Planner-Friedrich, 2008). Therefore, the most comprehensive chemical data should be used. For speciation of exchangeable or adsorbed species data on ion exchange capacity and adsorbing surfaces is required. In most other types of models (except, sometimes, transport models) the speciation is included.

Table 6.2. Distortion of the natural composition of deep pressurized groundwater at its sampling and analysis in the lab (results of the PHREEQC modeling).

	Hot and pressurized artesian groundwater at a depth of 3000 m	Discharging artesian water in equilibrium with atmospheric air	Sampled groundwater during its analysis in the lab
Temperature, °C	95	85	25
Pressure, atm	300	1	1
pH	5.15	7.76	7.44
Alkalinity, as $0.5CaCO_3$, meq kg_w^{-1}	65.8	0.3	0.3
Total Ca, mmol kg_w^{-1}	10.50	8.09	8.09
Ca^{2+}, mmol kg_w^{-1}	8.24	7.25	7.22
HCO_3^-, mmol kg_w^{-1}	46.8	0.17	0.20
CO_3^{2-}, mmol kg_w^{-1}	0.0023	0.0032	0.0008
CH_4, mmol kg_w^{-1}	218	0	0
CO_2 release, mmol kg_w^{-1}		474	−0.031
Total volume of gases released, L		10.6	−0.0007
$CaCO_3$ precipitation, mmol kg_w^{-1}		2.31	0
SI Calcite	0	0	−1.02
Ca^{2+}/Total Ca	0.79	0.90	0.89

6.3.2 *Data for reaction-path models*

For reaction-path models additional data on abundances of mineral and gaseous phases is needed. There should be a special accent on presence and characterization of solid solutions. In a case of kinetic reactions, rate equations and their parameters should be obtained in advance. A good compilation of the experimentally detected (in the lab) rate parameters of water-mineral interaction kinetics for application to geochemical modeling was prepared by Palandri and Kharaka (2004). When gaseous phase is involved it is important to know whether the system is open or closed. This should be previously grounded. The recent review of the reaction-path modeling has been done by Zhu (2009).

6.3.3 *Data for inverse (mass-balance) models*

Inverse (mass-balance) models, besides water chemistry data, require elementary knowledge about the water movement in a studied system, namely, knowledge of the existing flow paths. In addition, the complete mineralogical analysis should be performed. Preliminary information about possible hydrogeochemical processes is needed in order to make mass-balance calculations realistic.

6.3.4 *Data for reactive transport models*

For reactive transport models, in addition to a complete set of geochemical data, the most comprehensive knowledge of hydrogeology is necessary. Actually, each of these models includes a hydrogeological model, that is, requires definition of hydrological boundaries and setting the boundary conditions upon them, definition of sinks and sources of water, spatial spread of either hydraulic conductivity and specific storage/specific yield (for groundwater flow), or capillary conductivity and specific moisture capacity (for the unsaturated-zone flow).

6.4 SCHEMATIZATION AND CHOICE OF THERMODYNAMIC DATABASE

Schematization of a natural setting consists of defining a minimal set of its physical and geochemical features for adequate description of the nature (with required preciseness) in a framework of the chosen model. However, in many cases the sensitivity analysis for the applied data limitation and nature simplification is not performed.

Actually, the schematization includes definition and removal of uncertainties in our knowledge about the geological setting and studied processes. These uncertainties could be classified into three categories (Palenik *et al.*, 2004). Conceptual model uncertainty includes theoretical misconceptions about the modeling system, omission of relevant processes or inclusion of the non-existent processes. Parametric uncertainty refers to the quantitative thermodynamic parameters used in models. Scaling uncertainties arise due to heterogeneities inherent to natural systems over different length scales.

Some simplification of the nature is imposed by most of the existing models themselves (homogeneous equilibrium, ideal behavior of several components, instantaneous mixing, stepwise presentation of spatial heterogeneity, steady flow etc). Several assumptions about a state and properties of the system (open – closed, equilibrium – kinetic, homogeneous – double-porosity, ideal – non-ideal behavior of its components etc.) should be made by the modeler. Special attention should be paid to schematization of mineral phases either as pure ones, or as solid solutions. It is well known that the extent of reaction of minerals with a liquid phase strongly depends on this choice (Appelo and Postma, 2005).

In all cases, a degree and fidelity of the schematization, that predetermine the goodness of the modeling, are defined by our knowledge of the studied natural setting and/or acting processes. For example, a flow path for inverse modeling in many cases could be roughly defined from general considerations: from aquifer outcrops to a discharge area, but it can be much more precisely delineated from a detailed map of groundwater levels, if it is available. An imprecise schematization of water flow and wrong choice of the flow path could cause either unreliable results of the modeling or its impossibility at all. Another example, it is quite difficult to perform inverse modeling for transformation of rain water composition to this of groundwater in coastal aquifers if they are schematized as built of carboniferous sand only (without clay particles and/or accessory minerals).

Transport models, especially, 3D ones, require the hardest job of schematization since they combine the schematization for both hydrogeological and hydrogeochemical models.

In most of the cases a thermodynamic database (or several databases) is included in the existing hydrogeochemical software. However, in most of the programs it is possible to set new chemical species with their thermodynamic parameters and to correct the parameters for the species included.

The databases may differ by the use of parameters either of Debye-Hückel and Davies equations, or Pitzer's equations for calculation of activity coefficients. Strictly saying, the databases that employ Davies equations should be used for solutions with ion strength less than 0.5 (Appelo and Postma, 2005). However, practically, they are effectively used up to ionic strength of seawater, 0.7 (Parkhurst and Appelo, 1999). For non chloride-dominated solutions the limit of use of these databases can reach much lower ionic strengths, up to 0.1 (PHREEQC FAQ).

In the case of brines, databases that use Pitzer's specific interaction approach for modeling thermodynamic properties should be used. Their drawback is a lack of data for redox reactions or aluminosilicate reactions. In addition, a total number of species with known viral coefficients for Pitzer's equations in these databases is much less than a number of species with known thermodynamic properties in databases that use the Debye-Hückel and Davies equations.

Only several standard databases include a possibility of correction of thermodynamic data for temperatures above 50°C. However, temperature correction in many cases can be introduced manually or by using special software, SUPCRT92 (Johnson *et al.*, 1992), for example.

Thermodynamic properties of the same mineral and aqueous species in different databases can differ from each other. There may be several reasons for this (Oelkers *et al.*, 2009): (i) compilation of different sources of thermodynamic data detected using different methods, (ii) use of different estimation methods for missing species and (iii) updating a portion of the data. The absence of internal consistency in many of existing thermodynamic databases is a serious problem that has not been solved until today (Oelkers *et al.*, 2009).

A developing project "Common thermodynamic database for speciation models" (http://www.ctdp.org/#http://www.ctdp.org/) deals with "a compilation of a thermodynamic database with all exploitable data available, including organics, colloids and sorption reactions. The constants and values in this database are not forcedly certified or quality-assured: a confidence-level is attributed instead. The confidence attributed to a database item is subject to permanent discussion and may evolve in time". The most progress in development of thermodynamic databases for geochemical modeling occurs now in a sphere of performance assessment studies for radioactive or toxic waste disposal (Voigt *et al.*, 2007). Yucca Mountain project thermodynamic database comprises actinides, main inorganic elements, a few organics and about 2000 minerals related to the geology of Yucca Mountains (USA). Some kinetic data are also included. NEA-TDB project develops a database for radioactive waste that includes comprehensive, internally consistent, internationally recognized and quality assured thermodynamic data. Dortmund commercial databank (http://www.ddbst.de) for mostly organic compounds contains more than 100,000 data sets.

Results of the speciation-solubility modeling are strongly dependent on the thermodynamic database used by the model. The likely variability introduced into speciation calculation by the uncertainties (specified sometimes together with the respective formation constants) can be estimated by using LGUNGSKILE probabilistic software (Ekberg *et al.*, 2002). The highest uncertainty in thermodynamic properties exists for clay minerals and both minor and trace bearing minerals, phosphates and sulfides (Oelkers *et al.*, 2009).

The scaling uncertainties manifest themselves for most of model parameters by showing a strong discrepancy between values determined in the lab and in the field. The main reason for this is difficulty in performing a laboratory experiment that completely corresponds with field conditions including a difference in field heterogeneity at different scales. For example, hydraulic conductivities determined using Darcy tubes in the lab may sharply differ from pumping test results, owing to fracture or karst manifestation at a field local scale. On the other hand, pumping tests in shallow wells (that do not capture all the aquifer thickness) provide parameters only for the uppermost portion of the aquifer. They do not characterize both the deepest portions and whole thickness of the aquifer. Another well known problem is up-scaling the kinetics rate parameters determined through the lab experiments for their use in local or regional geochemical models. Reaction rates determined in field studies can be order of magnitude lower comparing to those determined in the lab. Several ways of the upscaling have been proposed in the literature, but the best way for estimation of representative values of these parameters is model calibration using field data, when it is possible.

When using regional reactive-transport models, spatial variability sometimes makes it difficult to set representative concentrations of species averaged over a large volume of rock in each cell or element of a model, especially, in fractured and karstic geological settings.

6.5 MODELING AND INTERPRETATION OF ITS RESULTS

After performing schematization of a natural setting and choice of a thermodynamic database, an input file for a chosen model should be built. This file concentrates all the relevant information about the schematized setting and chosen database. Analysis of the letter archive of questions about the use of PHREEQC (US Geological Survey, 1999), shows that most of problems in modeling originate from errors in the PHREEQC input file. Relatively few difficulties arise from convergence problems. The same situation exists, probably, for other geochemical codes.

Since, practically, the modeling consist in the running of an input file by using the chosen geochemical code, it is a responsibility of a modeler to build this file correctly. In most cases existing geochemical codes do not provide troubleshooting in input files, except of finding syntax errors and, sometimes, consistency errors in a geometry of hydrogeological settings. The running programs usually don't send any warning messages in case of an error in schematization and/or a wrong choice of thermodynamic database, but merely print wrong results to output files. The problem is that sometimes these results look realistic.

Analysis of plausibility of the received results and, if it is possible, their verification are important components of the modeling process. Verification of speciation-solubility models can seldom be performed, and only partially, by using data regarding development of mineral precipitation or dissolution. For reaction-path and reactive-transport models the verification is usually possible only at a short-time scale of lab or field experiments, or using short-term monitoring results. Results of application of inverse models can be verified by forward reactive-path modeling, and vice versa (Plummer, 1984). The preference of such check is in both involving thermodynamic considerations in analysis of results of inverse modeling and estimation of influence of unconsidered processes (in particular, kinetics, but also different reactions) on results of reaction-path modeling. However, since the inverse modeling provides results of the net mass change along a flow path, saturation indices calculated in the initial and final points could not reflect dissolution or precipitation that possibly exist in different intermediate points along the flow path.

Reporting of results is an important portion of the geochemical modeling. It is very convenient to present results of chemical speciation and inverse modeling in a form of Excel column-type charts (Merkel and Planer-Friedrich, 2005). In such a way results for several systems can be compared by putting them at the same chart.

Excel's AutoFilter facility enables effective analysis in a case of very numerous multi-variant results of inverse modeling. Apparently implausible scenarios can be filtered out without deletion. The possibility of hiding suspicious models makes it possible to come back and restore them in a case of need.

6.6 POSSIBLE ERRORS AND MISCONCEPTIONS IN MODEL ELABORATION

The main potential errors in elaboration of *speciation-solubility and reaction-path models* are:

1. Use of erroneous characteristics of alkalinity and/or redox potential basing on changed (not correctly measured in field) values of *pH* and *pe*;
2. Not including all the relevant chemical components, minerals and gases;
3. Neglecting presence of solid solutions;
4. Misconception in definition of open or closed character of the modeled system;
5. Disregarding kinetics;
6. Considering addition of minerals or gases in excessive amounts (above solubility thresholds);
7. Disregarding ion exchange and surface complexation.

In addition to the above, misconceptions in elaboration of *reactive transport models* are connected to an inappropriate schematization of a hydrogeological setting:

1. Wrong treatment of heterogeneity (either considering a homogenous system instead of a two-domain porous-fractured system or incorrect averaging for setting representative parameters of the media);
2. Wrong definition of the model boundaries;
3. Wrong setting the initial and boundary conditions.

The main misconceptions in elaboration of *inverse models* are:

1. Choice of an incorrect flow path;
2. Absence of the steady state along the chosen flow path (concentrations in both initial and final points are not constant in time);
3. A try to model cases when either dispersion or multi-component diffusion changes water chemistry;
4. Disregarding some of the contributing end-member solutions (for example, interstitial water of porous blocks in fractured or karstic aquifers; vertical fluxes of groundwater recharge or uprising deep waters along faults);
5. Considering phases (minerals or gases) that do not exist in the studied geological setting or neglecting the relevant existing phases;
6. Disregarding of ion exchange, sorption and uptake of nutrition elements by plants in the modeling of the unsaturated zone.

ACKNOWLEDGEMENTS

The author thanks to Dr. David Parkhurst of the USGS (U.S. Geological Survey, Denver Federal Center, P.O. Box 25046, MS 413, Denver, CO 80225, USA) for his valuable comments and to Mrs. Margarita Sapir for her thoughtful and constructive review of this chapter.

REFERENCES

Appelo, C.A.J. & Postma, D.: *Geochemistry, groundwater and pollution*. Balkema Publ., Amsterdam, The Netherlands, 2005.

Chen Zhu & Anderson, G.: *Environmental applications of geochemical modeling*. Cambridge University Press, Cambridge, UK, 2005.

Ding, K. & Seyfried, W.E.: *In situ* measurement of pH and dissolved H_2 in mid-ocean ridge hydrothermal fluids at elevated temperatures and pressures. *Chem. Rev.* 107 (2007), pp. 601–622.

Ekberg, C., Ödegaard-Jensen, A. & Meinrath, G.: LJUNGSKILE 1.0. A computer program for investigation of uncertainties in chemical speciation. SKI report 2003:03, Stockholm, Sweden, 2002.

EPA's Council for Regulatory Environmental Modeling (CREM), 2008 Internet site: http://www.epa.gov/crem/library/CREM-Guidance-Public-Review-Draft.pdf (accessed 10 April 2009).

Ii, H., Horie, Y., Ishii, T. & Shimada, J.: Development of an apparatus to measure groundwater qualities *in situ* and to sample groundwater using boreholes. *Environ. Geol.* 32:1 (1997), pp. 17–22.

Johnson, J.W., Oelkers, E.H. & Helgeson, H.C.: SUPCRT92: a software package for calculating the standard molal thermodynamic properties of minerals, gases, aqueous species, and reactions from 1 to 5000 bar and 0 to 1000°C. *Computers & Geosciences*, 18:7 (1992), pp. 899–947.

Meng, Z., Zhou, X.Y., Lvov, S.N. & Macdonald, D.D.: Development of advanced *in-situ* techniques for chemistry monitoring and corrosion mitigation in SCWO environments. Center for Advanced Materials, the Pennsylvania State University, Report to U. S. Department of Energy, 1997.

Merkel, B.J. & Planner-Friedrich, B.: *Groundwater geochemistry. A practical guide to modeling of natural and contaminated aquatic systems*. Springer, New York, NY, 2008.

Models in environmental regulatory decision making. Committee on Models in the Regulatory Decision Process, National Research Council. The National Academics Press, Washington, DC, 2007.

Oelkers, E.H., Benezeth, P. & Pokrovski, G.S.: Thermodynamic databases for water-rock interaction. *Reviews in Mineralogy and Geochemistry* 70 (2009), pp. 1–46.

Palandri, J.L. & Kharaka, Y.K.: A compilation of rate parameters of water-mineral interaction kinetics for application to geochemical modelling. U. S. Geological Survey, Open file report 2004-1068, Menlo Park, CA, 2004.

Palenik, C.S., Jensen, K.A. & Ewing, R.C.: Uncertainties in geochemical models of natural systems: implications for performance assessments. In: *Scientific Basis for Nuclear Waste Management, XXVIII Symp. Proc. of Material Research Society*, 824, pp. 543–548, 2004.

Parkhurst, D.L. & Appelo, C.A.J.: User's Guide to PHREEQC (Version 2)—A computer program for speciation, batch reaction, one-dimensional transport and inverse geochemical calculations. USGS Water-Resources Investigation Report 99-4259, Denver, CO, 1999.

Plummer, L.N.: Geochemical modelling: A comparison of forward and inverse methods. Reprinted from: *Procedings of the 1-st Canadian/American Conference on Hydrogeology. Practical applications of ground water geochemistry.* Banff, Alberta, Canada, June 22–26, 1984.

US Geological Survey, Parkhurst, D.L., 1999. PHREEQC FAQ. Internet site: http://wwwbrr.cr.usgs.gov/projects/GWC_coupled/phreeqc/faq.html (accessed 10 June 2010).

Voigt, W., Brendler, V., Marsh, K., Rarey, R., Wanner, H., Gaune-Escard, M., Cloke, P., Vercouter, Th., Bastrakov, E. & Hagemann, S.: Quality assurance in thermodynamic databases for performance assessment studies in waste disposal. *Appl. Chem.* 79:5 (2007), pp. 883–894.

CHAPTER 7

Advances in geochemical modeling for geothermal applications

Peter Birkle

"Properly constructed, a model is neither so simplified that it is unrealistic nor so detailed that it cannot be readily evaluated and applied to the problem of interest"
Craig M. Bethke (1996)*

7.1 INTRODUCTION

Following the aphorism of Albert Einstein: "*Everything should be made as simple as possible, but not simpler*", models need not to be completely realistic to be useful (Stumm and Morgan, 1996), but need to meet a successful balance between realism and practicality (Kühn, 2004). Properly constructed, a model is neither too simplified that it is unrealistic nor too detailed that it cannot be readily evaluated and applied to the problem of interest (Bethke, 2008). The results of a model have to be at least partially observable or experimentally verifiable (Zhu and Anderson, 2002).

The geochemical modeling of fossil hydrothermal systems allows an understanding of chemical processes through the circulation of water within the deep crust, such as seafloor circulation processes, the formation of hydrothermal ore deposits in fossil geothermal systems (e.g., porphyry copper deposits, Mississippi Valley type deposits, Western US uranium), diagenetic processes (e.g., clay alteration, dolomitization), and partially insights into maturation, migration and accumulation dynamics of petroleum in profound reservoirs. On the other hand, the evolution of recent geothermal systems is geochemically modeled during feasibility and exploitation stages to reconstruct thermodynamic conditions of the reservoir and potential water-rock interaction processes, which are decisive for the positioning of exploratory drillings and production boreholes. An additional application of geochemical modeling in geothermal reservoirs is the predictive modeling of chemical reactions during reservoir stimulation by acid injection. The latter method is commonly applied for low permeable reservoirs, where fluids are unable to circulate under natural conditions.

Recent geothermal systems can mostly be found in regions around plate margins where the existence of shallow located magma chambers provides the required heat for geothermal reservoir exploitation. The geothermal gradient of active geothermal areas can be more than ten times higher than the normal geothermal gradient of 2.5–3°C per 100 m. Therefore, enhanced temperature and pressure conditions (above 300°C and several hundreds of bars, respectively) are specific features to be considered for geochemical modeling of any geothermal reservoir system.

7.2 DEVELOPMENT OF GEOTHERMAL RESERVOIR TOOLS

Advances in modeling flow and transport processes: The starting point for the acceptance of computer simulation for the geothermal industry was the 1980 Code Comparison Study (Stanford Geothermal Program, 1980), where several simulators were tested for specific case

*Source: Bethke, C.M.: *Geochemical Reaction Modeling, Concepts and Applications*. Oxford University Press, New York, 1996.

studies. During the 1980s, most of the studies were developed by pioneering groups in the public sector (Lawrence Berkeley National Laboratory, Los Alamos National Laboratory, US Geological Survey, University of Auckland) and the private industry (i.e., GeothermEx, Unocal) (O'Sullivan *et al.*, 2001). Computer capacities limited the size of computational meshes to vertical slices (i.e., Lippmann and Bodvarsson, 1983; Ingebritsen and Sorey, 1985), to single layer models (Bodvarsson, *et al.*, 1982; Tulinius *et al.*, 1987) or to a radial symmetry (Lippmann and Bodvarsson, 1985) in one or two dimensions. Towards the end of the 1980s, a few quite-complex three-dimensional models were developed (i.e., Morris and Campbell, 1981; Pruess *et al.*, 1983; Aunzo *et al.*, 1989).

Since then, the incorporation of complex physical and chemical processes, such as the coupled heat and mass transfer in heterogeneous geothermal reservoirs and the presence of multi-phase system with additional chemical species (such as gases or dissolved salts), improved considerably the capabilities of geothermal reservoir simulation codes (e.g., Pruess *et al.*, 1998). Geothermal modelers adopted continuing improvements in numerical modeling, and models were diversified by extra features, such as a numerical representation of porous, fractured or double porosity systems. The common acceptance of numerical methods for geothermal exploration and exploitation has been reflected by a documented application of at least 125 global field simulation studies during the 1990s. An extensive list of modeled geothermal reservoirs can be found in O'Sullivan *et al.* (2001). Although the use of numerical models has become standard practice in the last 20–25 years, present and future efforts aim at the development of efficient coupled 3D mass and energy transport models with detailed interactions between aqueous fluids, gases and mineral assemblages, considering thermodynamic equilibrium as well as kinetic effects. Current efforts implement these models on massively parallel computers.

Advances in modeling chemical processes: After an initial period of pure water as idealized reservoir fluids (Stanford Geothermal Program, 1980), more sophisticated codes in the 1980s and 1990s included CO_2 (e.g., Pritchett *et al.*, 1981) and dissolved solids, typically represented as NaCl (Battistelli *et al.*, 1997) as input parameters. Further developments in the 1990s considered the interaction between different dissolved and gaseous chemical species in geothermal fluids (e.g., Kissling *et al.*, 1996; Battistelli *et al.*, 1997), as well as porosity and permeability changes through mineral dissolution and precipitation (i.e., White and Mroczek, 1998). Advanced multi-species chemical models describe reactions between aqueous, gaseous, and solid species, but are limited to zero-dimensional systems (i.e., "batch" systems) with no flow and transport considerations (e.g., Reed, 1982, 1998; Wolery, 1992; Helgeson, 1992; Moller *et al.*, 1998) and/or designed mostly for low temperature applications (e.g., Truesdell and Jones, 1974; Ball and Nordstrom, 1991; Parkhurst and Appelo, 1999; Allison *et al.*, 1991). A more realistic description of fully-coupled, fluid flow and mass transport with chemical interactions between aqueous, gaseous, and solid species is more difficult (O'Sullivan *et al.*, 2001) and has been the subject of more recent developments (e.g., Steefel and Lasaga, 1994; Lichtner *et al.*, 1996; Xu and Pruess, 2001; Xu *et al.*, 2011; Bethke 2007). Ongoing research is approaching coupled processes with different intrinsic time scales, and is addressing uncertainties in thermodynamic parameters, reactive surface areas and kinetic rate constants (e.g., Yabusaki *et al.*, 1998; Xu and Pruess; 2000; Palandri and Kharaka, 2004; Steefel *et al.*, 2005).

The following topics are more recent advances in the development of geochemical codes for geothermal reservoir simulation:

- Use of natural and man-made tracers such as soluble and volatile chemicals, noble gases, and isotopes for determining fluid flow paths and reservoir processes (i.e., Adams, 1995; Rose *et al.*, 1999).
- Modeling of reactive tracers subject to thermal degradation can anticipate cooling effects from injection (Plummer *et al.*, 2010).
- Isotopes as indicators for water-rock exchange processes (Wu and Pruess, 2000).
- Extending the thermodynamic range of fluid properties, especially to higher and super-critical temperatures for modeling deep reservoir zones (i.e., Hayba and Ingebritsen, 1997; Yano and Ishido, 1998).

- Coupled T-H-M (thermal-hydrologic-mechanical) processes in enhanced geothermal systems (EGS) and hot dry rock (HDR) geothermal reservoirs (i.e., Kohl *et al.*, 1995a,b; Bower and Zyvoloski, 1997).

7.3 TYPES OF GEOCHEMICAL MODELS FOR GEOTHERMAL SYSTEMS

A network of chemical reactions is the basic part of any geochemical model, without containing any temporal or spatial information ("zero-dimensional model"). Chemical reactions at elevated temperatures (>200 C) in hydrothermal systems often tend to operate near thermodynamic equilibrium, although this is generally the case only for secondary (alteration) minerals as opposed to primary rock-forming minerals. Near chemical equilibrium, the solute concentrations are governed by thermodynamic principles (e.g., Reed, 1982, 1998). Closed, static or so called "*batch*" systems delineate the distribution of mass among species and minerals (and/or gases), as well as the species activities, the saturation states of the solution with respect to various minerals phases, the fugacities of different gases that can exist in the chemical system, and the stable species distribution depending on temperature, fluid composition and mineral assemblage in contact with the water. In the case of geothermal reservoirs, this kind of geochemical modeling deciphers the initial equilibrium state of the geothermal system at the temperature of interest, and is primarily applied to delineate speciation-solubility characteristics of different geothermal phases (e.g., Reed, 1998).

As concentrations of reactants and products approach equilibrium after a certain time period, *kinetic processes* may have to be considered in certain chemical models. Reaction times can range from fractions of seconds or minutes (solute-solute), seconds to days (sorption-desorption), minutes to days (gas-solute) and hours to millions of years (crystalline solid-solute) (Langmuir and Mahoney, 1984). At low temperatures, calculations addressing chemical equilibrium are only possible if the geochemical system remains essentially closed long enough to approach equilibrium for the reactions of interest (Lasaga, 1998).

Reaction path models simulate the successive reactions steps of a system according to mass or energy fluxes (e.g., Reed, 1998; Wolery, 1992). Knowing the initial equilibrium state of a system, the reaction path of a model can be traced. For geothermal systems, chemical reaction paths are modeled to determine the effect of diagenetic reactions, especially temperature-accelerated water-rock interaction processes. In the case of polythermal reaction *models*, varying temperatures or a temperature gradient are commonly applied to closed systems (Kühn, 2004). The *titration model* is the simplest option for reaction modeling in open systems, with the gradual addition of a reactant to the solution in equilibrium. In practice, this type of model is used for (i) mixing between aqueous solutions, (ii) simulation of evaporation processes (i.e., formation of geothermal brine by evaporation of seawater), and (iii) mineral titration into formation water by pore fluid-rock reaction, i.e., the formation of skarn (Lu *et al.*, 1992).

The automatic calibration of the numerical model against field data by *inverse modeling* is a means to obtain model-related parameters that can be considered optimal for a given conceptual model (Finsterle *et al.*, 2000; Finsterle and Zhang, 2011).

Reactive transport modeling can provide opportunities to reduce costs and environmental impacts due to geothermal power generation. In detail, following chemical problems can be modeled:

- Chemical reactions by the injection of undersaturated, supersaturated or acidic brines into wells (e.g., André *et al.*, 2006a; Wagner *et al.*, 2005; Xu *et al.*, 2004b; Pham *et al.*, 2001).
- Enhanced reservoir management by modeling chemically reactive tracers (Pruess, 2002; Wu and Pruess, 2000).
- Recovery of precious minerals from geothermal brines (e.g., Pham *et al.*, 2001).
- Minimizing gas production and scaling through an optimized water injection and/or production strategy (e.g., Rodríguez, 2006; Xu *et al.*, 2004b).
- Effects of exploitation on CO_2 flux from geothermal systems (e.g., Ziegler 2011; Macini and Mesini, 2010; Pruess and Spycher, 2010; Xu and Pruess, 2010).

Further details on chemical modeling are given in Chapter 6 of this book, and practical applications of the geochemical models are demonstrated below in section 7.7.

7.4 REQUIREMENTS FOR GEOCHEMICAL SIMULATIONS OF GEOTHERMAL RESERVOIRS

The quality of any geochemical model depends on the availability of field specific input data as well as (and most particularly) on the thermodynamic and kinetic data required for the computations. In general, field sampling of surface and reservoir fluids and gases, the *in-situ* determination of certain parameters (*T*, *pH*, *Eh*, conductivity, dissolved oxygen and alkalinity), and a set of analytical data (Major, minor and trace elements, stable and radiogenic isotopes) are essential as primary parameters. Depending on the kind of simulator, additional data on thermodynamic properties of the reservoir, solid/gas phases and aqueous species, lithological and mineralogical composition of host rock, tectonic features, petrophysical and geophysical well log data are required to complete the input package. Besides, historical production data, such as mass discharge history of wells, wellhead pressure of production wells, enthalpy or temperature (liquid or dry steam), injection rates and wellhead pressures of injection wells, reservoir pressure in observation wells, reservoir temperatures through temperature logs in observation wells and wells status through caliper logs, injectivity tests and other methods are parameters to be selectively used for different computer codes.

The applied thermodynamic data is subject to error in experiment, chemical analysis and interpretation of the experimental results. The stabilities of many aqueous species have been determined only at room temperature, whereas mineral solubilities are generally measured at high temperatures where reactions approach equilibrium most rapidly (Bethke, 2008).

Application of kinetics to geochemical modeling is subject to a number of limitations especially by uncertainties in experimental methods and materials, as well as by the type of applied software. The kinetics of mineral dissolution and precipitation depend on a temperature-dependent kinetic rate constant k_n, the reactive surface area A_n of the mineral, the equilibrium constant K_n for the water-mineral reaction and a reaction quotient Q_n (Lasaga *et al.*, 1994). The reactive surface area is typically the parameter that is the least constrained and can vary by orders of magnitude. Equilibration rates are expected to be much slower than those predicted by the selected computer code, primarily because actual geochemical processes commonly involve flow through porous or fractured media (Palandri and Kharaka, 2004). Chemical equilibria among fluids, gases and alteration minerals can be expected for volcanic-hosted hydrothermal systems (Reed, 1982, 1998; Reed and Spycher, 1984), whereas low-temperature systems cause slower rates of chemical reaction. Equilibrium between aqueous fluids and high-temperature primary minerals is commonly not attained, and equilibrium between aqueous fluid and secondary minerals containing redox elements such as Fe and S generally requires higher temperatures than for those minerals with non-redox elements (Palandri and Kharaka, 2004).

Temperature and *pH* have greatest effect on reaction rates. Mostly experiments are conducted far from equilibrium conditions. Palandri and Kharaka (2004) compiled parameters of an Arrhenius-type rate equation for over 70 minerals, including phases from all the major classes of silicate, most carbonates, and many other non-silicates. The parameters are compiled for surface controlled reactions at conditions far from equilibrium, rather than diffusion controlled reactions. For many minerals, variations in composition, degree of cation disorder, degree of crystallinity, as well as the frequency and distribution of crystal defects contribute to a high degree of uncertainty in experimentally determined dissolution and precipitation rates.

7.5 POPULAR COMPUTER SOFTWARE FOR GEOTHERMAL SYSTEM MODELING

A variety of software tools were developed for the simulation of geothermal systems, to compute fluid flow, heat transfer and mass transport or fluid flow, mass transport and chemical reactions, but only a small number of codes are able to simulate all the processes for a comprehensive study of geothermal systems. At present, geothermal simulators include realistic descriptions of fluid properties and account for phase transitions and associated latent heat effects (Pruess, 1990; Hayba and Ingebritsen, 1997). A selection of most commonly used programs for geothermal systems and their capabilities are given in Table 7.1. All are specialized for particular applications with resulting inherent advantages and disadvantages.

The geochemical programs are applied for manipulating chemical reactions, calculating stability diagrams and the equilibrium states of natural waters, computing speciation in aqueous solutions, tracing reaction paths involving fluids, minerals, and gases, modeling reactive transport in one or two dimensions, sorption of species onto mineral surfaces including surface complexation and ion exchange, and plotting the result of these calculations. Thermodynamic data bases are included in many of the software packages (Table 7.1), including the properties of aqueous species, minerals, and gases, equilibrium constants for reactions to form these species, and data required to calculate activity coefficients. Extensive reviews of geochemical models can be found in the literature (Appelo and Postma, 2005; Bethke, 2008; Kühn, 2004; Melchior and Bassett, 1990; Nordstrom and Ball, 1984; Paschke and van der Heijde, 1996; Zhu and Anderson, 2002). O'Sullivan *et al.* (2001) presented the state of the art of geothermal reservoir simulation with a comprehensive compilation of more than 100 computer-modeled geothermal systems worldwide. Most of the standard models for geothermal fields are restricted to fluid flow and heat transfer with TOUGH2 (Pruess, 1998), STAR (Pritchett, 1995) and TETRAD (Vinsome and Shook, 1993) as most commonly used 3D simulators, whereas few cases incorporate the geochemical evolution and mineral recovery.

Table 7.1. Available software and thermodynamic databases* for geochemical modeling of geothermal reservoirs.

Software	References	Applications
3DHYDROGEOCHEM	Cheng and Yeh (1998)	Finite element (FE) program for saturated and unsaturated zone modeling including species transport, chemical reactions (Debye-Hückel, equilibrium and kinetics), and microbiological processes.
BRIGHT	Olivella *et al.* (1986)	Thermo-hydro mechanical model code for saline media.
CHEM-TOUGH2	White (1995) White and Mroczec (1998)	TOUGH2 code extended for transport of reactive species with Debye-Hückel activities and extended temperatures up to 800°C. Fixed and limited number of reactive components to calculate permeability changes due to chemical reactions.
CHILLER	Reed (1982, 1998) Reed and Spycher (2006a)	Multicomponent equilibrium reaction path program capable to simulate a wide range of water-rock-gas interaction processes, including solid titrations, mixing of fluids, heating/cooling, and boiling ("flash" calculations) following various temperature, pressure and enthalpy paths.
CRUNCH	Steefel and Lasaga (1994)	Multicomponent reactive transport model for chemically reactive (equilibrium or kinetic rates), non-isothermal flows in porous media, including the coupling of mineral reactions with changes in formation porosity and permeability.
EQ3/6	Wolery (1992)	Geochemical modeling of aqueous systems, including speciation-solubility and multicomponent equilibrium and kinetic reaction paths calculations.

(*Continued*)

Table 7.1. (Continued).

Software	References	Applications
FLOTRAN	Lichtner *et al.* (1996) Lichtner (2004)	Multicomponent reactive transport model for chemically reactive (equilibrium or kinetic rates), non-isothermal flows in porous media, including the coupling of mineral reactions with changes in formation porosity and permeability.
FRACHEM	Bächler (2003) Durst (2002)	FE system for modeling coupled flow, heat, transport, and reactions in fractured systems. Combination of CHEMTOUGH2 and FRACTure.
GEOFLUIDS	Duan *et al.* (1996)	Calculation of gas solubilities, phase equilibria of gaseous mixtures, breakout pressures, densities, fugacities and other thermodynamic properties.
HST3D	Kipp (1997)	A computer code for simulation of heat and solute transport in three-dimensional ground-water flow systems.
KINDIS	Madé (1991) Madé *et al.* (1994)	Thermodynamic and kinetic code for modeling of multicomponent kinetic mineral-solution interactions.
MINTEQA2	Allison *et al.* (1991) U.S. EPA (1999)	Multicomponent equilibrium speciation model, including redox, ion-exchange and surface complexation models to calculate the equilibrium composition of dilute aqueous solutions at low temperatures.
NETPATH	Plummer *et al.* (1994)	Modeling inverse mass balance reactions along a hydrologic flow path, including modeling of isotopic fractionation (^{2}H, ^{14}C, ^{18}O, ^{34}S).
PHREEQC	Parkhurst and Appelo (1999) Parkhurst (1995)	A computer program for multicomponent equilibrium/kinetic speciation, batch-reaction, one-dimensional reactive transport, and inverse mass balance calculations.
RST2D	Raffensberger and Garven (1995a,b)	Coupled simulation of fluid flow, heat transfer, transport, and chemistry with a fixed number of reactive components.
SHEMAT	Bartels *et al.* (2003) Clauser (2003)	Finite differences (FD) reactive transport code including coupled HEat and MAss transport. The chemistry is applicable for dilute solutions (Debye-Hückel) to highly saline brines (Pitzer equation) and flow reactions are coupled to a porosity-permeability relationship.
SOLMNEQ.88	Kharaka *et al.* (1988) Perkins (1992)	Multicomponent equilibrium model of speciation and water-rock interactions at ambient and elevated temperatures.
SOLTHERM*	Reed and Palandri (2006)	Data base of equilibrium constants for aqueous-mineral-gas equilibria (CHILLER and SOLVEQ).
SOLVEQ	Reed and Spycher, (2006b) Reed (1998, 1982)	Multicomponent equilibrium speciation program at ambient and elevated temperatures (also preprocessor for CHILLER).
SUPCTR92 SUPCRT98	Johnson *et al.* (1992)	Calculation of standard molal thermodynamic properties of minerals, gases, aqueous species and reactions from 1 to 5000 bar and 0 to 1000°C. Used to create thermodynamic data for input into geochemical computations.
THE GEOCHEMIST'S WORKBENCH™	Bethke (2007)	Collection of five geochemical modules comprising most geochemical modeling options (Stability and *Eh*/*pH* diagrams, multicomponent speciation, reaction paths, reactive transport in 1-D and 2-D).

(*Continued*)

Table 7.1. (Continued).

Software	References	Applications
TEQUIL	Duan *et al.* (1996)	Calculation of solubilities of various evaporite minerals and related thermodynamic properties.
THERMAL	Fritz (1981)	Calculation of the thermodynamic evolution of an aqueous solution between two temperatures.
TOUGHREACT	Xu (2006) Xu and Pruess (2001) Xu *et al.* (2004a, 2006, 2011)	Multicomponent reactive transport model for chemically reactive (equilibrium or kinetic rates), non-isothermal flows of multiphase fluids in porous and fractured media. Coupled precipitation and dissolution reactions with changes in formation porosity and permeability.
WATCH	Arnórsson *et al.* (1982)	Calculation of aqueous speciation in non-thermal and thermal water.
WATEQ4F WATEQ	Ball and Nordstrom (1991) Truesdell and Jones (1974)	Multicomponent equilibrium speciation model to predict species distributions, mineral saturation states, and gas fugacities from chemical analyses.

The development of fully coupled 3D mass and energy transport models (e.g., Steefel and Lasaga, 1994) with detailed chemical interactions between aqueous fluids, gases and mineral assemblages, considering thermodynamic equilibrium as well as kinetic effects, provide a more realistic description of the coupled physical and chemical processes inherent in geothermal reservoirs. The subsurface multiphase flow and multicomponent reactive transport code PFLOTRAN is an excellent example of more recent high-performance computation within an existing groundwater model (Hammond *et al.*, 2007).

7.6 FLOW AND GEOCHEMICAL MODEL CALIBRATION

The process of model calibration for natural state matching with past historical matching is difficult and time consuming, but is essential to verify the accuracy of designed reservoir models. Required corrections of the model can be performed by adjusting a few input parameters until the differences between model results and field data are at a minimum. Several techniques were proposed and applied for flow model calibration:

- Adjustments of permeabilities and porosity parameters to model *pressure decline* (O'Sullivan *et al.*, 2001).
- Calibration of reservoir porosity through comparison of *production enthalpies* between model and field data, which is especially useful in wells discharge with two-phase enthalpies (Pham *et al.*, 1996; O'Sullivan *et al.*, 1998; Mannington *et al.*, 2000; Sanyal *et al.*, 2000).
- Calibration of permeability in reservoirs where production temperatures are affected by *injection returns* (e.g., Aunzo *et al.*, 1991; Tokita *et al.*, 1995; Parini *et al.*, 1996).
- Calibration of highly fractured reservoirs by *tracer test results* (e.g., Parini *et al.*, 1996; Rose *et al.*, 1997) or, in general, by historical changes in *chemical composition* (Kissling *et al.*, 1996).
- Use of *geopysical data*, such as gravity measurements or electricity potentials (see complete list in O'Sullivan *et al.*, 2001) for model calibration.

In the case of geochemical modeling, calibrations should be addressed especially in the case of reactive chemistry. Some selected case studies are:

- Adjustment of *thermodynamic and physicochemical parameters* to approach the modeled geochemical behavior to field and experimental data, e.g., calibration of petrophysical

input parameters resulted in coherent velocity scenarios of well plugging by mineral precipitation for both simulated and *in-situ* cases (e.g., Birkle *et al.*, 2008).
- *Extension of the thermodynamic database* by adding stoichiometric and equilibrium constants for new species (Reed, 1998).
- *Matching of time scale* for the process being developed to the kinetics of aqueous and mineral reactions, e.g., quartz fails to equilibrate on a time scale of days to weeks in an ascending, cooling hydrothermal fluid, but the faster reaction rate of amorphous silica suggest to perform equilibrium calculations with the latter mineral (Reed, 1998).
- Depending on the ionic strength of the modeled solution, different equations with individual limitations (e.g., Debye-Hückel, Pitzer) can be used to derive the *species' activity coefficient* (Bethke, 2008; Reed, 1998).

7.7 SELECTION OF RECENT APPLICATIONS (2000–2010)—CASE STUDIES

7.7.1 *General applications*

The versatility of geothermal reservoir simulators provided a wide range of applications for heat and flow problems. The following list shows the principal areas, where numerical simulations have produced significant advances in the understanding of chemical and physical processes in geothermal systems:

- Pressure decline during the depletion of boiling reservoirs.
- Evaluation of boiling and condensation zones.
- Strategies for reservoir exploitation (sustainable management).
- Transition from liquid-dominated to vapor-dominated conditions.
- Evolution of natural convection systems.
- Fluid and heat transfer in fractured-porous media.
- Tracer modeling to prognosticate thermal drawdown.
- Effects of reinjection and natural recharge.
- Non-condensable gas effects.
- Scaling and corrosion processes by geothermal exploitation.
- Geological sequestration and storage of CO_2 in geothermal reservoirs.
- Enhanced geothermal systems (EGS) with CO_2 as heat transmission fluid.
- Fluid-rock interaction in hot dry rock (HDR) reservoirs.

Typical questions to be addressed by numerical simulations are related to (i) the generating capacity of a field, (ii) future rates, enthalphies, and chemical composition of well discharges, (iii) identification of drilling targets, (iv) optimal well spacing and completion and (v) design and impact of heat-depleted fluids by reinjection processes. Most of the numerical modeling cases for geothermal systems are generally focused on the reconstruction of historical production sequences (e.g., temperature, pressure, water level) by adjusting input parameters into a fitting thermodynamic flow model, whereas reactive chemical transport modeling is less applied for practical reservoir management.

In the following sections, a number of recent geochemical simulation studies from the open literature are presented for specific geothermal fields (Model description and location map in Table 7.2. and Fig. 7.1, respectively), although a considerably larger number of studies remains proprietary of engineering consulting firms and geothermal operators.

7.7.2 *Conceptual reservoir models*

As a prerequisite for any computer simulation of a geothermal reservoir, a *conceptual model* has to be developed. This model is normally presented as plan views and cross sections with incorporated technical data from a multidisciplinary team of geologists, mineralogists,

Table 7.2. Examples for recent applications of numerical geochemical models in geothermal systems.

Reservoir	Country	Numerical code	Target	Reservoir type	Model type	References
Shahe (Bejing)*	China	Lumped parameters	Prognostics for water-level drawdown without reinjection.	l.d.	1D	Axelsson and Stefansson (2003)
Berlín and Ahuachapán	El Salvador	TOUGHREACT	Prediction of well injectivity by mineral scaling.	l.d.	1D	Montalvo *et al.* (2005)
Berlín	El Salvador	SOLVEQ/ CHILLER	Modeling scaling of silica from reinjection waters at well head.	l.d.	1D	Castro *et al.* (2006)
Berlín	El Salvador	FRACHEM	Prediction of well injectivity by mineral scaling.	l.d.	2D	Portier *et al.* (2010)
HDR, Soultz	France	TOUGHREACT	Mineral alteration and changes in reservoir porosity by CO_2-injection in EGS systems.	l+g	1D	Xu and Pruess (2010)
Soultz-sous-Forêts	France	FRACHEM	Chemical and mechanical HDR performance during fluid injection.	l.d.	2D	Bächler (2003) Portier *et al.* (2007b)
Soultz-sous-Forêts	France	FRACHEM/ SHEMAT TOUGHREACT	Evolution of reservoir temperature, porosity, and mineral dissolution/ precipitation.	l.d.	2D	Portier *et al.* (2007a) André *et al.* (2006b)
Soultz-sous-Forêts	France	TOUGHREACT	Injection of CO_2 into EGS (Enhanced Geothermal System)	l.d.	1D	Xu *et al.* (2008)
Soultz-sous-Forêts	France	THERMAL/ KINDIS/BRIGHT	Multi-methodic approach to reconstruct geochemical processes of the hot-dry rock thermal loop.	l.d.	2D	Jacquot (2000) Fritz *et al.* (2010)
Svartsengi*	Iceland	iTOUGH2	Inverse modeling for assessment of production capacity.	l+v	2D	Ketilsson *et al.* (2008)
Darajat*	Indonesia (Java)	TOUGH2 autough2	Sensitivity study of reservoir parameters for reinjection process.	v.d.	2D	Kaya and O'Sullivan (2010)

(*Continued*)

Table 7.2. (Continued).

Reservoir	Country	Numerical code	Target	Reservoir type	Model type	References
Ogiri*	Japan	iTOUGH2	History matching simulation of reservoir conditions.	l+v	3D	Kumanoto *et al.* (2009)
Uenotai	Japan	CHEM-TOUGH2	Reactive chemical modeling of rock alteration, reservoir chemistry and effect of permeability structures.	l.d.	3D	White *et al.* (2005)
Kakkonda Taupo	Japan New Zealand	CHEM-TOUGH2	Permeability changes by dissolution and precipitation of quartz under super-critical reservoir conditions.	l.d.	3D	White and Mroczek (1998)
Los Azufres	Mexico	PHREEQC	Simulation of geochemical equilibrium conditions in evaporation ponds, including precipitation/dissolution processes of minerals and aqueous speciation of geothermal brines.	l.d. (shallow part) v.d. (deep part)	1D	Birkle and Merkel (2002)
Los Humeros	Mexico	SOLVEQ/ CHILLER	Geochemical modeling to reconstruct the origin of acid reservoir fluids.	l.d. (shallow part) v.d. (deep part)	1D	Bienkowski *et al.* (2005)
Pinatubo	Philippines	SOLVEQ	Chemical equilibrium between present day geothermal fluids and hydrothermal minerals.	l.d. (shallow part) l+v (deep part)	1D	Balangue and Collantes (2005)
Tiwi	Philippines	TOUGHREACT	Reactive transport modeling of scaling and acidizing of injection well.	l.d.	1D	Xu *et al.* (2004b)
Kızıldere	Turkey	SOLMINEQ.88	Reinjection modeling to predict scaling.	l.d.	1D	Giese and Cetiner (2006)
Imperial Valley Steamboat Spring	U.S.A. (CA) U.S.A. (NV)	GEOFLUIDS / TEQUIL	Prediction of boiling pressure and scale formation on geothermal wells.	l+g	1D	Duan *et al.* (1996)
Coso	U.S.A. (CA)	GWB (Xlt)	Reduction of silica precipitation through acid and alkali injection.	l.d.	1D	Park *et al.* (2006)
Coso	U.S.A. (CA)	TOUGHREACT	Prediction of scaling process through fluid reinjection.	l.d.	1D	Kovac *et al.* (2006) McLin *et al.* (2006)

Legend: v.d. vapor dominated; l.d. liquid dominated; l+v liquid + vapor; l+g liquid + gas; *Multi-phase fluid and heat flow.

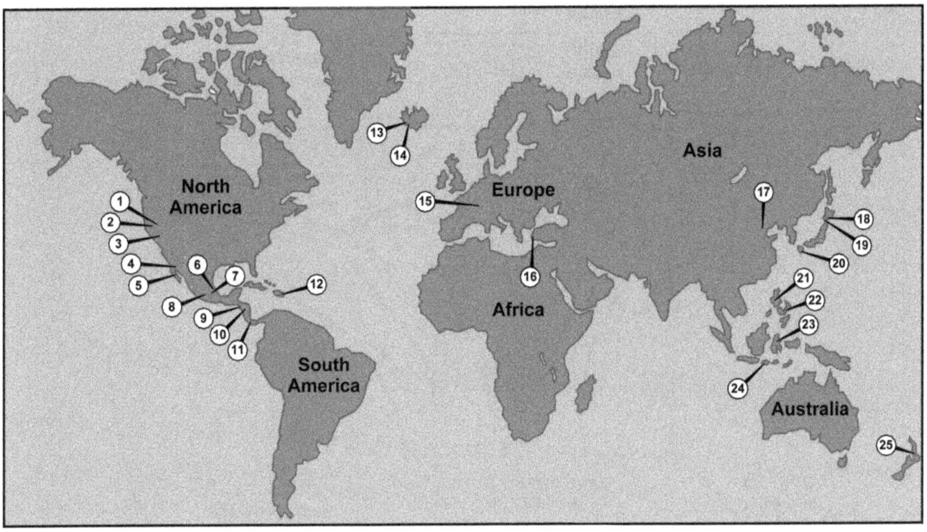

Figure 7.1. Location map of cited cases studies of global geothermal (petroleum*) fields with conceptual or numerical geochemical simulations.
Numbers in Figure: 1-Steamboat Springs (Nevada, U.S.A.); 2-Imperial Valley (The Geysers, CA, U.S.A.); 3-Coso (CA, U.S.A.); 4-Cerro Prieto (Mexico); 5-Las Tres Vírgenes (Mexico); 6-Poza Rica Altamira (Mexico)*; 7-Los Humeros (Mexico); 8-Los Azufres (Mexico); 9-Berlín and Ahuachapán (El Salvador); 10-Momotombo (Nicaragua); 11-Miravalles (Costa Rica); 12-Bouillante (Guadeloupe, French Lesser Antilles); 13-Hengill (Iceland); 14-Svartsengi (Iceland); 15-Soultz-sous-Forêts (France); 16-Kızıldere, Bayındır, Salihli (Turkey); 17-Shahe (China); 18-Kakkonda (Japan); 19-Uenotai (Japan); 20-Ogiri (Japan); 21-Pinatubo (Philippines); 22-Tiwi (Philippines); 23-Lahendong (Indonesia); 24-Darajat (Indonesia); 25-Taupo (New Zealand).

geophysicists, geochemists, hydrologists, reservoir engineers, and project managers. The profile includes information from surface manifestations (i.e., hot springs, mineral alteration zones, fault zones), as well as geological features (lithology, fault structures), petrophysical behavior (low and high porous and permeable strata), petrophysical data (resistivity boundaries, temperature logs, heat flow contours), hydraulic dynamics (flow boundaries, boiling zones, flow migration, pathways of convective-conductive systems) and hydrological conditions (recharge and discharge zones, infiltration rates, deep inflow zones) at the geothermal reservoir (Fig. 7.2). A major challenge for the computer modeler is the integration and synthesis of a partially incomplete or incoherent data set into a logical and sensible computer model.

A conceptual model represents the initial step of any reservoir modeling, incorporating all available data of a geothermal system in a qualitative manner (Bodvarsson *et al.*, 1986). Conceptual geochemical models have been developed for major geothermal production sites worldwide, i.e., for the Mexican geothermal fields of Cerro Prieto (e.g., Lippmann *et al.*, 1991), Los Azufres (e.g., Birkle *et al.*, 2001; Birkle and Verma, 2003; Zhanshi *et al.*, 2005), and Las Tres Vírgenes (e.g., Portugal *et al.*, 2000), the U.S. geothermal field of The Geysers (Bodvarsson *et al.*, 1992; Truesdell *et al.*, 1993), for the Indonesian Lahendong reservoir (Koestono *et al.*, 2010), Menderes Massif system in W-Anatolia (Özgur, 2010), and for the Icelandic Hengill geothermal system (Franzson *et al.*, 2010). A summary and re-interpretation of conceptual geochemical models for Central American geothermal fields is given by Birkle and Bundschuh (2007), such as for Berlín (El Salvador), Miravalles (Costa Rica), and Momotombo (Nicaragua) (Fig. 7.1).

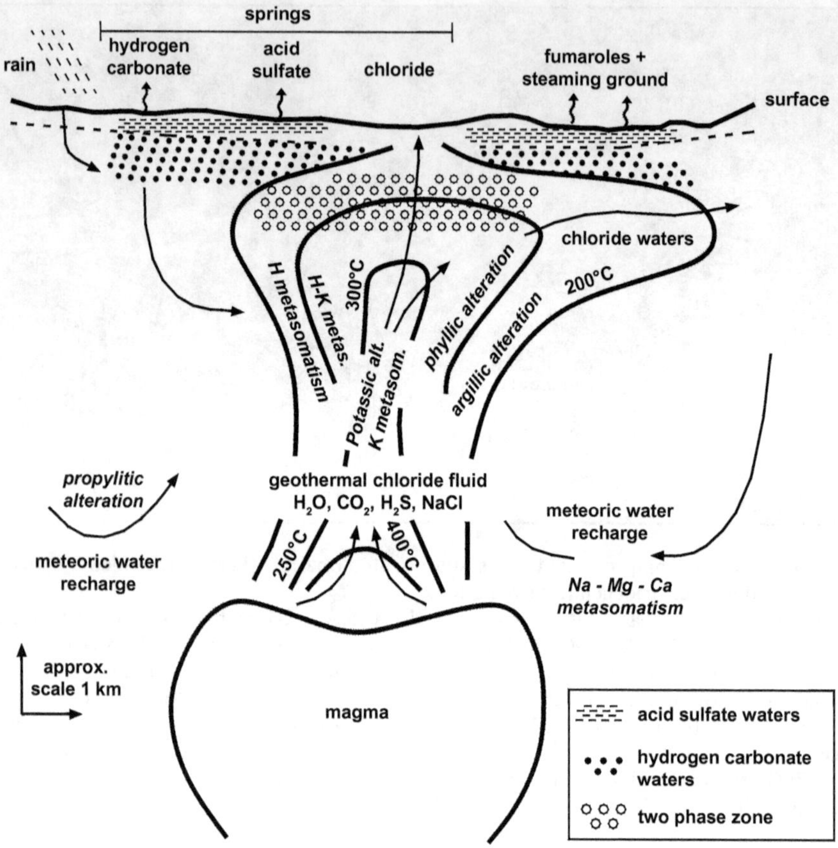

Figure 7.2. Conceptual model of liquid dominated geothermal systems in a low relief, typical for silicic volcanic terrain. Alteration and metasomatic process in italic letters, adapted from Kühn (2004), Nicholson (1993), and Giggenbach (1988) (reproduced with permission from Springer).

7.7.3 *Lumped parameter models*

Lumped parameter models can match production response with acceptable accuracy but geometrical structures of well-by-well models can not be modeled (Kaya and O'Sullivan, 2010). They are mostly used to estimate the production potential of a geothermal field and pressure responses to various exploitation strategies. Model parameters can be obtained by applying nonlinear least-square estimation techniques in which measured field pressure (or water level) data are matched to the corresponding model response. Then, by using history-matched models, the future performance of the reservoir can be predicted for different production/re-injection scenarios to optimize the management of a given low-temperature geothermal system (Sarak *et al.*, 2005).

Lumped models have been developed for low-temperature and high-temperature geothermal systems in Iceland, as well as in China, Turkey, Eastern Europe, Central America and The Philippines (Axelsson, 2003). Axelsson *et al.* (2005) presented a simple lumped parameter model for the general case of pressure changes in low-temperature geothermal reservoirs. Lopez *et al.* (2010) have designed a conceptual geological model of the Bouillante geothermal field (Guadeloupe, French Lesser Antilles) and a lumped parameter model as a preliminary exploration tool to test conceptual field schemes and to forecast pressure evolution.

7.7.4 *Advanced numerical modeling*

7.7.4.1 *Reservoir design and magnitude—Reconstruction of reservoir parameters*

White *et al.* (2005) designed a three-dimensional model of the Uenotai geothermal reservoir, Japan, with the reactive chemical transport model CHEM-TOUGH2 to calculate water-rock interaction products in the reservoir. A best match of simulated parameters (permeability) with registered reservoir conditions (temperatures, rock alteration) was obtained for a large region (2.5×1 km) with a permeability of 2×10.15 m^2 beneath the existing reservoir. White and Mroczek (1998) simulated the thermodyamic behavior (temperature, pressure, permeability, pore pressure) for the Kakkonda (Japan) and Taupo (New Zealand) geothermal systems for a time period of up to 500,000 years. As a result, parts of the lower reservoirs become impermeable during parts of their lifetime, mainly due to a minimum of silica solubility under super-critical reservoir conditions.

7.7.4.2 *Origin of acidity for reservoir fluids*

Some geothermal reservoirs are characterized by the presence of acid fluids with a *pH* as low as 2. In the case of the Los Humeros geothermal field, Mexico, multiphase modeling with SOLVEC/CHILLER confirmed that an underlying heat source (e.g., magma chamber) can provide enough amounts of HF or HCl to explain *pH* values as low as 3.5 for the deeper geothermal reservoir in Los Humeros (Bienkowski *et al.*, 2005).

7.7.4.3 *Mineral-fluid equilibria*

Using the speciation program SOLVEQ (Reed and Spycher, 1984) and the database SOLTHERM (Reed and Spycher, 1993), the activity of aqueous ions in geothermal water from the Pinatubo system, Philippines, was calculated from 300 to 50°C to determine the ion activity product (Q) for each mineral. The modeling of the mineral-fluid equilibria revealed that the hydrothermal mineral assemblage was partially formed by the present-day system, as indicated by the super saturation of three wells with acidic minerals, such as pyrite and pyrophyllite. In contrast, the occurrence of epidote at temperatures 40°C lower than the theoretical temperature of 200°C or more indicates that it is not in equilibrium with present PIN 1 geothermal fluids from the southeastern flank of Mt. Pinatubo. In general, chemical equilibrium between the present day fluids and hydrothermal minerals was not attained as indicated by a spread of temperatures and the occurrence of some minerals that have not been reported (Balangue and Collantes, 2005).

Geochemical simulations are also very useful to reconstruct mobility and solubility behavior of chemical compounds and precipitation/dissolution of minerals in geothermal brines stored temporally in surface evaporation ponds prior to reinjection. In the case of the Los Azufres geothermal field (Mexico), hydrochemical analysis of geothermal brine, extraction experiments with pond precipitates, mineralogical analysis of pond sediments and geochemical modeling of the mobility and solubility behavior of chemical compounds in evaporation ponds with PHREEQC (Parkhurst, 1995) confirmed the dominance of amorphous silica as major precipitate from saturated geothermal brines. A minor precipitation potential exists for other mineral species, such as iron oxides and hydroxides, manganese carbonate and selenium (Birkle and Merkel, 2002). Less than 1% of most brine solutes, except for Co, Cu, Mn, Pb, As, Fe and Si, are accumulated at the pond bottom, which indicates high solubilities for most aqueous species as part of an equilibrium-state between precipitation and dissolution processes. The mobility potential of several metals and non-metals requires the application of remediation technique for the final disposal of the pond sediments (Birkle and Merkel, 2002).

7.7.4.4 *Fluid reinjection—Scaling effects*

Geothermal reinjection is an essential part of any modern, environmentally friendly geothermal utilization, and has been an integral part of the operation of at least 50 geothermal fields in 20 countries. Besides a method for waste-disposal, it is used to counteract pressure draw-down and to extract more thermal energy from reservoir rocks (Axelsson, 2003). The injection into vapor dominated reservoirs, such as Larderello or The Geysers, involves complex fluid flow and heat transfer processes including boiling and condensation, vapor-liquid counter-flow and mixing of water with different temperatures. Generally, injection in this reservoir type has as an important impact in maintaining steam production (Cappetti *et al.*, 1995; Sanyal, 2000). Kaya and O'Sullivan (2010) carried out a sensitivity study of the Darajat reservoir (Java, Indonesia) to predict injection effects. The simulation results with the TOUGH2 code (Pruess *et al.*, 1998) showed that variations of some parameters (e.g., double porosity, vertical permeability, porous medium or fractured medium model, diagonal or parallel grids, grid size and orientation) have relative little effect on injection-induced steam flow. In contrast, a lower cut-off pressure in the deliverability formula increased the impact of injection on steam flow. A multitude of further case studies performed with the TOUGH2 geothermal reservoir simulator are given in O'Sullivan *et al.* (2001).

In the case of the Coso geothermal field in California, the simulated injection into a fracture zone with TOUGHREACT predicted the deposition of amorphous silica as dominant precipitating mineral, causing a decrease in porosity 1 to 2 years after initiation of injection (McLin *et al.*, 2006). Significant amounts of amorphous silica and minor amounts of calcite, quartz, and anhydrite are shown to precipitate while only calcite shows dissolution in the fractures (Kovac *et al.*, 2006). These predictions are consistent with current observations of injectivity losses by mineral deposition. Park *et al.* (2006) simulated with Xlt, a one dimensional reactive transport model (Bethke, 2008), three strategies to reduce long-term silica precipitation near Coso injection wells. An alternate injection of acid and alkali resulted to be the most effective method to reduce mineral precipitation as the kinetic rate of silica precipitation is reduced by acid injection, and alkali injection causes undersaturated conditions for silica, which dissolves as $NaHSiO_3$(aq). The loss of injectivity by scale deposition of precipitated amorphous silica within a 10 m radius from the well, as well as injectivity recovery by acid injection was reconstructed for the Tiwi geothermal field (Philippines) using a calibrated porosity-permeability relationship as part of the TOUGHREACT code (Xu *et al.*, 2004a). The simulated amounts for precipitated and dissolved amorphous silica during fluid reinjection and acid injection, respectively, captured very well the observed fluctuations of injectivity from field data (Xu *et al.*, 2004b).

In the case of the Berlín geothermal field in El Salvador, the increased exploitation of the field and the additional installation of a turbo-generated unit caused the decline of the capacity of most injection wells with time. A preliminary simulation with the coupled thermal-hydraulic-chemical FRACHEM code to study scaling problems resulted in a decline of porosity and permeability by the precipitation of amorphous silica and quartz (as principal precipitate) in the vicinity of the injection well at a maximum of 140 kg m^{-3} and 0.4 kg m^{-3}, respectively, after six years of simulation (Portier *et al.*, 2010). Similarly, reactive transport modeling with TOUGHREACT reproduces the decline in absorption capacity of the well TR-1A due to silica scaling, whereas other minerals maintain undersaturated conditions (Montalvo *et al.*, 2005). Geochemical simulations with SOLVEQ and CHILLER showed that the mass of reinjected water, its silica composition, and reinjection temperature determine the mass of potential quartz to precipitate. The temperature of the reservoir and water-rock interaction has only a minor effect, assuming chemical equilibrium with quartz in the original reservoir (Castro *et al.*, 2006).

Water-level prediction through lumped parameter modeling in the Shahe field in the northern part of Bejing, China, showed an increasing water-level draw-down, therefore future reinjection will be essential for sustainable utilization of the reservoir (Axelsson and Stefansson, 2003).

Reactive chemical transport modeling is also very useful in water flooding projects for enhanced oil recovery (EOR) in petroleum reservoirs, as the injected waters must be chemically compatible with connate reservoir waters, to avoid mineral dissolution-and-precipitation cycles that could seriously degrade formation permeability and injectivity. Birkle *et al.* (2008)

performed petrophysical laboratory experiments at core-scale and reactive chemistry modeling with TOUGHREACT to reconstruct potential chemical and physical processes during the injection of ambient-temperatured, untreated and treated formation water (30°C) into the Poza Rica-Altamira oil reservoir (Mexico) with initial temperatures of 80°C.

Similar to the experienced injectivity loss after five months of reinjection of formation waters into well AF-847 in 1999, numerical simulation confirmed a serious porosity and permeability reduction due to calcite precipitation after few months of injection, which is promoted by the retrograde solubility of this mineral. After 130 days, a relatively small decrease of the initial porosity from 16.3 to 13.5% caused a strong drop of the permeability from 50 mD to 0.93 mD (Fig. 7.3b) around the injection well. The further injection of fluids is inhibited by the maximum filling of 2.24 and 0.85% of the total formation volume with secondary calcite (Fig. 7.3a) and coelestine, respectively. In contrast, the use of treated water that performed well in the laboratory flow experiments was found to avoid excessive precipitation, and allowed injection to proceed.

Due to the heterogeneous lithological composition of the Poza Rica reservoir units, different probabilistic scenarios had to be simulated with variations of the primary permeability (K_L = 5–500 mD) and effective porosity input values (ϕ_{eff} = 12.3–16.0%) (Table 7.3.). As a principal result, the injection of untreated formation water into the oil reservoir will cause plugging of the injection interval, independent of the initial petrophysical composition. Variations of the primary permeability and porosity alter the time scale of plugging. Assuming a low permeability of 5 mD (*Case 3*), the reservoir pressure rises to 300 bar within 24 hours of injection, whereas a complete plugging of the reservoir by mechanical obstruction and mineral precipitation (mainly by calcite precipitation) is predicted to occur after 30 days. Larger K_L values of 50 mD (*Case 1*) or 500 mD (*Case 2*) can extend the time period for formation plugging to between 130 to 150 days.

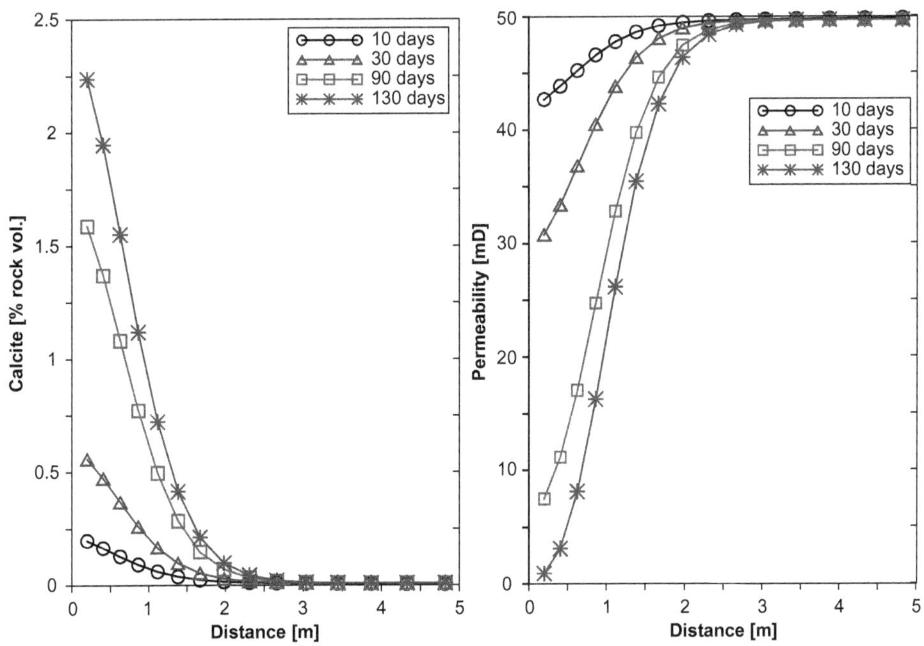

Figure 7.3. Numerical simulation with TOUGHREACT of the accumulation of precipitated calcite (a) and permeability loss (b) during the injection of untreated formation water at different time steps (10, 30, 90, and 130 days) to a distance of 5 m from the injection well at the Poza Rica-Altamira oil field (modified from Birkle *et al.*, 2008).

Table 7.3. Comparison of simulation parameters for different scenarios of water re-injection at the Poza Rica—Altamira oilfield, Mexico (taken from Birkle *et al.*, 2008).

Parameter	Case 1: Reference case	Case 2: Optimistic case	Case 3: Worst case
Initial K_L [mD]	50.0	500.0	5.0
Initial Φ_{eff} [%]	16.3	16.3	12.0
1. Pressure rise from 240 bar close to 300 bar:			
Time period [days]	130	145	1
Final K_L [mD]	0.93	1.01	4.78
Final ϕ_{eff} [%]	13.50	13.19	11.97
Final pressure [bar]	296.2	265.0	293.9
Calcite precipitation [rock vol.%]	2.24	2.47	0.02
2. Complete clogging of injection well:			
Time period [days]	145	150	30
Final K_L [mD]	0.09	1.04	1.50
Final ϕ_{eff} [%]	13.18	13.09	11.46
Final pressure [bar]	497.9	418.9	347.4
Calcite precipitation [rock vol.%]	2.48	2.55	0.43

7.7.4.5 *Hot-Dry Rock (HDR) systems (Soultz-sous-Forêts, France)*

Bächler (2003) developed the numerical code FRACHEM to simulate thermal, hydraulic and chemical processes, accounting for the specific reservoir conditions of the Soultz-sous-Forêts HDR site (Alsace, France). Modeling geochemical reactions during the fluid injection into a fracture zone at a depth of 3500 m during a two-year-period, calcite resulted as the most reactive mineral. Although a total of 990 kg (3.6% of the initial amount) is dissolved near the injection point and 1180 kg (4.3% of the initial amount) are precipitated close to the production zone, little variations of the porosity and permeability evolution suggest that mineral dissolution and precipitation do not represent a limiting factor on the HDR performance at the Soultz site.

Portier *et al.* (2007a) applied two geochemical transport codes, FRACHEM and SHEMAT to model a forced fluid circulation of five years for the complex Enhanced Geothermal System of the Soultz reservoir. Although significant differences were found in model results, primarily due to differences in kinetic models and their parameters, changes of the hydraulic system are equally well described by both models. FRACHEM and SHEMAT calculate the same temperature development and fluid velocities as well as resulting porosities from mineral reactions. Similar code comparisons were presented by André *et al.* (2006b) for simulations of the Soultz system using FRACHEM and TOUGHREACT, in this case also examining the effect of the activity coefficient models (Debye-Hückle versus Pitzer) on simulation results.

Jacquot (2000) used the code KINDIS to model the geochemical equilibrium stages of circulating fluid from the injection well to the production well, where the fluid is re-heated from 65 to 165°C. The results of the modeling of the sequence of dissolution and precipitation processes during the re-heating of the re-injected geothermal fluid are shown in Figure 7.4.

Carbonates are responsible for the main part of the mineral volume change, despite their minor abundance in the granitic rock reacting with the hydrothermal fluid. This is due to the differences in reactivities between carbonates and silicates, as the latter one is several orders of magnitude slower to dissolve or precipitate (Fritz *et al.*, 2010). On a laboratory scale, tube-in-tube petrophysical experiments with powdered granite and subsequent SEM, X-Ray and TEM mineralogical analysis were compared with the modeling of fluid-rock interactions with the codes THERMAL and KINDIS. Both methods indicate that the crystallization of feldspars at high temperatures (around 200°C) and clays at lower temperatures (60–150°C) will contribute to the control of the porosity in the granitic reservoir. As a further step, petrophysical data was

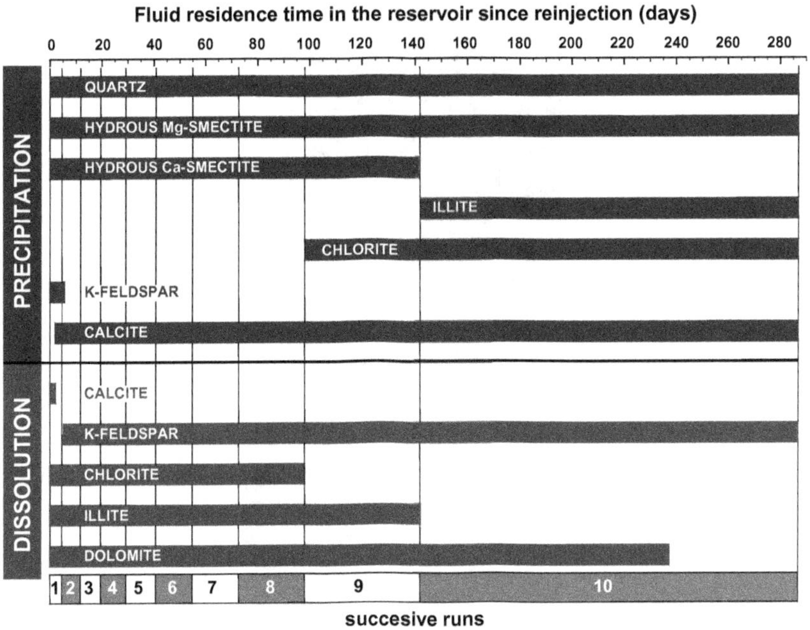

Figure 7.4. Qualitative sequence of dissolution and precipitation processes occurring within the geothermal reservoir during the re-heating of re-injected geothermal fluid, as simulated for the summer 1997 circulation test of the Soultz-sous-Forêts EGS (taken from Fritz *et al.*, 2010) (reproduced with permission from Elsevier).

incorporated into fault zone models with the thermo-hydro mechanical model code BRIGHT (Olivella *et al.*, 1986). Different geometries of fault zones with an increasing complexity from a single gauge zone to a complex fault zone, including a damage zone with a secondary fracture network, were modeled. Concerning mass transfer, the injected fluid had a different composition than the original one (Fig. 7.5), as the matrix of the damage zone will work as buffer, and the secondary fracture network will allow exchanges between the matrix and the gauge zone. Depending on the density of the secondary fracture network, the flow could be deviated in the damage zone and pass over the plug. As a conclusion, the geothermal loop with fluid temperature variations between 200 and 65°C may be responsible for dissolution/precipitation of minerals, which subsequently alter the porosity and permeability of the granite, similar to fossil hydrothermal circulation in the Rhine Graben system (Fritz *et al.*, 2010).

7.7.4.6 *CO_2 injection into geothermal reservoirs*

Recently, there is a growing interest in the novel concept of operating *Enhanced Geothermal Systems* (EGS) with CO_2 instead of water as heat transmission fluid. Numerical simulations of fluid dynamics and heat transfer indicate that CO_2 may be superior to water in its ability to mine heat from hot fractured rock (Fouillac *et al.*, 2004; Pruess, 2006, 2008). A fully developed EGS would consist of three zones, (i) a central zone in which all aqueous phase has been removed by dissolution into the flowing CO_2 stream, (ii) a surrounding intermediate zone with a two-phase mixture of water and CO_2, and (iii) a peripheral zone with reservoir fluid as a single aqueous phase and dissolved CO_2 (Fouillac *et al.*, 2004; Xu and Pruess, 2010).

Initial studies from Xu *et al.* (2008) have suggested that CO_2 will achieve larger rates of extraction, and can offer geologic storage as a benefit. A combination of mineral dissolution and precipitation effects in the peripheral zone of the system suggest impacts on reservoir growth and longevity. Xu and Pruess (2010) performed chemical reactive transport modeling

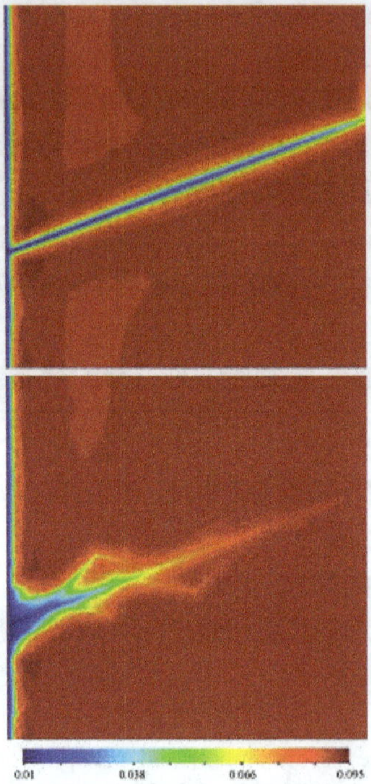

Figure 7.5. Distribution of total dissolved matter (kg/kg) after 30 days of fresh water injection in a single gauge zone (top) and in a complex fault zone (bottom) (taken from Fritz *et al.*, 2010; reproduced with permission from Elsevier).

with TOUGHREACT (Xu *et al.*, 2004a; Xu and Pruess, 2001) to investigate potential fluid-rock interaction in the peripheral zone of an EGS operated with CO_2. Temperature and pressure conditions, mineralogical composition and water chemistry was taken from the Soultz EGS site. As a main result, dissolved CO_2 may diffuse to the peripheral zone of the EGS and can induce dissolution of primary minerals (calcite, K-feldspar, chlorite) and precipitation of secondary carbonate (dolomite, siderite, ankerite) and clay minerals (Na-smectite, Ca-smectite, illite). Significant amounts of CO_2 could be fixed through precipitation of carbonate minerals, which causes reduced reservoir porosity, but can offer geological storage of carbon as an ancillary benefit (Xu and Pruess, 2010). Krupka *et al.* (2010) give an extended literature overview of the actual advances in geochemical modeling of carbonate reactions associated with CO_2 sequestration.

7.8 CONCLUSIONS—FUTURE CHALLENGES

Geochemical and reactive transport modeling is an essential tool during the exploration and exploitation stages of a geothermal field to optimize production and maximize reservoir lifetime, but its application has to be coupled and verified by real field and experimental data, and to be tested against field experience. For example, the comparison of reactive transport codes done by André *et al.* (2006b) revealed that differences in thermodynamic equilibrium constants, activity coefficients (or ion-interaction processes) and kinetic models can result in significant discrepancies in predicted mineral precipitation and reservoir-porosity and permeability evaluation. Existing thermodynamic data is limited for high-ionic strength and CO_2/water mixtures, and reaction kinetics and the dependence of reaction parameters on spatial scale are not fully understood.

The ability to correctly predict the saturation (supersaturation or undersaturation) of the reservoir fluids with respect to the mineralogy of the reservoir formation is a critical step in understanding and controlling permeability, as the dissolution and precipitation of minerals in the pore and fracture structure of a hydrothermal formation has a significant influence on the flow volume (e.g., Moller *et al.*, 2007). Especially for geothermal systems, the prediction of solubilities for a range of ionic strengths under high temperature conditions (e.g., >150°C) represents still a challenge for the scientific community. As an example, Moller *et al.* (2007) developed a thermochemical model of aluminum silicate mineral solubilities in aqueous solutions containing H^+, Na^+, K^+, Al^{3+}, Cl^-, $Si(OH)_4$, $SiO(OH)_3^-$, OH^-, $Al(OH)^{2+}$ $Al(OH)_2^+$, $Al(OH)_3^0$, $Al(OH)_4^-$ for 100°C and 300°C systems.

In the new field of CO_2–EGS systems, the time frame to form a lower barrier and the fixation of CO_2 is a function of mineral kinetics as well as mineral dissolution and precipitation, which requires future studies (Xu and Pruess, 2010). Sensitivity studies on primary rock mineralogy, the use of natural analogues of high-pressure CO_2 reservoirs, refinements on thermodynamic, kinetic and physical data, experimental studies to describe reaction kinetic are essential studies to understand chemical changes in geothermal CO_2-EGS systems and should be systematically addressed in the future (Xu and Pruess, 2010). Extensive thermodynamic databases exist for CO_2 and CH_4, gases, carbonate aqueous species and carbonate minerals, but dissociation constants and/or heat capacity data at temperatures above 25°C exists for less than approximately one-third of the mentioned compounds (Krupka *et al.*, 2010).

As shown for the Soultz-sous-Forêts HDR systems, short-term experiments and equilibrium geochemical modeling are able to reconstruct mass transfers for fast-equilibrating minerals (e.g., calcite), but the long-term behavior of some minerals (e.g., silicates) require modeling of dissolution and precipitation processes under kinetic constraints. In the future, fully coupled hydrologic-thermal-mechanical-chemical (H-T-M-C) simulators will be required to predict fluid flow, heat extraction and temperature drawdown, rock-mechanical processes, and chemical reactions that affecting EGS reservoir operation and sustainability.

ACKNOWLEDGEMENTS

The author is grateful to Nicolas F. Spycher (LBNL) for his very constructive review comments. The author apologizes for any not cited reference, as a complete discussion of available information about geochemical modeling in geothermal systems would exceed the permitted chapter size.

REFERENCES

Adams, M.: Vapor, liquid, and two-phase tracers for geothermal systems. *Proc. World Geothermal Congress'95*, Florence, Italy, 1995, pp. 1875–1880.

Allison, J.D., Brown, D.S. & Novo-Gradac, K.J.: MINTEQA2/PRODEFA2, A geochemical assessment model for environmental systems: Version 3.0. U.S. Environmental Protection Agency Report EPA/600/3-91/021, Athens, GA, 1991.

André, L., Rabemanana, V. & Vuataz, F.-D.: Influence on water-rock interactions on fracture permeability of the deep reservoir at Soultz-sous-Forêts, France. *Geothermics* 35:5–6 (2006a), pp. 507–531.

André, L., Spycher, N., Xu, T., Pruess, K. & Vuataz, F.-D.: Comparing FRACHEM and TOUGHREACT for reactive transport modeling of brine-rock interactions in enhanced geothermal systems (EGS). *Proc. 31th Workshop on Geothermal Reservoir Engineering*, 30 Jan.–1 Feb. 2006, Stanford University, Stanford, CA, 2006b.

Appelo, C.A.J. & Postma, D.: *Geochemistry, groundwater and pollution.* A.A. Balkema, Rotterdam, The Netherlands, 2005.

Arnórsson, S., Sigurdsson, S. & Svavarsson, H.: The chemistry of geothermal waters in Iceland. I. Calculation of aqueous speciation from 0°C to 370°C. *Geochim. Cosmochim. Acta* 46 (1982), pp. 1513–1532.

Aunzo, Z., Steingrimsson, B., Bodvarsson, G.S., Escobar, C. & Quintanilla, A.: Modeling studies of the Ahuachapán geothermal field, El Salvador. *Proc. 14th Workshop on Geothermal Reservoir Engineering*, 24–26 Jan. 1989, Stanford University, Stanford, CA, 1989, pp. 287–295.

Aunzo, Z., Laky, C., Steingrimsson, B., Bodvarsson, G.S., Lippmann, M.J., Truesdell, A.H., Escobar, C., Quintanilla, A. & Cuellar, G.: Pre-exploitation state of the Ahuachapán geothermal field, El Salvador. *Geothermics* 20 (1991), pp. 1–22.

Axelsson, G.: Essence of geothermal resource management. *Geothermal Training Programme*, United Nations University, IGC Shourt Course 2003, pp. 129–151.

Axelsson, G. & Stefansson, V.: Sustainable management of geothermal resources. *Proc. Int. Conference on Multiple Integrated Use of Geothermal Resources*, 14–17 Sept. 2003, Reykjavik, 2003.

Axelsson, G. & Bjornsson, G. & Quijano, J.E.: Reliability of lumped parameter modeling of pressure changes in geothermal reservoirs. *Proc. World Geothermal Congress*, Antalya, Turkey, 2005.

Bächler, D. *Coupled thermal-hydraulic-chemical modeling at the Soultz-sous-Forêts HDR reservoir (France)*. PhD thesis, Swiss Federal Institute of Technology Zurich, Zurich, Switzerland, 2003.

Balangue, Ma.I.R. & Collantes, Ma.G.C. Modeling mineral equilibrium in the Pinatubo geothermal system. *Proc. World Geothermal Congress*, 24–29 April 2005, Antalya, Turkey, 2005.

Ball, J.W. & Nordstrom, D.K.: User's manual for WATEQ4F, with revised thermodynamic data base and test cases for calculating speciation of major, trace and redox elements in natural waters. USGS Open-file Report 91-183, 1991.

Bartels, J., Kühn, M. & Clauser, C.: Numerical simulation of reactive flow using SHEMAT. In: C. Clauser (ed): *Numerical simulation of reactive flow in hot aquifers – SHEMAT and Processing SHEMAT*. Springer Publishers, Heidelberg, 2003.

Battistelli, A., Calore, C. & Pruess, K.: The simulator TOUGH2/EWASG for modeling geothermal reservoirs with brines and non-condensible gas. *Geothermics* 26:4 (1997), pp. 437–464.

Bethke, C.M.: *Geochemical and biochemical reaction modeling.* 2nd edition, Oxford University Press, Inc., NY, 2008.

Bethke, C.M. & Yeakel, S.: The Geochemist's Workbench™, Version 7.0, GWB Essentials Guide, Hydrogeology Program, University of Illinois, Urbana, IL, 2007.

Bienkowski, R., Torres-Alvarado, I. & Hinderer, M.: Geochemical modeling of acid fluids in Los Humeros geothermal field, Mexico. *Proc. World Geothermal Congress*, 24–29 April 2005, Antalya, Turkey, 2005.

Birkle, P. & Bundschuh, J.: Hydrogeochemical and isotope composition of geothermal fluids. In: J. Bundschuh & G. Alvarado (eds): *Central America—Geology, Resources and Hazards*, Chapter 27. Taylor & Francis / Balkema Publishers, The Netherlands, 2007, pp. 777–838.

Birkle, P. & Merkel, B.: Mineralogical-chemical composition and environmental risk potential of the evaporation pond sediments at the geothermal field of Los Azufres, Michoacán, México. *Environ. Geol.* 41 (2002), pp. 583–592.

Birkle, P. & Verma, M.: Geothermal development in Mexico. In: D. Chandrasekharam & J. Bundschuh (eds): *Geothermal energy resources for developing countries*. A.A. Balkema Publishers, Lisse, The Netherlands, 2003, pp. 385–404.

Birkle P., Merkel B., Portugal, E. & Torres, I.: The origin of reservoir fluids at the geothermal field of Los Azufres, México—Isotopic and hydrological indications. *Appl. Geochem.* 16 (2001), pp. 1595–1610.

Birkle, P., Pruess, K., Xu, T., Hernández, R.F., Díaz, M.L. & Contreras, E.L.: Using laboratory flow experiments and reactive chemical transport modeling for designing waterflooding of the Agua Fría reservoir, Poza Rica-Altamira field, Mexico. *SPE Reservoir Evaluation & Engineering* 11:6 (2008), pp. 1029–1045.

Bodvarsson, G.S., Vonder Haar, S., Wilt, M. & Tsang, C.F.: Preliminary studies of the reservoir capacity and generating potential of the Baca geothermal field, New Mexico. *Water Resour. Res.* 18:6 (1982), pp. 1713–1723.

Bodvarsson, G.S., Pruess, K. & Lippmann, M.J.: Modeling of geothermal systems. *J. Pet. Tech. 38* (1986), pp. 1007–1021.

Bodvarsson, G.S., Lippmann, M.J., Majer, E.L. & Pruess, K.: LBL research on The Geysers: Conceptual models, simulation and monitoring studies. LBL report No. 32139, 1992.

Bower, K.M. & Zyvoloski, G.: A numerical model for thermo-hydro-mechanical coupling in fractured rock. *Int. J. Rock Mechan. Mining Sci.* 34: 8 (1997), pp. 1201–1211.

Cappetti, G., Parisi, L., Ridolfi, A. & Stefani, G.: Fifteen years of reinjection in the Lardarello-Valle Secolo area: analysis of the production data. *Proc. World Geothermal Congress*, 18–31 May 1995, Florence, Italy, 1995.

Castro, M.R., López, D.L., Reyes-López, J.A., Montalvo, F.E., Romero, R., Ramírez-Hernández, J. & Lázaro-Mancilla, O.: Modeling scaling of silica from reinjection waters at well head conditions in the Berlín geothermal field, El Salvador, Central America. *Proc. 31th Workshop on Geothermal Reservoir Engineering*, 30 Jan.–1 Feb. 2006, Stanford University, Stanford, CA, 2006.
Cheng, H.P. & Yeh, G.T.: Development and demonstrative application of a 3-D numerical model of subsurface flow, heat transfer, and reactive chemical transport: 3DHYDROGEOCHEM. *J. Contam. Hydrol.* 34 (1998), pp. 47–83.
Clauser, C.: Numerical simulation of reactive flow in hot aquifers. Springer Publishers, Heidelberg, 2003.
Duan, Z., Møller, N., DeRocher, T. & Weare, J.H.: Prediction of boiling, scaling and formation conditions in geothermal reservoirs using computer programs TEQUIL and GEOFLUIDS. *Geothermics* 25:6 (1996), pp. 663–678.
Durst, P.: *Geochemical modelling of the Soultz-sous-Forêts Hot Dry Rock rest site: coupling fluid-rock interactions to heat and fluid transport*. PhD thesis, University of Neuchatel, Neuchatel, Switzerland, 2002.
Finsterle, S. & Zhang, Y.: Solving iTOUGH2 simulation and optimization problems using the PEST protocol. *Environmental Modelling and Software*, 26 (2011): pp. 959–968, doi:10.1016/j.envsoft.2011.02.008.
Finsterle, S., Björnsson, G., Pruess, K. & Battistelli, A.: Evaluation of geothermal well behavior using inverse modeling. In: B. Faybishenko, P. Adams Witherspoon & S. Benson (eds): *Dynamics of fluids in fractured rock*. American Geophysical Union, 2000, pp. 377–388.
Fouillac, C., Sanjuan, B., Gentier, S. & Czernichowski-Lauriol, I.: Could sequestration of CO_2 be combined with the development of enhanced geothermal systems? *Third Annual Conference on Carbon Capture and Sequestration*, 3–6 May 2004, Alexandria, VA, 2004.
Franzson, H., Gunnlaugsson, E., Árnason, K., Sæmundsson, K., Steingrímsson, B. & Harðarson, B.S.: The Hengill geothermal system, conceptual model and thermal evolution. *Proc. World Geothermal Congress*, 25–29 April 2010, Bali, Indonesia, 2010.
Fritz, B.: Etude thermodynamique et modélisation des réactions hydrothermales et diagénétiques. *Sci. Géol. Mém.* 65, 1981.
Fritz, B., Jacquot, E., Jacquemont, B., Baldeyrou-Bailly, A., Rosener, M. & Vidal, O.: Geochemical modelling of fluid-rock interactions in the context of the Soultz-sous-Forets geothermal system. *Comptes Rendus Geosci.* 342:7–8 (2010), pp. 653–667.
Giese, L.B. & Cetiner L.: Geochemical modelling of geothermal fluids—Application of the computer program SOLMINEQ.88. In: A. Baba, K. Howard & O. Gunduz (eds): Groundwater and ecosystems. Springer, printed in The Netherlands, 2006, pp. 93–106.
Giggenbach, W.F.: Geothermal solute equilibria. Derivation of Na-K-Mg-Ca indicators. *Geochim. Cosmochim. Acta* 52 (1988), pp. 2749–2765.
Hammond, G., Lichtner, P. & Lu, C.: Subsurface multiphase flow and multicomponent reactive transport modeling using high-performance computing. *J. Phys. Conf. Series* 78(012025), 2007.
Hayba, D.O. & Ingebritsen, S.E.: Multiphase groundwater flow near cooling plutons. *J. Geophys. Res.* 102:12 (1997), pp. 235–252.
Helgeson, H.C.: Effects of complex formation in flowing fluids on the hydrothermal solubilities of minerals as a function of fluid pressure and temperature in the critical and supercritical regions of the system H_2O. *Geochim. Cosmochim. Acta* 56:8 (1992), pp. 3191–3207.
Ingebritsen, S.E. & Sorey, M.L.: Quantitative analysis of the Lassen hydrothermal system, north-central California. *Water Resour. Res.* 21:6 (1985), pp. 853–868.
Jacquot, E.: *Modélisation thermodynamique et cinétique des réactions géochimiques entre fluides de bassin et socle cristallin: application au site expérimental du programme Européen de recherche en géothermie profonde (Soultz-Sous-Forêts, Bas-Rhin, France)*. PhD thesis, University Louis Pasteur, Strasbourg, France, 2000.
Johnson, J.W., Oelkers, E.H. & Helgeson, H.C.: SUPCRT92, a software package for calculating the standard molal thermodynamic properties of minerals, gases, aqueous species and reactions from 1 to 5000 bars and 0° to 1000°C. User's manual, 1992.
Kaya, E. & O'Sullivan, M.: Modelling of injection into vapour-dominated geothermal systems. *Proc. World Geothermal Congress*, 25–29 Abril 2010, Bali, Indonesia, 2010.
Ketilsson, J., Axelsson, G., Palsson, H. & Jonsson, M.T.: Production capacity assessment: Numerical modeling of geothermal resources. *Proc. 33rd Workshop on Geothermal Reservoir Engineering*, 28–30 Jan 2008, Stanford University, Stanford, CA.

Kharaka, Y.K., Gunter, W.D., Aggarwal, P.K., Perins, E.H. & DeBraal, J.D.: SOLMNEQ.88, a computer program for geochemical modeling of water-rock interactions. U.S. Geological Survey Water Resources Investigation Report 88-4227, 1988.

Kipp, K.L. Jr.: Guide to the revised heat and solute transport simulator: HST3D—Version 2. U.S. Geological Survey, Water-Resources Investigations Report 97-4157, 1997.

Kissling, W.M., Brown, K.L., O'Sullivan, M.J., White, S.P. & Bullivant, D.P.: Modeling chloride and CO_2 chemistry in the Wairakei geothermal reservoir. *Geothermics* 25 (1996), pp. 285–305.

Koestono, H., Siahaan, E.E., Silaban, M. & Franzson, H.: Geothermal model of the Lahendong geothermal field, Indonesia. *Proc. World Geothermal Congress*, 25–29 April 2010, Bali, Indonesia, 2010.

Kohl, T., Evans, K.F., Hopkirk, R.J. & Rybach, L.: Coupled hydraulic and mechanical considerations for the simulation of hot dry rock reservoirs. *Geothermics* 24:3 (1995a), pp. 333–343.

Kohl, T., Evans, K.F., Hopkirk, R.J. & Rybach, L.: Coupled hydraulic, thermal, and mechanical considerations for the simulation of hot dry rock reservoirs. *Geothermics* 24:3 (1995b), pp. 345–359.

Kovac, K.M., Xu, T., Pruess, K. & Adams, M.C.: Reactive chemical flow modeling applied to injection in the Coso EGS experiment. *31th Workshop on Geothermal Reservoir Engineering*, 30 Jan.–1 Feb. 2006, Stanford University, Stanford, CA, 2006.

Krupka, K.M., Cantrell, K.J. & McGrail, B.P.: Geochemical modeling of carbonate reactions associated with CO_2 sequestration—Literature review. Pacific Northwest National Laboratory, Report PNNL, 19766, prepared for U.S. Department of Energy, 2010.

Kühn, M.: Reactive flow modeling of hydrothermal systems. *Lecture Notes in Earth Sciences*, No. 103, Springer, 2004.

Kumamoto, Y., Itoi, R., Tanaka, T. & Hazama, Y.: Modeling and numerical analysis of the two-phase geothermal reservoir at Ogiri, Kyushu, Japan. *34th Workshop on Geothermal Reservoir Engineering*, 9–11 Feb. 2009, Stanford University, Stanford, CA, 2009.

Langmuir, D. & Mahoney, J.: Chemical equilibrium and kinetics of geochemical processes in groundwater studies. In B. Hitchon & E. Walleck (eds): *First Canadian-American Conference on Hydrogeology*, National Water Well Association, Dublin, OH, 1984, pp. 69–75.

Lasaga, A.C.: Kinetic theory in the earth sciences. *Princeton Series in Geochemistry*. Princeton University Press, Princeton, NJ, 1998.

Lasaga, A.C., Soler, J.M., Ganor, J., Burch, T.E. & Nagy, K.L.: Chemical weathering rate laws and global geochemical cycles. *Geochim. Cosmochim. Acta* 58 (1994), pp. 2361–2386.

Lichtner, P.C.: Flotran User's Manual: Two phase, non-isothermal coupled thermal-hydrologic-chemical (THC) reactive flow and transport code. LANL Report LA-UR-01-2349, Los Alamos National Laboratory, Los Alamos, NM, USA, 2004.

Lichtner, P.C., Steefel, C.I. & Oelkers, E.H.: Reactive transport in porous media. *Reviews in Mineralogy* 34, Mineralogical Society of America, Washington, DC, 1996.

Lippmann, M.J. & Bodvarsson, G.S.: Numerical studies of the heat and transport in the Cerro Prieto geothermal field, Mexico. *Water Resour. Res.* 19:3 (1983), pp. 753–767.

Lippmann, M.J. & Bodvarsson, G.S.: The Heber geothermal field, California: natural state and exploitation modeling studies. *J. Geophys. Res.* 90:B1 (1985), pp. 745–758.

Lippmann, M.J., Truesdell, A.H., Halfman-Dooley, S.E. & Mañón, A.: A review of the hydrogeologic-geochemical model for Cerro Prieto. *Geothermics* 20:1–2 (1991), pp. 39–52.

Lopez, S., Bouchot, V., Lakhssassi, M., Calcagno, P. & Grappe, B.: Modeling of Bouillante geothermal field (Guadeloupe, French Lesser Antilles). *35th Workshop on Geothermal Reservoir Engineering*, 1–3 Feb. 2010, Stanford University, Stanford, CA, 2010.

Lu, C., Reed, M.H. & Misra, K.C.: Zinc-lead skarn mineralization at Tin Creek, Alaska: Fluid inclusions and skarn-forming reactions. *Geochim. Cosmochim. Acta* 56 (1992), pp. 109–119.

Macini, P. & Mesini, E.: Assessing the underground CO_2 storage potential in a highly populated and industrialized area: The case of Lombardia region (Italy). *SPE Annual Technical Conference and Exhibition*, 19–22 Sept. 2010, Florence, Italy, 2010.

Madé, B.: *Modélisation thermodynamique et cinétique des réactions géochimiques dans les interactions eau-roche*. PhD thesis, University Louis Pasteur, Strasbourg, 1991.

Madé, B., Clément, A. & Fritz, B.: Modelling mineral-solutions interactions: The thermodynamic and kinetic code KINDISP. *Computer Geosci.* 20: 9 (1994), pp. 1347–1363.

Mannington, W., O'Sullivan, M.J. & Bullivant, D.P.: An air/water model of the Wairakei-Tauhara geothermal system. *Proc. World Geothermal Congress*, 28 May–June 10 2000, Kyushu-Tohuku, Japan, 2000, pp. 2713–2718.

McLin, K.S., Kovac, K.M., Moore, J.N., Adams, M.C. & Xu, T.: Modeling the geochemical effects of injection at Coso geothermal field, CA; Comparison with field observations. *31th Workshop on Geothermal Reservoir Engineering*, 30 Jan.–1 Feb. 2006, Stanford University, Stanford, CA, 2006.

Melchior, D.C. & Bassett, R.L. (eds): Chemical modeling of aqueous systems II. *ACS Symposium Series* Vol. 416, American Chemical Society, Washington, DC, 1990.

Moller, N., Greenberg, J.P. & Weare, J.H.: Computer modeling for geothermal systems: predicting carbonate and silica scale formation, CO_2 breakout and H_2S exchange. *Transp. Porous Media* 33 (1998), pp. 173–204.

Moller, N., Christov, C. & Weare, J.: Thermodynamic model for predicting interactions of geothermal brines with hydrothermal aluminum silicate minerals. *32th Workshop on Geothermal Reservoir Engineering*, 22–24 Jan. 2007, Stanford University, Stanford, CA, 2007.

Montalvo, F., Xu, T. & Pruess, K.: TOUGHREACT code applications to problems of reactive chemistry in geothermal production-injection wells. First exploratory model for Ahuachapán and Berlín geothermal fields. *Proc. World Geothermal Congress*, 24–29 April 2005, Antalya, Turkey, 2005.

Morris, C.W. & Campbell, D.A.: Geothermal reservoir energy recovery—a three-dimensional simulation study of the East Mesa field. *J. Petrol. Technol.* 33:4 (1981), pp. 735–742.

Nicholson, K.: Geothermal fluids—Chemistry and exploration techniques. Springer Verlag, Berlin, Germany, 1993.

Nordstrom, D.K. & Ball, J.W.: Chemical models, computer programs, and metal complexation in natural waters. In: C.J.M. Kramer & J.C. Brinker (eds): *Complexation of trace metals in natural waters.* Martinus Nijhoff/Dr. W. Junk Publishers, The Haque, The Netherlands, 1984.

Olivella, S., Gens, A., Carrera, J. & Alonso, E.E.: Numerical formulation for a single simulator (code_bright) for the coupled analysis of saline media. *Eng. Comput.* 13 (1986), pp. 87–112.

O'Sullivan, M.J., Bullivant, D.P., Follows, S.E. & Mannington, W.I.: Modeling of the Wairakei-Tauhara geothermal system. *Proc. of the TOUGH Workshop 1998*, 4–6 May 1998, Berkeley, California, 1998, pp. 1–6.

O'Sullivan, M.J., Pruess, K. & Lippmann, M.J.: The state of the art of geothermal reservoir simulation. *Geothermics* 30 (2001), pp. 395–429.

Özgur, N.: Hydrogeological, hydrogeochemical and isotope geochemical modeling of the thermal waters in the contintental rift zones of the Menderes Massif, Western Anatolia, Turkey. *Proc. World Geothermal Congress*, 25–29 April 2010, Bali, Indonesia, 2010.

Palandri, J.L. & Kharaka, Y.K.: A compilation of rate parameters of water-mineral interaction kinetics for application to geochemical modeling. U.S. Geological Survey, Open file report 2004–1068, 2004.

Parini, M., Acuna, J.A. & Laudiano, M.: Reinjected water return at Miravalles geothermal reservoir, Costa Rica: numerical modeling and observations. *Proc. 21st Workshop on Geothermal Reservoir Engineering*, 22–24 Jan 1996, Stanford University, Stanford, CA, 1996, pp. 127–134.

Park, J., Norman, D., McLin, K. & Moore, J.: Modeling amorphous silica precipitation: a strategy to reduce silica precipitation near Coso injection wells. *31th Workshop on Geothermal Reservoir Engineering*, 30 Jan–1 Feb 2006, Stanford University, Stanford, CA, 2006.

Parkhurst, D.L.: User's guide to PHREEQC—a computer program for speciation, reaction-path, advective transport, and inverse geochemical calculations: U.S. Geological Survey Water-Resources Investigations Report 95-4227, 1995.

Parkhurst, D.L. & Appelo, C.A.J.: User's guide to PHREEQC (version 2)—a computer program for speciation, batch-reaction, one-dimensional transport, and inverse geochemical calculations: U.S. Geological Survey Water-Resources Investigations Report 99-4259, 1999.

Paschke, SS. & van der Heijde, P.K.M.: *Overview of chemical modeling in groundwater and listing of available geochemical models.* IGWMC Publications, International Ground Water Modeling Center, Colorado School of Mines, Golden, CO, 1996.

Perkins, E.H.: Integration of intensive variable diagrams and fluid phase equilibrium with SOLMNEQ.88 pc/shell. In: Y.K. Kharaka and A.S. Maest (eds): *Water-Rock Interaction*, Balkema, Rotterdam, The Netherlands, 1992, pp. 1079–1081.

Pham, M., Menzies, A.J., Sanyal, S.K., Lima, E., Shimada, K., Juarez, J. & Cuevas, A.: Numerical modeling of the high-temperature geothermal system of Amatitlan, Guatemala. *GRC Transactions* 20 (1996), pp. 833–838.

Pham, M., Klein C. & Sanyal, S.: Reducing costs and environmental impact of geothermal power through modeling of chemical processes in the reservoir. *Proc. 26th Workshop on Geothermal Reservoir Engineering*, 20–31 Jan 2001, Stanford University, Stanford, CA, 2001.

Portugal, E., Birkle, P., Barragán, R.M., Arellano, V.M., Tello, E. & Tello, M.: Hydrochemical-isotopical and hydrogeological conceptual model of the Las Tres Vírgenes geothermal field, Baja California, Mexico. *J. Volcanol. Geotherm. Res.* 101:3–4 (2000), pp. 223–244.

Plummer, M.A., Palmer, C.D., Mattson, E.D., Redden, G.D. & Hull, L.C.: Advancing reactive tracer methods for monitoring thermal drawdown in geothermal enhanced geothermal reservoirs. *Geothermal Resource Council Meeting*, Idaho National Laboratory, INL/CON-10-18881, 2010.

Plummer, L.N., Prestemon, E.C. & Parkhurst, D.L.: An interactive code (NETPATH) for modeling net geochemical reactions along a flow path. Version 2.0. U.S. Geological Survey Water Resources Investigation Report 94-4169, Reston, VA, 1994.

Portier, S., Kühn, M. & Vuataz, F.-D.: Comparing FRACHEM and SHEMAT for the modelling of brine-rock interactions in Enhanced Geothermal Systems. *Proc. European Geothermal Congress*, 30 May–1 June 2007, Unterhaching, Germany, 2007a.

Portier, S., Laurent, A., Vuataz, F.-D. & Kohl, T.: Modelling the impact of forced fluid-rock interactions on reservoir properties at Soultz-sous-Forêts EGS geothermal site. *Proc. European Geothermal Congress*, 30 May–1 June 2007, Unterhaching, Germany, 2007b.

Portier, S., Vuataz, F.D., Barrios Martínez, A.L. & Siddiqi, G.: Preliminary modelling of the permeability reduction in the injection zone at Berlín geothermal field, El Salvador. *Proc. World Geothermal Congress*, 25–29 April 2010, Bali, Indonesia, 2010.

Pritchett, J.W.: STAR: A geothermal reservoir simulation system. *Proc. World Geothermal Congress 1995*, 18–31 May 2005, Florence, Italy, 2005, pp. 2959–2963.

Pritchett, J.W., Rice, M.H. & Riney, T.D.: Equation-of-state for water-carbon dioxide mixtures: Implications for Baca reservoir. Report DOE/ET/27163-8, Systems, Science and Software, La Jolla, CA, 1981.

Pruess, K.: Modeling of geothermal reservoirs: Fundamental processes, computer simulation and field applications. *Geothermics* 19 (1990), pp. 3–15.

Pruess, K. (ed): Proceedings of the TOUGH Workshop'98, Berkeley, California, Lawrence Berkeley National Laboratory report LBNL-41995, 1998.

Pruess, K.: Numerical simulation of multiphase tracer transport in fractured geothermal reservoirs. *Geothermics* 31:4 (2002), pp. 475–499.

Pruess, K.: Enhanced geothermal systems (EGS) using CO_2 as working fluid—a novel approach for generating renewable energy with simultaneous sequestration of carbon. *Geothermics* 35 (2006), pp. 351–367.

Pruess, K.: On production behavior of enhanced geothermal systems with CO_2 as working fluid. *Energy Convers. Manage.* 49 (2008), pp. 1446–1454.

Pruess, K. & Spycher, N.: Enhanced geothermal systems (EGS) with CO_2 as heat transmission fluid—A scheme for combining recovery of renewable energy with geologic storage of CO_2. Lawrence Berkeley National Laboratory, LBNL Paper LBNL-1949E, http://escholarship.org/uc/item/3v70v3ns, 2010.

Pruess, K., Weres, O. & Schroeder, R.: Distributed parameter modeling of a producing vapor-dominated geothermal reservoir: Serrazzano, Italy. *Water Resour. Res.* 19:5 (1983), pp. 1219–1230.

Pruess, K., Oldenburg, C. & Moridis, G.: An overview of TOUGH2, Version 2.0. *Proc. of the TOUGH Workshop '98*, 4–6 May 1998, Berkeley, California, 1998, pp. 307–314.

Raffensberger, J.B. & Garven, G.: The formation of unconformity-type uranium ore deposits 1. Coupled groundwater flow and heat transport modeling. *Am. J. Sci.* 295 (1995a), pp. 581-636.

Raffensberger, J.B. & Garven, G.: The formation of unconformity-type uranium ore deposits 2. Coupled hydrochemical modeling. *Am. J. Sci.* 295 (1995b), pp. 639-696.

Reed, M.H.: Calculation of multicomponent chemical equilibria and reaction processes in systems involving minerals, gases, and an aqueous phase. *Geochim. Cosmochim. Acta* 46 (1982), pp. 513–528.

Reed, M.H.: Calculation of simultaneous chemical equilibria in aqueous-mineral-gas systems and its application to modeling hydrothermal processes. In J.P. Richards and P.B. Larson (eds): Techniques in hydrothermal ore deposits geology. *Reviews in Economic Geology* 10, 1998, pp. 109–124.

Reed, M. & Palandri, J.: SOLTHERM.H06, a database of equilibrium constants for minerals and aqueous species. Available from the authors, University of Oregon, Eugene, OR, 2006.

Reed, M.H. & Spycher, N.F.: Calculation of pH and mineral equilibria in hydrothermal waters with application to geothermometry and studies of boiling and dilution. *Geochim. Cosmochim. Acta* 48 (1984), pp. 1479–1492.

Reed, M.H. & Spycher, N.F.: SOLTHERM: Data base of equilibrium constants for aqueous-mineral-gas equilibria. Available from the authors, University of Oregon, Eugene, OR, 1993.

Reed, M.H & Spycher, N.F.: CHILLER: A program for computing water-rock reactions, boiling, mixing and other reaction processes in aqueous-mineral-gas systems. 3rd edition, Dept. Geol. Sci., University of Oregon, Eugene, OR, 2006a.

Reed, M.H. & Spycher, N.F.: SOLVEQ: A computer program for computing aqueous-mineral–gas equilibria. 3rd edition. Department of Geological Sciences, University of Oregon, Eugene, OR, 2006b.

Rodríguez, A.: Amorphous iron silicate scales in surface pipelines: Characterization and geochemical constraints on formation conditions in the Miravalles geothermal field, Costa Rica. The United Nations University, Geothermal Training Programme, Reports 2006, No. 19, 2006, pp. 429–452.

Rose, P.E., Apperson, K.D., Johnson, S.D. & Adams, M.C.: Numerical simulation of a tracer test at Dixie Valley, Nevada. Proc. 22nd Workshop on Geothermal Reservoir Engineering, 27–29 Jan 1997, Stanford University, Stanford, CA, 1997, pp. 169–176.

Rose, P.E., Goranson, C., Salls, D. & Kilbourn, P.: Tracer testing at Steamboat Hills, Nevada, using fluorescein and 1,5-naphthalene disulfonate. *Proc. 24th Workshop on Geothermal Reservoir Engineering*, 25–27 Jan. 1999, Stanford University, Stanford, CA, 1999, pp. 17–23.

Sanyal, S.K.: Forty years of production history at The Geysers geothermal field, California—The lessons learned. *Geothermal Resources Council Transactions* 24, 2000, pp. 317–323.

Sanyal, S.K., Pham, M., Iwata, S., Suzuki, M., Inoue, T., Yamada, K. & Futagoishi, M.: Numerical simulation of the Wasabizawa geothermal field, Akita Prefecture, Japan. *Proc. World Geothermal Congress*, 28 May–June 10 2000, Kyushu-Tohuku, Japan, 2000, pp. 2189–2194.

Sarak, H., Onur, M. & Satman, A.: Lumped-parameter models for low-temperature geothermal reservoirs and their application. *Geothermics* 34 (2005), pp. 728–755.

Stanford Geothermal Program (ed): Proceedings of the Special Panel on Geothermal Model Intercomparison Study. Report SGP-TR42. *6th Workshop on Geothermal Reservoir Engineering*, 16–18 Dec. 1980, Stanford University, Stanford, CA, 1980.

Steefel, C.I. & Lasaga, A.C.: A coupled model for transport of multiple chemical species and kinetic precipitation/dissolution reactions with application to reactive flow in single phase hydrothermal systems. *Amer. J. Sci.* 294 (1994), pp. 529–592.

Steefel, C.I., DePaolo, D.J. & Lichtner, P.C.: Reactive transport modeling: An essential tool and a new research approach for the Earth Sciences. *Earth Plan. Sci. Lett.* 240:3–4 (2005), pp. 539–558.

Stumm, W. & Morgan, J.J.: *Aquatic chemistry—Chemical equilibria and rates in natural waters*. John Wiley & Sons, NY, 1996.

Tokita, H., Yahara, T. & Kitakoga, I.: Cooling effect and fluid behavior due to reinjected hot water in the Hatchobaru geothermal field, Japan. *Proc. World Geothermal Congress* 1995, 18–31 May 1995, Florence, Italy, 1995, pp. 1869–1874.

Truesdell, A.H. & Jones, B.F.: WATEQ, a computer program for calculating chemical equilibria of natural waters. *Jour. Res. U.S. Geological Survey* 2 (1974), pp. 233–248.

Truesdell, A.H., Walters, M., Kennedy, M. & Lippmann, M.: An integrated model for the origin of The Geysers geothermal field. *GRC Transactions* 17 (1993), pp. 273–280.

Tulinius, H., Spencer, A.L., Bodvarsson, G.S., Kristmannsdottir, H., Thorsteinsson, T. & Sveinbjornsdottir, A.E.: Reservoir studies of the Seltjarnarnes geothermal field, Iceland. *Proc. 12th Workshop on Reservoir Engineering*, 22–27 Jan. 1987, Stanford University, Stanford, CA, 1987, pp. 67–75.

U.S.EPA (United States Environmental Protection Agency): MINTEQA2/PRODEFA2. A geochemical assessment model for environmental systems: User manual supplement of Version 4.0. HydroGeoLogic, Inc. & Allison Geoscience Consultants, Inc., 1999.

Vinsome, P.K.W. & Shook, G.M.: Multi-purpose simulation. *J. Petrol. Sci. Eng.* 9:1 (1993), pp. 29–38.

Wagner, R., Kühn, M., Meyn, V., Pape, H., Vath, U. & Clauser, C.: Numerical simulation of pore space clogging in geothermal reservoirs by precipitation of anhydrite. *Int. J. Rock Mech. Min. Sci.* 42:7–8 (2005), pp. 1070–1081.

White, S.P.: Multiphase non-isothermal transport of systems of reacting chemicals. *Water Resour. Res.* 31 (1995), pp. 1761–1772.

White, S.P. & Mroczek, E.K.: Permeability changes during the evolution of a geothermal field due to dissolution and precipitation of quartz. *Transp. Porous Media* 33 (1998), pp. 81–101.

White, S.P., Okabe, T., Sato, T., Sato, M., Shiga, T. & Takahashi, Y.: Modelling of the deep geothermal system of the Uenotai reservoir. *Proc. World Geothermal Congress*, 24–29 April 2005, Antalya, Turkey, 2005.

Wolery, T.J.: EQ376, A software package for geochemical modeling of aqueous systems: Package overview and installation guide (Version 7.0). Report UCRL-MA-110662 PT I, Lawrence Livermore National Laboratory, CA, 1992.

Wu, Y.S. & Pruess, K.: Numerical simulation of non-isothermal multiphase tracer transport in heterogeneous fractured porous media. *Adv. Water Resour.* 23:7 (2000), pp. 699–723.

Xu, T.: Incorporation of aqueous and sorption kinetics and biodegradation into TOUGHREACT. *Proc. TOUGH Symposium 2006*, 15–17 May 2006, Lawrence Berkeley National Laboratory, Berkeley, CA, 2006.

Xu, T. & Pruess, K.: Hydrothermal fluid flow and mineral alteration in a fractured rock under multiphase H_2O-CO_2 mixture conditions. *Proc. World Geothermal Congress*, 28 May–June 10 2000, Kyushu-Tohuku, Japan, 2000, pp. 2983–2988.

Xu, T. & Pruess, K.: Modeling multiphase non-isothermal fluid flow and reactive geochemical transport in variably saturated fractured rocks: 1. Methodoloy. *Amer. J. Sci.* 301 (2001), pp. 16–33.

Xu, T., Sonnenthal, E., Spycher, N. & Pruess, K.: TOUGHREACT user's guide: A simulation program for non-isothermal multiphase reactive geochemical transport in variably saturated geologic media. Lawrence Berkeley National Laboratory Publication, LBNL-55460, 2004a.

Xu, T., Ontoy, Y., Molling, P., Spycher, N., Parini, M. & Pruess, K.: Reactive transport modeling of injection well scaling and acidizing at Tiwi field, Philippines. *Geothermics* 33 (2004b), pp. 477–491.

Xu, T., Sonnenthal, E., Spycher, N. & Pruess, K.: TOUGHREACT: A simulation program for non-isothermal multiphase reactive geochemical transport in variably saturated geologic media. *Comp. Geosci. 32* (2006), pp. 145–165.

Xu, T., Pruess, K. & Apps, J.: Numerical studies of fluid-rock interactions in Enhanced Geothermal Systems (EGS) with CO_2 as working fluid. *Proc. 33rd Workshop on Geothermal Reservoir Engineering*, Stanford University, Stanford, CA, 2008.

Xu, T. & Pruess, K.: Reactive transport modeling to study fluid-rock interactions in enhanced geothermal systems (EGS) with CO_2 as working fluid. *Proc. World Geothermal Congress*, 25–29 April 2010, Bali, Indonesia, 2010.

Xu, T., Spycher, N., Sonnenthal, E., Zhang, G., Zheng, L. & Pruess, K.: TOUGHREACT Version 2.0: A simulator for subsurface reactive transport under non-isothermal multiphase flow conditions. *Comp. Geosci.* 37 (2011), pp. 763–774.

Yabusaki, S.B., Steefel, C.I. & Wood, B.D.: Multidimensional, multicomponent, subsurface reactive transport in nonuniform velocity fields: code verification using an advective reactive streamtube approach. *J. Contam. Hydrol.* 30:3–4 (1998), pp. 299–331.

Yano, Y. & Ishido, T.: A reservoir engineering study on production behaviour of deep-seated geothermal reservoirs. *Geothermal Resources Council Transactions* 22, 1998, pp. 503–506.

Zhanshi, Z., Mangen, L. & Wenbin, Z.: Experimental and modeling study of water-rock interaction in active geothermal fields: Los Azufres, Mexico. *Proc. World Geothermal Congress*, 24–29 April 2005, Antalya, Turkey, 2005.

Zhu, C. & Anderson, G.: *Environmental applications of geochemical modeling*. Cambridge University Press Publication, 2002.

Ziegler, M.: *The future of hot dry rock systems with CO_2 as a production fluid.* Bachelor's thesis. Freie Universität Berlin & University of Iceland, 2011.

Section 2: Cases studies

CHAPTER 8

Integrating field observations and inverse and forward modeling: application at a site with acidic, heavy-metal-contaminated groundwater

Pierre Glynn & James Brown

> *"'I had,' said he, 'come to an entirely erroneous conclusion which shows, my dear Watson, how dangerous it always is to reason from insufficient data."*
>
> Sir Arthur Conan Doyle (1891)

> *"Remember that all models are wrong; the practical question is how wrong do they have to be to not be useful."*
>
> George Edward Pelham Box (1987)

8.1 INTRODUCTION

Historical note: This chapter presents a revised version of Glynn and Brown (1996), a study that discussed in detail the theory and application of inverse and forward geochemical modeling with the computer codes PHREEQC, PHREEQM, and NETPATH, and the application of these codes in modeling contaminated groundwaters in the Pinal Creek basin (Arizona, USA). The modeling study focuses on the evolution and transport of acidic, reducing, waters because those conditions control the transport of metal contaminants in the groundwaters of the Pinal Creek basin. In addition to presenting the essential results and conclusions of Glynn and Brown (1996), this chapter provides a 15-year perspective on the original study. It discusses follow-up work done in the Pinal Creek basin, comments on lessons learned from investigations at the site, and provides suggestions that may be useful to geochemical and modeling investigations at other sites.

The construction of a multispecies reactive transport model used to predict the future evolution and movement of groundwater contaminants requires, at a minimum, three separate but related elements: (i) an understanding of the groundwater flow system and its possible transients, (ii) an understanding of dispersion and other processes causing observed dilution or "mixing" of different water types, and (iii) an understanding of the primary reactions controlling the distribution of various contaminants, not only among solid and phases, but also within the groundwater itself. The degree of understanding of all three of these elements, and perhaps more importantly an appreciation for the remaining knowledge gaps, will be essential in determining the usefulness of the constructed model. Indeed, even though a groundwater model may not adequately predict the future evolution of a contaminant plume, the process of constructing and using a model often results in an improved understanding of contaminant transport at the site.

Sources: *Sir Arthur Conan Doyle: The Adventure of the Speckled Band', The Adventures of Sherlock Holmes, 1891; **George Edward Pelham Box; In: George E.P. Box and Norman R. Draper: Empirical Model-Building and Response Surfaces, 1987.

Most groundwater contamination sites have less geochemical and hydrogeologic information known about them than is desirable for predictive modeling of reactive contaminant transport. Detailed hydrogeologic and geochemical studies are usually much too expensive to consider.[1] The resulting lack of knowledge on the operative chemical and hydrologic processes at a given site means that investigators should try to use, as efficiently as possible, all tools and knowledge available. We believe that a combination of inverse and forward modeling of groundwater flow, contaminant transport, and geochemical evolution, may often provide the greatest knowledge gains for the least amount of money and time. In particular, geochemical inverse modeling should be used first, prior to forward geochemical modeling, both to explain the currently observed groundwater chemistry in the aquifer system, and to make predictions on the future chemical evolution of the groundwaters.

This chapter focuses on geochemical modeling and will show how both inverse and forward geochemical modeling approaches were used to better understand the evolution of acidic, heavy-metal-contaminated groundwaters in the Pinal Creek basin, near Globe, Arizona. The Pinal Creek basin is a site with sparse spatial information (30 wells distributed in a 15 km long and 10^2 to 10^3 m wide sulfate plume) and with significant temporal variations in both chemical and hydrological characteristics (water-table movements of more than 15 m during a three month period, groundwater velocities on the order of 3 to 5 m day^{-1}; Brown and Harvey, 1994). The Pinal Creek site is well suited to test our modeling philosophy.

8.2 GEOCHEMICAL MODELING: COMPUTER CODES, THEORY AND ASSUMPTIONS

8.2.1 *Inverse geochemical modeling*

8.2.1.1 *Principles, codes and theory*

Inverse geochemical modeling uses existing groundwater chemical and isotopic analyses, which are assumed to be representative of the groundwater along a given groundwater flowpath, and attempts to identify and quantify the reactions that are responsible for the chemical and isotopic evolution of the groundwater. Although an aqueous speciation code may be used to identify thermodynamically possible (or impossible) reactions and to determine the dissolved inorganic carbon content and the redox state (RS) of the groundwaters, the inverse modeling approach does not require that reactions proceed to thermodynamic equilibrium. Indeed, mass-balance constraints and the judgment of the user concerning the possible reactions are the only constraints posed in the inverse modeling approach.

Inverse geochemical modeling codes (BALANCE, Parkhurst *et al.*, 1982; NETPATH, Plummer *et al.*, 1991, 1994; PHREEQC,[2] Parkhurst, 1995, 1997, Parkhurst and Appelo, 1999; PHREEQCi,[3] Charlton *et al.*, 1997 and Charlton and Parkhurst, 1999) essentially solve a system of algebraic mass-balance equations. These relate the masses of elements, isotopes, redox oxidation states, and water found in a "final" sampled water along a flowpath to the masses found in contributing "initial" waters along the same flowpath, and to the

[1]Studies at the Cape Cod (LeBlanc, 1984) and Borden sites (Mackay *et al.*, 1986) are examples of what we would consider detailed studies. On the order of 10^3 to 10^4 sampling points were installed to study plumes on the order of 10^2 meters to a few kilometers long. However, even at these sites, after three decades, many questions remain regarding the operative geochemical and hydrogeologic processes, and studies continue to refine and improve existing knowledge.

[2]The PHREEQC series of numerical codes (PHREEQCi, PHREEQC versions 1 and 2) have both inverse and forward geochemical modeling capabilities.

[3]PHREEQCi is an interactive version of PHREEQC with a windows-based graphical user interface.

masses lost or gained through heterogeneous reactions between the aqueous phase and other phases (solid, gaseous, exchange or surface sorption phases). Each "inverse model" calculated by an inverse geochemical modeling code consists of a set of reaction mass transfers and "initial" solution fractions that satisfies the algebraic mass balance constraints for the observed "initial" and "final" solution compositions. The list of possible reactions and contributing "initial" solutions is postulated by the user. Usually, more than one model, or set of mass transfers, can be calculated for a given problem, and the user then tries to add additional constraints, for example additional information regarding the chemical and/or isotopic compositions of the initial and final solutions, to limit the number of models.

Mass-action equations and thermodynamic equilibrium speciation of the "initial" and "final" aqueous solutions are also solved in the NETPATH and PHREEQC codes (but not in the earlier BALANCE code). These speciation calculations primarily serve to inform the user as to the reactions that might be thermodynamically feasible. The calculations are also used to establish the redox and charge balance states and to determine the total dissolved inorganic carbon balance for the various aqueous solutions. Significant differences exist between the NETPATH and the PHREEQC inverse modeling capabilities, for example in the treatment of redox balances and solution charge balances. Perhaps, the most important difference between NETPATH and PHREEQC is that PHREEQC allows each analytical input datum for each of the aqueous solutions to be adjusted within an uncertainty range specified by the user. PHREEQC then calculates sets of phase mass transfers, solution mixing fractions, and adjustments to the analytical data that satisfy the mass-balance constraints, are consistent with the specified uncertainty ranges, and minimize the sum of the adjustments to the analytical data. The mathematical descriptions of the NETPATH and PHREEQC (and PHREEQCi) codes will not be discussed here. The reader is referred to Glynn and Brown (1996), Parkhurst (1995, 1997), Parkhurst and Appelo (1999), and Plummer *et al.* (1994) for further information. More recently, in their guide on radiocarbon dating of groundwater systems, Plummer and Glynn (in press) review the unique capabilities of the NETPATH code to simulate Rayleigh fractionation processes and assess groundwater residence times through inverse geochemical modeling.

8.2.1.2 *Assumptions used in inverse modeling*

By definition, a model is a construct of assumptions that is meant to help understand some facet(s) of reality. Inverse geochemical modeling of groundwaters requires the user to make assumptions concerning (i) the types of geochemical reactions postulated to be present, (ii) the rates of reaction relative to the movement of the water and its mobile constituents, and (iii) the present distribution of chemical constituents in the aquifer system studied and the prior evolution of this distribution. The last 2 sets of assumptions require that the user have some presumptive knowledge of the groundwater flow and transport system and of its prior evolution.

Knowledge of flowpaths and the assumption of a steady-state groundwater flow field

These are the most important and possibly the most tenuous assumptions used in inverse modeling of the chemical and isotopic evolution in a groundwater system. The user often does not have enough hydrologic knowledge to precisely determine the flowpaths in a groundwater system. Furthermore, even if there is sufficient knowledge of the hydrogeologic system, existing wells must often be used. One rarely has the luxury of emplacing new sampling wells. However, when analyses are available from several wells, the spatial array of chemical and isotopic information may itself be used to decide the most likely flowpath. In most cases, the user will pick a direction that shows the least amount of dilution for the most conservative solutes to deduce the direction of flow.

Most groundwater systems are likely to experience some seasonal and interannual fluctuations in hydraulic heads. Therefore, flowlines and groundwater velocities are likely to change at least seasonally, and steady-state conditions may not apply during the time scale of interest. The time scale of interest will normally be the time required for groundwater flow between the wells used in the inverse modeling simulation. The user typically assumes a steady-state groundwater flow field over the time scale of interest, or at the least assumes that the groundwater flow field observed at the time of sampling represents the long-term average velocity field.

The assumption of chemical steady-state

The groundwater analyses used in an inverse model usually represent samples taken concurrently or near-concurrently. The inverse modeling approach generally assumes that the parcel of water sampled from a "final", downgradient, well (well B in Fig. 8.1) used to have the same composition as that of the water sampled concurrently at an "initial", upgradient, well (well A in Fig. 8.1). This assumption will certainly be reasonable if the groundwater system has remained in chemical (and isotopic) steady-state at least during the travel time required for the water to move from the initial well to the final well. The assumption of chemical and isotopic steady-state simply states that although chemical and isotopic compositions may vary spatially, they do not vary in time at any given point in the groundwater system. In groundwater systems with spatially varying chemical and isotopic compositions, the assumption of chemical steady state implies a steady-state groundwater flow field (i.e., hydrologic steady-state), that is, flow lines and groundwater velocities that have not varied in time.

Most groundwater contamination cases involve dynamically evolving contaminant plumes, for which there can be no assumption of chemical steady-state. (Steady-state plumes, in which the rate of diffusive/dispersive and reactive loss of solutes balances the rate of their influx and/or production are uncommon). Fortunately, the assumption of chemical and

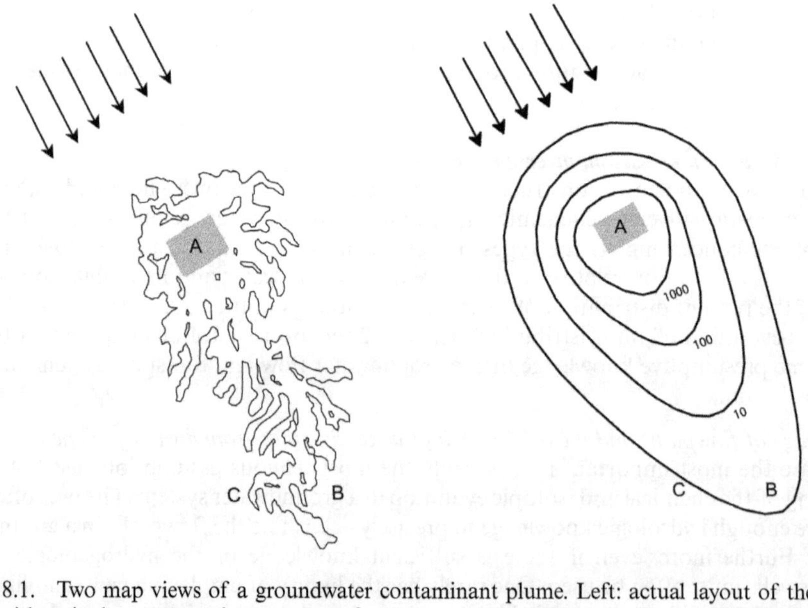

Figure 8.1. Two map views of a groundwater contaminant plume. Left: actual layout of the plume, drawn with a single concentration contour of concern. A is a well emplaced near or in the source of the contamination (stippled) and B and C are wells further downgradient. Right: results of a transport model for the same groundwater contaminant plume based on a fit of concentration data obtained from several observation wells. Additional concentration contours are drawn. The large transverse and longitudinal dispersion of the modeled plume results not only from the mixing that actually occurs in the ground but also from mixing that occurs during pumping at the observation wells. Uncertainty in simulation results is also caused by the inability to obtain a sufficiently detailed time-dependent representation of the contaminant plume and of the transient groundwater velocity field.

isotopic steady-state is not required for inverse geochemical modeling to be meaningful. The less stringent constraint required for inverse modeling is that the "initial", upgradient, water composition used in inverse modeling (well A, sampled ideally at time t_0, or more commonly at time t_s) should be representative of the composition that the "final" water sampled down-gradient at well B (at time t_s) used to have, prior to its chemical evolution and movement from A to B. This less stringent constraint allows the groundwater compositions at points *in between* the "initial" and "final" wells (A and B) to have varied with time, as long as the chemical composition of waters from the "initial" well (A) has remained invariant, or the appropriate sample from well A at time t_0 can be obtained. Strictly from a mass-balance point of view, it could be argued that the constraint could be reduced further to require only that the *changes* in composition between waters from well A (at time t_0) and B (at time t_s), rather than the *actual* compositions of waters from wells A (at time t_0) and B (at time t_s) should have remained constant. (A uniform dilution, or concentration, of the waters sampled at the "initial" and "final" wells, however, could lead the user to conclude from his inspection of the mineral saturation indices and general speciation of the waters that some other set of reactions was responsible for the evolution of water A into water B.) Finally, although the locations of the "initial" and "final" waters used in inverse geochemical modeling do not enter in the calculation of the reaction sets and reaction extents responsible for the overall evolution of water A into B, location information is important in assessing whether the reactions are reasonable from a kinetic, mineralogic and hydrologic perspective.

How does "mixing" occur in groundwater systems?

The U.S. Geological Survey inverse geochemical modeling codes (BALANCE, NETPATH, and the general geochemical code PHREEQC) calculate "mixing" fractions of initial waters and extent of reactions with solid and gas phases that account for the chemical composition of the final water. Clearly in most groundwater environments, "mixing" of groundwaters should be modeled as a continuous process rather than as a discrete process where a small number of specified water compositions are mixed together. Unfortunately, the inverse geochemical codes presented here cannot replicate a "continuous" mixing process. Forward transport modeling codes can replicate a continuous mixing process such as dispersion, but even then, their results and the very basis of their conceptual models are usually fraught with uncertainty. In using a set of discrete "initial" water compositions, inverse geochemical models inherently assume that the "initial" waters chosen encompass the range of intermediate waters that are actually involved in the real, continuous mixing process.

Although the location, timing, and sequence of the mixing and reaction processes is of no mathematical significance in the solution of the mass-balance equations, the user should try to determine where, when, and why such mixing processes have occurred in the groundwater environment. The "mixing" of initial waters by dispersion, for example, may well have led to heterogeneous mass transfers in areas that are not directly on the flowpath between a principal "initial" water and the "final" water. The inverse models will, nevertheless, implicitly incorporate those mass transfers in their solution of *net* mass transfer amounts.

The premise of inverse geochemical modeling is that the "final" and "initial" groundwaters used in a model should be related to each other. Ideally, they represent very small volumes of water sampled from a unique flowline or pathline. If it were indeed possible to do so (it is not), then the "final" groundwater sampled could only have experienced "mixing" as the result of the following processes:

1. Diffusion of chemical and isotopic constituents (and possibly of water) to or from the flowline (or pathline) to neighboring flow lines or to stagnant water zones.
2. Sampling from the pathline for more than an infinitesimally small amount of time, in the case of a system not in chemical steady-state. (Although most groundwater analyses do not require large samples, and the samples are therefore typically collected over small time periods, the length of sampling time may be an important consideration for some analytes.) If the system is not in chemical steady-state, the concentration of various constituents

may be changing as a function of time at the final well and therefore a sample may in fact represent some groundwater composition averaged out in time and therefore in space.
3. The sampling of multiple flow lines that have undergone different chemical and isotopic evolution. Sampling-induced mixing of a diversity of solution chemistries may occur particularly in regions of converging flow, and may also occur when sampling from wells screened across large or multiple intervals. Flow convergence may occur naturally or may be the direct result of pumping.

Using the "mixing" option in an inverse geochemical model may also be needed because the "initial" and "final" groundwaters may not be chemically related despite the best efforts of the user. Just as excessively high values of dispersivity are often used in groundwater transport modeling because of a lack of precise spatial and temporal information (see Fig. 8.1), the use of the "mixing" option in NETPATH or in PHREEQC can often be the result of insufficient information on a groundwater system. For example, in Figure 8.1, if the "final" well used in the inverse model was off to the side of the path line of heaviest contamination (well C for example) and if well A was used as the "initial" well, the inverse model defined by the user would probably require a significant contribution of "background" water to explain the extent of "dilution" between well A and the "final" well. Similarly pumping a large amount of water from the "final" well chosen (well B or C) and using the average composition of this water as the "final" water composition in the inverse model could also lead to a serious misrepresentation of the amount of mixing. An error in the mixing fractions of "initial" waters could result in significant errors in the reaction mass transfers calculated by the inverse model. Furthermore, using water compositions averaged over a large volume by the sampling process could mislead the modeler into thinking that certain reactions were thermodynamically impossible, when in fact proper sampling, and location, of the "initial" and "final" waters would have indicated that these reactions were in fact possible.

8.2.2 *Forward geochemical modeling*

8.2.2.1 *Principles and codes*

Unlike inverse geochemical modeling, which attempts to explain observations and quantify the reactive mass transfers and mixing processes responsible for the observed chemical evolution along a flowpath in a groundwater system, forward modeling and reactive-transport modeling attempt to predict the evolution of a groundwater given certain postulated reactions and hydrodynamic processes. In their study, Glynn and Brown (1996) used the PHREEQM (Appelo and Willemsen, 1987) code for their reactive-transport simulations, rather than PHREEQC, because version 1 of PHREEQC (Parkhurst, 1995), the version that was available at the time, only simulated advective transport and reactions. PHREEQC version 2 (Parkhurst and Appelo, 1999) implemented advective-dispersive transport, by using an algorithm similar to the one used in PHREEQM. Because the concepts of forward geochemical modeling and reactive-transport modeling are widely known, this discussion will be limited to a brief description of the reactive transport capabilities of the PHREEQM and PHREEQC geochemical codes.

The PHREEQM code, and later the PHREEQC code, added significant capabilities to earlier geochemical speciation and mass-transfer codes, such as the U.S Geological Survey code PHREEQE (Parkhurst *et al.*, 1980). PHREEQM and PHREEQC simulated the transport of aqueous solutions by advection, dispersion, and diffusion in a 1-dimensional column (made up of a series of "cells") and simulated the reaction of those solutions with minerals and surfaces inside the column. The codes also allowed simulation of cation-exchange processes. PHREEQM typically used the local equilibrium assumption in its modeling of reactive transport. PHREEQC, starting with its second version (Parkhurst and Appelo, 1999), implemented a wide range of capabilities that were not available in PHREEQM. These included modeling of kinetically limited reactions, modeling of additional surface speciation and sorption reactions, modeling of solid-solution aqueous-solution reactions as presented in Glynn (1991a, 2000), transverse diffusion, and many other capabilities.

The transport algorithms in PHREEQM and PHREEQC use an operator splitting technique, where advection is modeled by shifting cell contents from one cell to the next at every time step or "shift". Dispersion and/or diffusion is simulated by mixing the aqueous contents of each cell with that of adjacent cells. This algorithm gives the codes the advantage (over most typical finite-difference and finite-element codes) of being able to simulate not only an advective-dispersive transport process or a diffusive transport process, but also a purely advective, albeit one-dimensional, transport process. The mixing factors f calculated are a function of the aquifer dispersivity α and the molecular diffusivity D^*:

$$f_i = \frac{\alpha_i + \alpha_{i+1}}{l_i + l_{i+1}} + 4D^* \frac{\Delta t}{(l_i + l_{i+1})^2} \tag{8.1}$$

where i is the cell number, l is the cell length, and Δt is the time step.

The above equation can be derived from a finite difference approximation (ignoring advection) centered in space and forward in time. Because the simulation of dispersion is centered in space, PHREEQM and PHREEQC show no numerical dispersion error for conservative constituents when simulating advection-dispersion processes. Numerical dispersion does occur, however, in dispersion of non-conservative constituents and is dependent on the amount of retardation experienced by each constituent and the cell lengths (the maximum numerical dispersivity equals 1/2 the cell length; *cf.* Herzer and Kinzelbach, 1989). The codes do not show any numerical dispersion in simulations with only diffusion as a transport process. The lack of sequential iterations between the solution of the chemical equilibrium equations and the simulation of the transport processes at every time step can theoretically generate errors, although comparisons (Glynn *et al.*, 1991; or Figs. 9 and 10 in Glynn and Brown, 1996) of PHREEQM with the sequential iteration finite difference code MSTID (Engesgaard and Kipp, 1992) suggest that the error is typically small as long as an appropriate discretization is used. Operator splitting can also generate error, although, again, comparisons of PHREEQM with the MSTID code and results by Steefel and MacQuarrie (1996) suggest that this error is usually minor. A much more complete description of the PHREEQM and PHREEQC codes and their capabilities can be found in Appelo and Postma (1993 and 2005, respectively). Unless mentioned otherwise, all PHREEQC simulations referred to in this chapter were performed with the first version of the code (Parkhurst, 1995).

Forward geochemical modeling is conceptually different from inverse geochemical modeling. Indeed, forward modeling seeks to *predict* the chemical composition of an aqueous solution, given the composition of an initial solution and given certain specified reactions, some of which are usually considered to reach thermodynamic equilibrium (or have a prescribed kinetic progression). Forward modeling is most suitable and most useful, when the amount of chemical and isotopic data available for a given groundwater system is limited, and when the objective is to predict the future evolution of the system. In contrast, inverse modeling is most useful when abundant chemical, isotopic, hydrologic and mineralogic data are present and all that is desired is an explanation of the past chemical evolution of the groundwater system. Of course, just as understanding the past is a key to understanding the future, inverse modeling can provide insight regarding the reactions that control the future chemical evolution of a groundwater system.

Forward modeling codes can also be used for the purpose of inverse modeling. A series of trial and error simulations or automated parameter estimation can be used to adjust reaction extents or reaction rates to align forward simulation results with actual observations (VanCappellen and Gaillard, 1996; Steefel and MacQuarrie, 1996). Although this latest approach can be time consuming, it does have the advantage over simpler inverse geochemical codes of a more accurate representation of groundwater mixing as a continuous, rather than discrete, process. This parameter-estimation approach does not necessarily require the assumption of chemical steady-state, although it is subject to the same constraints as simpler inverse geochemical codes in choosing initial and final water compositions (see discussion in previous section), their location and time of sampling. The disadvantages of this approach, relative to non-transport-oriented inverse geochemical modeling, are the computer time requirements, the significantly greater number of adjustable parameters, including flow and transport parameters, and the consequently greater number of solutions that may explain the observations. Further references in this chapter to inverse modeling will generally not refer to the use of forward codes as part of an inverse modeling approach, although many of the statements made may apply equally well to this latter more sophisticated approach.

The Pinal Creek Toxics Program investigation site, a site of groundwater contamination by acidic, metal-laden, and sulfate-rich wastewater near Globe, Arizona provides a good example of the improved understanding of the chemical reaction and transport processes that may be gained through the combined use of both inverse modeling and forward modeling approaches. The primary purpose of this chapter is to describe some of the insights that can be gained through this combined approach, and through the examination of available field observations. A brief description of the site is first needed.

8.3 THE PINAL CREEK BASIN SITE: BRIEF DESCRIPTION

The Pinal Creek basin is located in central Arizona, about 100 km east of Phoenix (Fig. 8.2). The surface drainage area of the basin occupies 516 km^2, of which 170 km^2 is covered by alluvium and basin fill, which form the regional aquifer; 27 km^2 of the basin's area is covered by mine

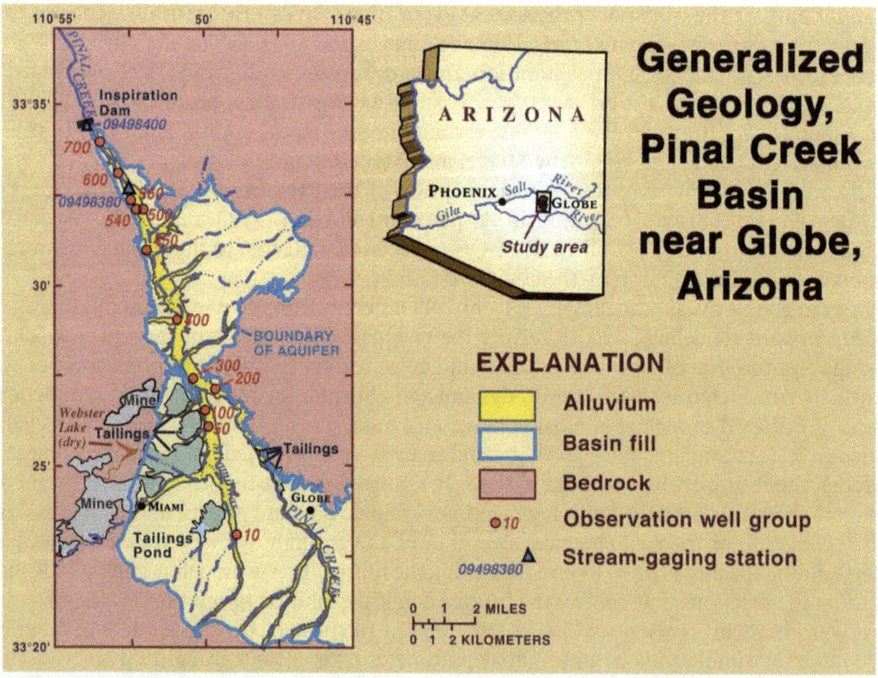

Figure 8.2. Location and generalized geology of the Pinal Creek basin study area. Several wells screened at different depths exist at each well group site. Well names are related to site names by their first, or first and second, digits. For example: site 400 includes well 402; site 500 includes wells 503 and 504; site 50 includes well 51; site 450 includes wells 451 and 452. Dashed dotted stream lines represent intermittent streams. All stream reaches are intermittent, except Pinal Creek itself to the north of well group 500.

tailings. Mining, mainly for copper, began in the late 1870s and has been the largest economic activity in the basin. Because of the long history of mining, there are several potential sources of contamination to the regional aquifer (Eychaner, 1991; Brown and Harvey, 1994). For example, pyrite is thought to be the most abundant sulfide mineral in the tailings. Following significant rainfall, oxidation of pyrite in the tailings and subsequent transport of this contaminated water into the alluvial aquifer could represent a significant source of acidity, iron, and sulfate in the aquifer. A similar phenomenon could also occur in the relatively undisturbed naturally mineralized areas in the basin, although the amount of this flow is expected to be small in relation to flow from the disturbed mining areas and tailings. Finally, the unlined water impoundments that were used during copper refining and mineral processing operations are also a likely source of groundwater contamination in the basin. The water in the impoundments is rich in sulfuric acid (used in copper refining) and heavy metals. The largest suspected single source of contamination in the basin was Webster Lake, an unlined surface-water impoundment that existed from 1940 to 1988. In 1988 it was drained at the order of the U.S. Environmental Protection Agency (Tolle and Arthur, 1991). Maximum volume of the lake was more than 7 million m^3. In 1988, a sample of water from the lake had a *pH* of 2.7; concentrations of iron and sulfate were 6 g L^{-1} and 20 g L^{-1}, respectively; and aluminum, copper, cobalt, nickel, and zinc were present at concentrations greater than 20 mg L^{-1}.

8.3.1 *Geology*

Peterson (1962) describes the complex and highly diverse geology and mineral deposits of the Pinal Creek basin area. Older Precambrian age rocks in the basin include schist, diorite, granite, conglomerate, quartzite, limestone, and basalt. Paleozoic age rocks include quartzite, limestone and shale. Younger Mesozoic- and Cenozoic-age rocks are mainly intrusive and include granite, granodiorite, diabase, and monzonite. These are exposed in the hills north of Globe and Miami. Finally and most importantly from an economic perspective, the igneous and metamorphic rocks include a major body of copper porphyry ore. Chalcocite, chalcopyrite, and pyrite predominate in the deeper parts of the ore body, while chrysocolla, malachite, and azurite predominate in its upper, oxidized zone.

The Pinal Creek basin's configuration was created by high-angle block faulting associated with basin subsidence that began 19 to 15 million years ago and continued until about 8 million years ago. The basin fill, which is derived from rocks of the surrounding mountains and forms the deeper aquifer in the basin, has a highly variable lithology, ranging "from completely unsorted and unconsolidated rubble of angular blocks as much as 4.5 m in diameter, to well-stratified deposits of firmly cemented sand, silt, and gravel containing well-rounded pebbles and cobbles" (Peterson, 1962). Carbonate content of the basin fill is about 1.5% (Eychaner, 1989).

Unconsolidated stream alluvium overlies the basin fill along Miami Wash, Pinal Creek, and other drainages. The alluvium is from 300 to 800 m wide and is less than 50 m thick. The alluvium contains cobble- to clay-sized material, although sand- to gravel-sized material is most abundant. Sand-sized particles contain quartz, feldspar, and lesser amounts of mica and a variety of rock fragments. Gravel-sized materials consist mainly of rock fragments of granite, volcanic rocks, and schist. The alluvium also contains interbedded clays and lenticular clay layers that were found to be as much as 12 m thick at Nugget Wash (Hydro Geo Chem, Inc., 1989). A sample of alluvium collected in 1985 (Eychaner and Stollenwerk, 1985) contained 0.34 percent calcite by weight. This concentration is equivalent to 0.18 moles of carbonate material per liter of water, using the bulk density of 1.65 g cm^{-3} and porosity of 0.316 determined for alluvium used in a column experiment (Stollenwerk, 1994).

Manganese cycling and transport is of interest in our study. Indeed, secondary manganese oxide deposits are highly visible in the perennial Pinal Creek streambed, frequently forming a black hardpan layer. Estimates of the concentration of primary manganese oxide minerals were based on samples of alluvium not affected by acidic contamination. At well site 500, in the neutralized part of the plume, the depth-averaged content of manganese oxides was 0.079 mol L^{-1}, based on sequential extractions done by Ficklin *et al.*(1991). Stollenwerk (1994) estimated that 0.0449 mol L^{-1} of manganese could be available for reaction in a sample of alluvium obtained from a gravel quarry just east of well site 200.

8.3.2 *Hydrology and groundwater flow*

Groundwater flow in the Pinal Creek basin is the result of past geologic events, the past and present climate, and human activities. Because the area climate is semiarid, most of the drainages in the basin are usually dry but convey large amounts of runoff during and after severe storms. Streams that drain the Pinal Mountains also flow during and following snowmelt in late winter and early spring. Most of the groundwater recharge in the basin occurs in the winter and spring, with lesser recharge events during the summer monsoonal period. The amount and the distribution of precipitation controls the size, frequency, and duration of streamflow, and therefore controls the quantity and distribution of water that infiltrates the permeable stream beds and recharges the unconsolidated alluvial and consolidated basin-fill aquifers.

Groundwater in the basin fill flows generally northward from the flanks of the Pinal Mountains, and westward from the Apache Peak alluvial fan. Most groundwater in the basin fill eventually flows upward into the alluvium and then flows generally north to the perennial reach of Pinal Creek. A greater quantity of water, however, recharges the alluvium directly and flows north, mixing with the water from the basin fill. In the northern part of the basin, the aquifer is constricted by the impermeable basement rocks. This constriction forces groundwater to the land surface, generating perennial flow from about 6 km above Inspiration Dam to the Salt River, which is a major source of drinking water for the Phoenix Metropolitan area.

8.4 INVERSE GEOCHEMICAL MODELING AT THE PINAL CREEK SITE

Like most sites with point-source groundwater contamination, the chemistry of groundwaters in the Pinal Creek basin exhibits both spatial and temporal variation. The most heavily contaminated groundwaters are typically found near the base of the unconsolidated alluvial aquifer, where a zone of coarser (and possibly less carbonate-rich) material is suspected to be present. The wells with the most contaminated waters at each well site are 51, 101, 302, 402, 503, 601 and 702 in a down-gradient direction (Fig. 8.2). Although other wells at each site also show the presence of contaminated water, wells with the most contaminated waters (as measured by total dissolved solids, chloride, or any other relatively conservative constituent) present the most logical choice for inverse modeling. To further narrow the scope of the inverse-modeling study, we focus on the significant changes in the chemical characteristics between samples from well 402 and 503. The two water samples chosen represent an acidic contaminated water sampled from well 402 in January 1989 and a neutralized contaminated water sampled from well 503 in November 1991. The two wells are 5.6 kilometers apart. From the difference in sampling times and the distance between wells, we calculate that a parcel of water leaving well 402 would have to travel at a linear groundwater velocity of about 5.2 m day^{-1}. This velocity is in the range of 4.2 to 5.6 m day^{-1} estimated by using Darcy's law, assuming an effective porosity of 0.3 and a hydraulic conductivity of 200 m day^{-1} (Brown *et al.*, 1995; Brown, 1996).

Glynn and Brown (1996) initially used the NETPATH inverse modeling code. Unlike the first version of PHREEQC, the only one available at the time, the NETPATH code was interactive and allowed the authors to quickly determine primary sources and sinks in their inverse simulations. The NETPATH code (i) helped identify some of the possible reaction mechanisms responsible for the chemical evolution of the groundwaters between wells 402 and 503, and (ii) quantified some of the reaction mass-transfers involved.

8.4.1 *Examination of end-member waters and their conservative constituents*

The first step in an inverse modeling study is to examine the chemical composition and thermodynamic state of the waters that will be used in the model. The chemical analyses for the waters chosen for the inverse modeling study are given below in Table 8.1. The charge imbalances calculated by the speciation code WATEQFP (embedded in NETPATH) for the well 402, 504 and 503 samples were −3.1, 1.5 and 4.4%, respectively.

As can be seen, the most significant differences in groundwater chemistry between the well 402 and 503 samples are the increase in *pH* from 3.9 to 5.6, the 25% increase in calcium and magnesium, the complete removal of 590 mg L^{-1} of dissolved iron, the 90% removal of 18.4 mg L^{-1} of dissolved aluminum, the 60% increase in manganese, and the 30% decrease in sulfate. Silica concentrations are nearly constant. As is the case in any geochemical modeling analysis, however, conservative (i.e., nonreactive) constituents are perhaps the most important constituents to examine because they give information on physical flow and transport processes. Any groundwater sampling and analysis program should ensure measurement of at least one, but preferably two or more, relatively non-reactive tracers, such as chloride, bromide, ^{18}O and ^{2}H. Sodium may also be relatively conservative although it may increase in solution due to cation exchange, feldspar dissolution, or evaporite dissolution processes. Sodium is rarely taken out of solution by reaction processes other than cation exchange. In the case of the Pinal Creek groundwaters, the high Ca/Na ratio in the acidic waters (Ca/Na = 2.4 mol/mol in well 402) and the even higher ratio in the neutralized waters (Ca/Na = 4.2 mol/mol in well 503) suggest that removal of sodium by cation exchange is not likely.

Table 8.1. Chemical composition of three groundwaters from the Pinal Creek basin: an acidic contaminated water (well 402), a background uncontaminated water (well 504), and a neutralized contaminated water (well 503). Concentrations in mg L^{-1}. Concentration changes are expressed relative to well 402. ND: not determined. *TDIC*: total dissolved inorganic carbon.

Concentration or value	Well 402 1989/1/12	Well 504 1991/11/22	Well 503 1991/11/22	Change: 402 to 503	Change: Reaction only[1]
pH	4.13	7.05	5.59		
Eh (in mV)	420 est.[2]	350	410		
Temperature	18°C	20.5°C	18.2°C	1.1%	
Dissolved oxygen	0.3	6.64	<0.1		
Calcium	502	44.6[3]	634	26%	57%
Magnesium	161	15.6[3]	200	24%	54%
Sodium	121	19.8[3]	86	−29%	−13%
Potassium	7 est.[2]	2.1	5 est.[2]	−29%	−16%
Iron	591	0.004	<0.1	−100%	−100%
Manganese	71.6	<0.001	116	62%	106%
Aluminum	18.4	<0.01	2.3 est.[2]	−88%	−84%
Strontium	2.29	0.335	2.7	18%	44%
Silica (as SiO_2)	85.6	27[3]	91.8	7.2%	26%
Chloride	140	9.7[3]	112	−20%	0%
TDIC (as C)	50	ND	ND	ND	ND
Alkalinity (as HCO_3)	ND	227	66	ND	ND
Sulfate	3260	14.2[3]	2350	−28%	−8%
Fluoride	10 est.[2]	0.3	1.5 est.[2]	−85%	−81%

[1] Assumes chloride is conservative. The relative change expressed represents the relative difference in concentration between a mixture of waters from wells 402 and 504, determined on the basis of chloride concentrations, and well 503 water. [2] Values were estimated by inspecting earlier and later analyses. [3] Average of two analyses.

The decrease in both Cl and Na between wells 402 and 503 suggests that dilution is occurring. This dilution may be caused either by longitudinal and transverse dispersion along the flowpath or by the well-sampling process. It is also important to recognize that the groundwater sample taken from well 503 in November 1991 was probably not exactly on the pathline originating from well 402 in January 1989. Well 503 may be "off" (to the side or above) the most contaminated pathline, and the well 503 sample might therefore be more diluted than a hypothetical sample taken from the pathline. In recognition of the difficulty in determining the causes and the exact proportions of the various groundwaters responsible for the dilution of the well 503 water relative to the well 402 water, an uncontaminated water sampled in November 1991 from below the plume at well 504 was used as the source of diluting water in our inverse geochemical model.

Although chloride undergoes a 20% decrease between wells 402 and 503, sodium undergoes an even greater decrease of about 29%. If the decrease in chloride is used to calculate the fraction (0.21) of water from well 504 diluting the water from well 402, the observed sodium concentration in well 503 is still 13% lower than the calculated diluted sodium concentration (Table 8.1, last column). This greater observed decrease in sodium may be at least partly due to a greater Cl/Na ratio in the average diluting water relative to that of the background water (well 504) used in the calculation. Indeed, although the average Cl/Na ratio in the uncontaminated waters below the plume (wells 404, 504) or upgradient (well 010) from the plume is 0.44 mg/mg (±0.10), the average Cl/Na ratio for the most contaminated waters along the flowpath is close to three times higher (well 51:1.48 ± 0.74, well 101: 1.35 ± 0.44, well 302: 1.32 ± 0.36, well 402: 1.29 ± 0.44, well 503: 1.17 ± 0.43; all ratios in mg/mg). [Note the decreasing Cl/Na ratio with distance downgradient, i.e., with increasing neutralization and dilution of the contaminated waters.] Dilute (only slightly contaminated) groundwaters sampled from wells on the side of the plume (wells 201, 202) also have a much higher average Cl/Na ratio (0.91 ± 0.30) than that of the uncontaminated groundwaters. An argument can therefore be made that these slightly contaminated waters should have been used as diluting waters in the NETPATH modeling, instead of (or in addition to) the uncontaminated water chosen here. The discrepancy in the chloride and sodium dilution factors is, however, a measure of the uncertainty inherent in trying to model the groundwater mixing process with a simple inverse geochemical model.

8.4.2 *The thermodynamic state of the end-member waters*

After examining the conservative constituent concentrations of the groundwaters, the next step is to examine the aqueous-speciation results, in particular the mineral saturation indices (Table 8.2) calculated for the three end-member waters chosen in our model. The speciation calculations were performed with the WATEQFP code incorporated in the database management code DB distributed with the NETPATH code. The thermodynamic database used in WATEQFP is a subset of the database described for WATEQ4F (Nordstrom *et al.*, 1990), and is also the basis for the phreeqc.dat and wateq4f.dat databases included with the PHREEQC code (Parkhurst and Appelo, 1999).

The background water (well 504) is predominantly rich in $CaHCO_3$, and is typical of the uncontaminated groundwaters in the Pinal Creek basin. These waters are usually near saturation with calcite, dolomite and chalcedony, have near to slightly above neutral *pH* values, and have equilibrium CO_2 partial pressures between $10^{-1.5}$ and $10^{-2.0}$. The uncontaminated groundwaters are typically rich in dissolved oxygen and other dissolved atmospheric gases (Glynn and Busenberg, 1994a; Robertson, 1991, Winograd and Robertson, 1982).

In comparison, the acidic water from well 402 is highly undersaturated with respect to calcite, dolomite, siderite ($FeCO_3$, *SI*: −2.6) and rhodochrosite ($MnCO_3$, *SI*: −3.3) and is near saturation with amorphous silica, kaolinite, and gypsum. The water is undersaturated

Table 8.2. Saturation indices and carbon dioxide equilibrium partial pressures for an acidic groundwater (well 402), an uncontaminated groundwater (well 504) and a neutralized contaminated groundwater (well 503) from the Pinal Creek basin alluvial and basin-fill aquifers. NC: could not be calculated.

Mineral	Well 402 (acidic) 1989/1/12	Well 504 (background) 1991/11/22	Well 503 (neutralized) 1991/11/22
Calcite	−5.0	−0.5	−1.8
Dolomite	−10.2	−1.1	−3.9
Siderite	−2.6	−11.5	NC
Rhodochrosite	−3.3	NC	−0.0
Gypsum	0.0	−2.6	0.1
Fluorite	−3.2	−2.2	−2.5
SiO_2 (am)	−0.1	−0.6	−0.0
Chalcedony	0.8	0.3	0.8
$Al(OH)_3$(am)	−4.0	NC	−0.8
Gibbsite	−1.3	NC	1.9
Kaolinite	0.7	NC	7.2
Alunite	1.8	NC	6.6
$Fe(OH)_3$(am)	−1.0	0.5	NC
Goethite	4.9	6.4	NC
K-Jarosite	0.0	−8.5	NC(<−3)
Log p_{CO_2} (in atm)	−1.0	−1.7	−0.9

with respect to amorphous $Fe(OH)_3$ but supersaturated with respect to goethite, and highly undersaturated with respect to all manganese oxides in the thermodynamic database used by NETPATH: pyrolusite (MnO_2), hausmanite (Mn_3O_4), manganite (MnOOH) and pyrochroite ($Mn(OH)_2$). These speciation results, based on the relatively high measured *Eh* (420 mV), are consistent with the high Fe and Mn contents of the water and the lack of any evidence of sulfate reduction. Surprisingly, the calculated equilibrium CO_2 partial pressure, $10^{-1.0}$ for well 402, is close to that of the neutralized water (well 503).

In comparison to the acidic water from well 402, the partially neutralized water from well 503 is not as highly undersaturated with respect to calcite and dolomite, and remains close to saturation with respect to both gypsum and amorphous silica. Unlike its more acidic precursor, the water is highly supersaturated with respect to kaolinite and is likely near saturation with respect to some $Al(OH)_3$ phase: it is undersaturated with respect to amorphous $Al(OH)_3$ but supersaturated with respect to the more stable form of $Al(OH)_3$, gibbsite. Although most of the dissolved iron has been removed, manganese has increased, and the water is at saturation with rhodochrosite. The water is still undersaturated with respect to several manganese oxides (pyrolusite *SI*: −9.1, hausmanite *SI*: −13.4, manganite *SI*: −4.7, pyrochroite *SI*: −7.2), although the uncertainty in these saturation indices is high, given the poor knowledge of manganese oxide thermodynamics and the dependence of the calcu lated saturation indices on the measured *Eh*. Indeed, lack of data on the vanishingly small dissolved Mn(IV) and Mn(III) concentrations makes any saturation index calculations for the Mn(IV) and Mn(III) minerals (pyrolusite, hausmanite, manganite) almost meaningless, because the calculations assume that the measured *Eh* values are representative of the Mn(IV)/Mn(II) and Mn(III)/Mn(II) aqueous activity ratios. Finally, the equilibrium CO_2 partial pressure ($10^{-0.9}$) is close to that of the acidic water from well 402, and is more than an order of magnitude higher than are expected from equilibrium with unsaturated zone CO_2 partial pressures (Glynn and Busenberg, 1994b).

8.4.3 *NETPATH inverse modeling: simulation results*

The first NETPATH inverse modeling simulation considered the following 11 mass balance constraints: Cl, Ca, Mg, Na, Al, Si, RS (redox state), Fe, Mn, C, and S. The simulation also considered 14 phases for possible reactions. In addition, some phases were "forced" to be included (NETPATH model results without the phase were excluded from consideration), and some phase reactions were specified either as dissolution only, or as precipitation/exsolution only. The phases specified are listed below along with any associated constraints:

Calcite (forced inclusion; dissolution only),
Goethite (forced inclusion; precipitation only),
Gypsum (forced inclusion; precipitation only),
Kaolinite (precipitation only),
SiO_2,
Dolomite (dissolution only),
MnO_2 (dissolution only),
Rhodochrosite ($MnCO_3$),
Anorthite ($CaAl_2Si_2O_8$; dissolution only),
Gibbsite,
$Mn(OH)_3$ (precipitation only),
O_2 gas (dissolution only),
CO_2 gas (exsolution only),
Pure Na phase.

The last phase in the list was added to keep track of the Na imbalance. In the first NETPATH simulation, the mixing fractions of well 402 and well 504 waters were determined through the chloride concentrations because no Cl phases were specified. The 11 element mass-balance constraints allow one mixing fraction and 10 phase mole transfers to be calculated. Because of the many models that include 10 of the 14 possible phases, additional mass-transfer limitations were necessary to minimize the number of possible models. NETPATH checked 330 models or possible solutions and found 12 that did not violate the limitations placed (i.e., whether a phase was forced to be included in all models, and whether it was allowed to dissolve only, precipitate/exsolve only, or both). Of the 12 models, 6 are given here (Table 8.3). They adequately represent the range of possible solutions given by the NETPATH code and will be further discussed. It is noted that linear combinations of models also form possible models.

The phases in the simulation were chosen, based on knowledge of the mineralogy of the basin-fill and alluvial aquifer materials, and the examination of saturation indices of the well 402 and well 503 water. In addition to selecting plausible phases, the saturation indices were used to determine which reactions might be thermodynamically feasible (dissolution-only and precipitation-only constraints). Although gypsum is not present in the uncontaminated aquifer, the contaminated waters are consistently slightly supersaturated with respect to gypsum. In fact, samples brought back from the field precipitate gypsum over the course of several months. Calcite and dolomite are known to be present in the aquifer materials and were therefore included in the model. Similarly, there is no lack of manganese oxides in the alluvial materials. Manganese oxides form at the contact between the Mn(II) rich-groundwaters and oxygenated groundwaters, and are also widely disseminated in the uncontaminated sand and gravel (Ficklin *et al.*, 1991). Lind and Stollenwerk (1994) conducted an elution experiment reacting acidic iron- and manganese-rich groundwater from well 101 with alluvial material from well 601, which is downgradient from the manganese-contaminated groundwaters. Based on X-ray diffraction results, Lind and Stollenwerk (1994) found that pyrolusite (β-MnO_2) and a solid resembling kutnahorite ($CaMn(CO_3)_2$) were present before, but not after the elution of the alluvial materials.

Although goethite was the Fe(III) phase chosen (for precipitation only), choosing any other Fe(III) oxide would have resulted in the same Fe mass-transfer values. Similarly, we could

Table 8.3. Models from the first NETPATH simulation. Results in millimoles per kilogram of H_2O. Positive numbers indicate dissolution, negative numbers precipitation or degassing.

Reaction or addition	Model 1 anorthite gibbsite	Model 2 gibbsite SiO_2	Model 3 kaolinite SiO_2	Model 4 electron transfer	Model 5 O_2 gas	Model 6 O_2 gas rhodo. diss.
Well 504	0.216	0.216	0.216	0.216	0.216	0.216
Pure Na	–0.579	–0.579	–0.579	–0.579	–0.579	–0.579
Dolomite +	2.899	2.899	2.899	2.899	2.899	2.899
Gypsum –F	–2.219	–2.219	–2.219	–2.219	–2.219	–2.219
Goethite –F	–8.339	–8.339	–8.339	–8.339	–8.339	–8.339
Calcite +F	4.929	5.086	5.086	4.929	4.929	4.929
Anorthite	0.157			0.157	0.157	0.157
Kaolinite			–0.226			
Gibbsite	–0.766	–0.452		–0.766	–0.766	–0.766
SiO_2		0.314	0.766			
Rhodochrosite	–2.972	–2.972	–2.972			1.092
MnO_2 +	4.064	4.064	4.064	7.036	1.092	
$Mn(OH)_3$–				–5.944		
O_2 gas +					1.486	2.032
CO_2 gas –	–6.033	–6.190	–6.190	–9.005	–9.005	–10.097
Net protons consumed	5.707	5.707	5.707	5.707	5.707	5.707

Notes: In the first column, F indicates a forced phase, + a dissolution only phase, – a precipitation only or an exsolution only phase.

have picked amorphous $Al(OH)_3$ instead of gibbsite. Whereas thermodynamic stability might distinguish among the potential iron phases, thermodynamic stability is not incorporated explicitly into NETPATH calculations. In fact, Fe and Al are most likely precipitating as fairly amorphous precipitates, that may recrystallize to more stable crystalline forms with time. The WATEQFP speciation results suggest that the waters near well 503 may be precipitating some $Al(OH)_3$ phase. The speciation results also suggest that kaolinite may be forming near well 402, but probably is not forming quickly near well 503 (as evidenced by the very high supersaturation with respect to kaolinite). The precipitation of amorphous forms of Al and Fe(III) minerals upon reaction of the alluvial sediments with acidic waters is also suggested by the elution experiments of Lind and Stollenwerk (1994) and by the selective extractions performed by Ficklin *et al.* (1991) on core materials from wells 107 (acidic), 451 (partially neutralized) and 505 (neutralized). Ficklin *et al.* (1991) also report no visible association between Al and SO_4 and argue against the formation of an $AlOHSO_4$ phase. Stollenwerk and Eychaner (1987) had earlier argued that this phase controlled aluminum concentrations in the acidic groundwaters. Furthermore, in his column elution studies, Stollenwerk (1994) found that he could best simulate the behavior of dissolved aluminum by using amorphous $Al(OH)_3$ as the solubility-limiting phase at *pH* values above 4.7 and $AlOHSO_4$ at lower *pH* values. Stollenwerk, however, changed the solubility product of the $AlOHSO_4$ phase to best fit his experimental results (from $\log K = -3.23$ to $\log K = -2.2$).

Considering the available evidence, Glynn and Brown (1996) considered that the issue of $AlOHSO_4$ precipitation was not resolved and required further research. More recently, Kirk Nordstrom (U.S. Geological Survey, personal communication, 2011), and a review by Bigham and Nordstrom (2000), have provided good arguments suggesting that an $AlOHSO_4$ phase probably did not occur in groundwaters at the Pinal Creek site. Because the water from well 402 is close to saturation with kaolinite, and because kaolinite is known to form in acidic waters with high dissolved silica (Blair Jones, U.S. Geological Survey, personal communication, 1996), Glynn and Brown (1996) preferred the hypothesis of Al control by kaolinite in the more acidic waters from the site. However, these authors also investigated the effect of possible $AlOHSO_4$ reactions in their inverse and forward geochemical modeling for the site. The results were interesting and therefore will be presented here, despite the recent knowledge arguing further against $AlOHSO_4$ reaction control at the Pinal Creek site.

Because well 503 water is close to saturation with respect to rhodochrosite, we chose this mineral as a possible Mn sink. We believe that reductive dissolution of MnO_2 is the primary process causing dissolved Fe(II) to oxidize and precipitate from solution. The only problem with this reaction mechanism is that the increase in dissolved Mn(II) is too small relative to the decrease in Fe(II). Several other possible reactions could explain the missing dissolved manganese:

1. Mn(II) may be precipitating out as rhodochrosite;
2. An electron transfer process may be taking place during which the oxidation state of the Mn oxide simply decreases while only partially releasing Mn into the solution.
3. Mn(II) may be sorbing onto the freshly precipitated Fe-oxyhydroxides.
4. Oxygen is known to be diffusing through the unsaturated zone into the groundwaters near the water table. However, because the sampling depths for wells 402 and 503 are substantially below the water table, the addition of significant amounts of oxygen is considered unlikely.

Although O_2 ingassing was considered in the first NETPATH simulation, this reaction will be discarded in the second simulation. For similar reasons, the possibility of CO_2 exsolution from a deep flowpath is unlikely. Glynn and Busenberg (1994b) estimated, based on their measurements of dissolved gases in the Pinal Creek basin groundwaters, that only groundwaters within 2 m of the water table could be exsolving dissolved gases and CO_2. Significant CO_2 exsolution would also cause exsolution of other dissolved gases such as N_2 and Ar. For example, exsolution of CH_4 and CO_2 from an hydrocarbon contaminant plume has been held responsible for the low dissolved Ar and N_2 concentrations measured in groundwaters from the U.S. Geological Survey Bemidji Toxics site (Revesz *et al.*, 1995). Instead, groundwaters from the Pinal Creek site show high concentrations of both dissolved N_2 and Ar because of the large amounts (often above 20 mL L^{-1}) of excess air entrained during groundwater recharge (Glynn and Busenberg, 1994b).

Interestingly, the results of the first NETPATH simulation suggest that other Ca and Mg sources (in addition to calcite and dolomite) are needed if CO_2 is disallowed as a sink for the excess carbon provided by the dissolution of the carbonates. We initially thought that rhodochrosite ($MnCO_3$) would provide an additional carbon sink, but found that, given the Mn mass-balance constraints, the rhodochrosite sink would not be sufficient to account for the excess carbon. The presence of another Mn sink (such as Mn(II) sorption) instead or in addition to rhodochrosite precipitation would only exacerbate this problem. Therefore, because no other carbon sinks are likely to be present (the waters are undersaturated with respect to siderite), the next approach was to incorporate another Ca source, specifically a Ca silicate, so as to reduce the amount of carbon coming into solution from calcite. Although anorthite was chosen, it is likely that any silicate mineral dissolution accelerated by the acidic groundwaters would also act as a source of Mg, Na, and K (and probably Fe and Mn) to the solution. However, the dissolution of Ca-rich silicates (and perhaps Mg-rich silicates) is expected to be faster than that of Na- and K-rich silicates. On the basis of their observations and elution experiments, Lind and Stollenwerk (1994) suggest that tremolite ($Ca_2Mg_5Si_8O_{22}(OH)_2$) dissolution may be a source of both Ca and Mg to the Pinal Creek groundwaters. Indeed, amphiboles, such as tremolite, and pyroxenes are expected to have faster reaction rates than feldspar minerals, although their abundance in the alluvial materials is minor compared to that of the feldspar minerals. The presence of CO_2 degassing in all the models found by the first NETPATH simulation suggests that some Mg-silicate phase (such as tremolite) must be included if models without CO_2 degassing are to be found.

The last row in Table 8.3 gives an estimate of the net number of millimoles of protons consumed by the various reaction models. The number of protons consumed in each reaction model was calculated by estimating the number of protons consumed by the *dissolution* of one millimole of each solid or gaseous phase multiplied by the mole transfer in the reaction model. The number of protons consumed is dependent on the degree of protonation or hydroxylation of the various aqueous species produced by the dissolution reactions. For example, a calcite dissolution reaction will show consumption of two protons per mole of calcite dissolved if the reaction is written to produce $H_2CO_3^0$ (or equivalently aqueous CO_2; henceforth, $H_2CO_3^0$ refers to the sum of the $H_2CO_3^0$ and the much more dominant aqueous CO_2 species), but will show consumption of only one proton if the reaction is written to produce HCO_3^-. The proton consumption calculations shown here assume that the reaction byproducts are the dominant aqueous species determined from the speciation of the well 402 water, such as: AlF^{2+}, AlF_2^+, $AlSO_4^+$, $Al(SO_4)_2^-$ and Al^{+3} for Al; $Mn^{2+}$and $MnSO_4^0$ for Mn; Fe^{2+} and $FeSO_4^0$ for Fe; H_2CO3^0 for *TDIC* (species listed in order of decreasing predominance). Using this assumption, the number of moles of protons consumed per mole of phase dissolved are: 14 for tremolite; 8 for anorthite; 6 for kaolinite; 4 for dolomite, MnO_2 and O_2 gas; 3 for goethite (or $Fe(OH)_3$), gibbsite (or $Al(OH)_3$) and $Mn(OH)_3$; 2 for calcite and rhodochrosite; 1 for $AlOHSO_4$. All other phases mentioned in Tables 8.3 and 8.4 are assumed not to consume protons upon dissolution. The consumption of protons by the heterogeneous mass-transfer reactions must necessarily be matched by an increase in solution *pH* and also by the release of protons from homogeneous deprotonation reactions (e.g., $H_2CO_3^0 \rightleftarrows HCO_3^- + H^+$; $HSO_4^- \rightleftarrows SO_4^{2-} + H^+$). Given (i) that the increase in *pH* between wells 402 and 503 corresponds to approximately a 0.1 millimole decrease in H^+ concentration, (ii) that the difference in $H_2CO_3^0$ concentrations in well 402 and 503 waters is less than 1 millimolal (and HCO_3^- is always at least 5 times lower than the $H_2CO_3^0$ concentration), (iii) that the concentration of HSO_4^- in well 402 water is close to 0.1 millimolal, and (iv) that there are no other major homogeneous deprotonation reactions, it appears that the 5.7 millimoles of proton consumption calculated for the various reaction models presented in Table 8.3 are at least 5 times too high. Unaccounted surface deprotonation or proton exchange reactions offer one possible reason for this discrepancy. Erroneous reaction models and analytical uncertainty in the basic data collected are other possible reasons.

The most interesting results of the first NETPATH inverse modeling simulation are the following. Gas exsolution or dissolution were found necessary in all models, even though anorthite dissolution and rhodochrosite precipitation were included. Of all the models found by the first NETPATH simulation, we prefer the 3 models that considered MnO_2 dissolution and rhodochrosite precipitation, rather than an electron-transfer mechanism ($MnO_2 \rightleftarrows Mn(OH)_3$) or O_2 ingassing (with or without accompanying rhodochrosite dissolution). Of those three models, we also prefer the two models (models 1 and 2 in Table 8.3) that did not involve kaolinite precipitation. Although possible, the very high supersaturation of well 503 water with respect to kaolinite suggests that the mineral does not undergo fast precipitation, at least at *pH* values > 4. Instead, we favor aluminum control by $Al(OH)_3$ precipitation (with possible recrystallization to gibbsite).

The second and third NETPATH simulations

A second NETPATH simulation used Na as the conservative constituent, instead of chloride, and resulted only in "invalid" models that required the dissolution, rather than the precipitation, of 2.22 millimoles of gypsum per kg of water. Because both the acidic and neutralized groundwaters at Pinal Creek are supersaturated with respect to gypsum, slightly but consistently, a model with gypsum dissolution was not plausible. A pure chloride source (0.484 millimoles) was used in this second simulation. The calculated mixing fraction of background water from well 504 was 0.347 (instead of 0.216). A third NETPATH simulation used an intermediate mixing fraction of 0.281 (instead of 0.216 or 0.347) and resulted in 12 models that were similar to those of the first NETPATH simulation, but had different mass-transfer amounts. Gypsum precipitation was small (–0.003 millimoles). All models required CO_2 mass transfer, but in somewhat smaller amounts (e.g., −5.4 instead of −6.0 millimoles for Model 1, Table 8.3).

The fourth NETPATH simulation

A fourth NETPATH simulation was used to explore the effects of including tremolite [$Ca_2Mg_5Si_8O_{22}(OH)_2$], biotite [$KMg_{1.5}Fe_{1.5}AlSi_3O_{10}(OH)_2$], forsterite [$Mg_2SiO_4$], a pure Mn sink (to simulate Mn(II) sorption) and $AlOHSO_4$, while excluding some of the reactions considered unrealistic in the previous simulations, namely CO_2 exsolution, O_2 dissolution, and kaolinite precipitation. Forsterite was included for numerical exploration because it is a pure Mg-silicate with a high Mg/Si ratio. Tremolite and biotite were included because those minerals are commonly found in the Pinal Creek basin sediments. Although K is present in biotite, no mass-balance for K was included in the NETPATH simulations because of the large uncertainties in our estimated K data. The fourth NETPATH simulation resulted in 19 possible models. Of the 19 models, 6 included more than 1 Mg-silicate phase and are not presented here (Table 8.4) for reasons of space and simplicity. Five models included tremolite as the only Mg-silicate phase, and differed from each other in their treatment of the Mn and Al mass-balances (Models 7 and 10 through 13 in Table 8.4). Five other similar models included biotite instead of tremolite (Model 8 in Table 8.4) and 3 remaining models included forsterite but did not include $AlOHSO_4$.

Table 8.4. Models from the fourth NETPATH simulation. Same phases included as in first three simulations, except for the following changes: (i) tremolite, biotite, forsterite, $AlOHSO_4$, and a pure Mn sink included as possible phases; (ii) kaolinite and gases excluded.

Reaction or addition	Model 7 tremolite	Model 8 biotite	Model 9 forsterite	Model 10 tremolite electron transfer	Model 11 tremolite Mn sink	Model 12 tremolite $AlOHSO_4$	Model 13 tremolite no anorth. $AlOHSO_4$
Well 504	0.216	0.216	0.216	0.216	0.216	0.216	0.216
Pure Na	−0.579	−0.579	−0.579	−0.579	−0.579	−0.579	−0.579
Dolomite +							
Gypsum −F	−2.219	−2.219	−2.219	−2.219	−2.219	−0.648	−0.089
Goethite −F	−8.339	−11.238	−8.339	−8.339	−8.339	−8.339	−8.339
Calcite +F	4.696	6.146	4.696	1.724	1.724	4.696	4.696
Anorthite	2.130	1.840	3.290	5.102	5.102	0.559	
$AlOHSO_4$–						−1.571	−2.130
Gibbsite	−4.712	−6.066	−7.032	−10.656	−10.656		1.678
SiO_2	−8.587	−9.166	−7.717	−14.530	−14.530	−5.445	−4.326
Rhodochrosite	−2.972	−4.422	−2.972			−2.972	−2.972
MnO_2 +	4.064	5.514	4.064	7.036	4.064	4.064	4.064
$Mn(OH)_3$–				−5.944			
Mn sink –					−2.972		
Tremolite	0.580			0.580	0.580	0.580	0.580
Biotite		1.933					
Forsterite			1.450				
Net protons consumed	5.711	7.642	5.711	5.711	11.655	5.708	5.711

Notes: Results in millimoles per kilogram of H_2O. Positive numbers indicate dissolution, negative numbers precipitation or degassing. In the first column, F indicates a forced phase, + a dissolution only phase, – a precipitation/exsolution only phase.

None of the models found included dolomite, a mineral present in the alluvium as rock fragments and in dolomite formations in the surrounding hills. Dolomite should certainly be as reactive with the acidic waters as some of the silicate minerals. The problem appears to be that dolomite dissolution requires a carbon sink in addition to rhodochrosite. This carbon sink has not yet been identified. Of all the models presented in Table 8.4, we prefer the models (Models 7 through 9) that considered MnO_2 dissolution and rhodochrosite precipitation and that did not include $AlOHSO_4$. The models with electron transfer combined with any of the Mg-silicate phases are also plausible. In reality, the reactions occurring and responsible for the evolution of well 402 water to well 503 water are likely to be some linear combination that may include, but will not be restricted to, the mass-transfer models that we found using the NETPATH code. Many more models could have been found had we included other phases (silicates in particular), but their description and classification would not add substantive insight into the predominant reactions occurring in the basin. Table 8.4 also gives the millimoles of protons consumed for each reaction, using the assumptions discussed earlier (see discussion of Table 8.3). Once again it appears that the millimoles of proton consumption calculated for the various reaction models presented in Table 8.4 are about 3 to 10 times too high. Unaccounted surface deprotonation and proton-exchange reactions and uncertainty in the analytical data offer possible reasons for the discrepancy. This problem will be circumvented in the PHREEQC inverse modeling demonstration discussed later, because PHREEQC always includes alkalinity mass-balance and charge-balance equations and also considers possible uncertainties in the analytical data.

Fifth and sixth NETPATH simulations used the phases considered in the fourth simulation, but explored the effects of using either (i) Na (instead of Cl) as the conservative constituent to determine the mixing fractions of the waters from wells 402 and 504, or (ii) an average mixing fraction of well 504 water (similar to the third NETPATH simulation). The results of these simulations are discussed in Glynn and Brown (1996), but the main results of the simulations are that they either produce unrealistic models (e.g., requiring gypsum dissolution rather than precipitation), or they require a Na sink. Loss of Na to an ion-exchange mechanism is unlikely given the relatively high and increasing Ca/Na ratio along the flowpath. It is possible that accelerated weathering of silicate minerals by the acidic waters may be causing a significant increase in the cation exchange capacity of minerals exposed to the groundwaters and could thereby be responsible for a net removal of Na (and other cations) from solution (Blair Jones, U.S. Geological Survey, verbal communication). However, it is more likely that analytical error or unrepresentative concentrations in the background water chosen to simulate the downgradient dilution process are responsible for the inconsistency in the Cl and Na dilution results.

Conclusions from the NETPATH inverse modeling simulations

The most important conclusion provided by the NETPATH simulations is that Ca- and Mg-silicate mineral dissolution must be a significant process. Many researchers at the Pinal Creek site originally believed that calcite and dolomite dissolution was responsible for the most of the acid neutralization. However, Glynn (1991b) demonstrated that the increase in Sr concentrations between wells 51 and 402 must have been caused by silicate mineral dissolution—the amount of Sr present in limestone and dolomite formations contributing to the carbonate content of the alluvial materials is too small relative to the amount of Sr that precipitates as an impurity in gypsum. Sr is a significant impurity in Ca-silicate minerals and is released to the solution during their dissolution. Similarly, dissolved inorganic carbon $\delta^{13}C$ data (Glynn, Busenberg and Brown, unpublished data collected in June 1993) ranges from −9.15 to −12.90 per mil for the acidic or neutralized contaminated groundwaters, and from −10.95 to −14.00 per mil for the uncontaminated waters, suggesting that neutralization of the acidic plume by silicate minerals must indeed be important. [All $\delta^{13}C$ values are expressed relative to the Vienna PDB standard.] If the calcite and dolomite $\delta^{13}C$ values are near 0 per mil, as is reasonable for marine carbonates, closed system dissolution of those carbonates caused by acid neutralization reactions deep within the aquifer should result in higher $\delta^{13}C$ values.

8.4.4 *Inverse geochemical modeling with PHREEQC*

PHREEQC provides additional capabilities for modeling the chemical evolution between the well 402 and well 503 waters because it considers uncertainties associated with individual element analyses and also solves alkalinity-balance, water mass-balance, and charge-balance equations. PHREEQC allows each analytical datum for each aqueous solution to be adjusted within an uncertainty range that is specified by the user. PHREEQC determines sets of phase mass transfers, solution mixing fractions, and adjustments to the analytical data that satisfy mass-balance constraints and are consistent with the specified uncertainties. As an option, PHREEQC will also determine mass-transfer models (later referred to as "minimal" models) that minimize the number of phases involved. Most constraints used by PHREEQC inverse modeling are automatically specified by providing a list of potentially reactive phases. For example, if tremolite is identified as a potential reactant, PHREEQC will automatically include mass-balance constraints on Ca, Mg, and Si (NETPATH does not do this). In addition to the mass-balance constraints defined by specifying a list of potential reactants, PHREEQC also lets the user specify additional mass-balance constraints that may be used in determining the mixing fractions for two or more solutions that mix to form a final solution.

Unlike NETPATH, PHREEQC includes a charge-balance constraint, which specifies that the sum of the deviations from the analytical data for a given solution must equal the charge imbalance present in that solution. PHREEQC also uses a water mass-balance constraint to account for mixing, water derived from mineral reactions, and water evaporation or dilution (PHREEQC and NETPATH are not limited to groundwater problems). The charge-balance and water mass-balance constraints used by PHREEQC are equivalent to including a mass balance on hydrogen or oxygen. During the inverse modeling simulation, PHREEQC will adjust not only the analytical element concentrations, it will also adjust the *pH* of the waters. The adjustment to total dissolved inorganic carbon is constrained to be consistent with the adjustments to *pH* and alkalinity. Finally, in addition to including a general electron balance constraint (such as is done in NETPATH), PHREEQC also includes mole-balance equations for individual valence states of redox-active elements.

A PHREEQC inverse-modeling simulation was constructed by including all the phases used in the NETPATH simulations, except for kaolinite and CO_2 and O_2 gases, which were excluded. Including biotite [$KMg_{1.5}Fe_{1.5}AlSi_3O_{10}(OH)_2$] in PHREEQC forced the code to account for the K mass-balance. However, K is not expected to accumulate in the Pinal Creek basin groundwaters and should also not be used to determine the mixing fractions of well 402 and well 504 waters. Therefore, a pure potassium-montmorillonite [$K_{0.33}Al_{2.33}Si_{3.67}O_{10}(OH)_2$] was also included in the simulations with biotite and was only allowed to precipitate. Both sodium and chloride were specified as mass-balance constraints. Pure Na, pure Cl and pure Mn sinks were not specified, because PHREEQC requires charge-balanced phases. Adding charge imbalance would prevent PHREEQC from correctly adjusting the analytical data within the user-specified uncertainties. A ±5% relative uncertainty was chosen for all elements, except for K (±20%) and for Cl for which an uncertainty of ±10% was initially chosen but later was reduced to ±5%. The lower uncertainty for Cl did not affect the number of mass transfer models calculated by PHREEQC, minimal or otherwise, did not significantly affect the calculated mixing fractions of well 504 water, and did not result in any changes in the phases included in the mass-transfer models. The uncertainty in the *pH* of the three waters from wells 402, 504 and 503 was ±0.05 *pH* units. The "minimal" option was initially chosen to reduce the number of possible models to those minimizing the number of phases involved. Some additional precipitation-only and dissolution-only constraints were added for the phases. The models shown in Table 8.5 are a representative selection of all the models found by PHREEQC. Table 8.5 shows most of the tremolite-containing models, but other

Table 8.5. PHREEQC inverse modeling simulation results. Amounts of mass transfer and net proton consumption are reported in millimoles per kilogram of H_2O. Only mass-transfer sets (models) with the minimum significant number of phases are shown.

Reaction or addition	Model 1 tremolite non-min.	Model 2 tremolite	Model 3 tremolite no calcite	Model 4 tremolite e^- transf.	Model 5 tremolite $AlOHSO_4$	Model 6 tremolite $AlOHSO_4$ e^- transf.	Model 7 biotite
504mf.	0.258	0.258	0.258	0.258	0.258	0.258	0.258
504 mf. min.	0.258	0.258	0.258	0.258	0.258	0.258	0.258
504 mf. max.	0.277	0.277	0.277	0.277	0.264	0.264	0.263
Dolomite +	0.398		2.191		0.290		2.291
Gypsum –							
Goethite –	–8.292	–8.292	–8.292	–8.291	–8.292	–8.292	–9.074
Calcite +	3.588	4.383		1.626	3.817	3.991	
Anorthite +			2.512	2.202			2.691
$AlOHSO_4$–					–0.423	–0.423	
Gibbsite–	–0.423	–0.423	–5.448	–4.827			–2.167
SiO_2–	–3.649	–3.649	–5.803	–8.687	–3.165	–3.629	
Rhodochrosite	–2.757	–2.757	–2.756		–2.769	–2.364	–3.253
MnO_2 +	3.903	3.903	3.903	6.659	3.903	4.309	4.294
$Mn(OH)_3$–				–5.512		–0.811	
Tremolite +	0.500	0.500	0.142	0.580	0.440	0.498	
Biotite +							0.521
K-mont.–							–1.796
Net protons consumed	–0.279	–0.281	–0.272	–0.266	–0.271	–0.270	–0.298
Alk. change	0.214	0.213	0.213	0.213	0.213	0.213	0.190
Difference	0.493	0.494	0.485	0.479	0.484	0.483	0.487

Abbreviations: 504 mf., well 504 mixing fraction; +, dissolution only; –, precipitation only; Alk., alkalinity change; non-min., not a "minimal" model; K-mont., K-montmorillonite; e^- transf., electron transfer.

models found (with biotite/K-montmorillonite, forsterite or tremolite) were essentially variations on the combination of reactions shown in Table 8.5. It should be remembered that, similar to the NETPATH models, linear combinations of PHREEQC inverse models also represent possible models. With the exception of Model 1[4], all models reported in Table 8.5 are "minimal" models, i.e., had the minimum number of phases.

PHREEQC calculates, as an option, the minimum and maximum mass transfers associated with any given phase in any given model. These minimum and maximum mass transfers are constrained by the uncertainty ranges specified for the various element and *pH* analyses. Only the minimum and maximum mixing fractions of well 504 water are given in Table 8.5. The optimal mixing fractions of well 504 water reported by PHREEQC are close to the minimum possible values and do not change significantly between the various models. The mixing fraction of 0.258 reported in Table 8.5 is between the values of 0.216 and 0.347 determined by NETPATH respectively assuming conservative dissolved Cl or conservative dissolved Na.

The absence of gypsum precipitation in all of the minimal or non-minimal models is one of the most important results of the PHREEQC inverse modeling. PHREEQC revealed that SO_4 could be considered a conservative entity given the ±5% uncertainty associated with the analyses. Although most of the reasonable NETPATH mass-transfer models

[4] Model 1 can be compared with its equivalent "minimal" model, Model 2.

based on chloride conservation precipitated 2 millimoles of gypsum, this mass-transfer was insignificant given the ±5% relative uncertainty on the SO_4 concentrations and the high SO_4 concentrations in the well 402 and 503 waters. All of the mass-transfer models reported by NETPATH included 10 phases. In contrast, given the uncertainties in the analytical data, inverse modeling with PHREEQC showed that only 7 or 8 phases were required.

The second major conclusion, which confirmed the earlier NETPATH simulation results, is that Ca- and Mg-silicate phases are needed in addition to calcite and dolomite to explain the Ca, Mg and C mass balances. Another conclusion, not revealed by the previous NETPATH simulations, was the presence of dolomite dissolution in some of the minimal models. This conclusion was satisfying, because dolomite is present in the aquifer and should react in acidic environments. The PHREEQC results presented in Table 8.5 show a net production of close to 0.3 millimoles of protons. This proton balance was calculated, as previously done for the NETPATH model results (see discussion of Table 8.3), from the mass transfers in the PHREEQC models. The net proton consumption calculated for the PHREEQC models is an order of magnitude smaller than calculated for the NETPATH models. Because of alkalinity-balance and charge-balance equations are included in PHREEQC calculations, the net amount of protons released and consumed by heterogeneous and homogeneous reactions is consistent with the *pH* values of the initial and final waters, given the uncertainties specified by the user. The proton consumption can be checked against the change in alkalinity between the final well 503 water and the mixture of the well 402 and 504 waters. The alkalinity changes reported in Table 8.5 use the adjusted alkalinities calculated by PHREEQC for each inverse model. The difference between the net proton consumption and the net alkalinity change is caused by the lack of accounting for the consumption of dissolved oxygen (6.64 mg L^{-1} in well 504) in our proton consumption calculations. The difference of 0.49 millimoles is close to 4 times the difference between the oxygen content of the well 503 water and a mixture of the well 504 and well 402 waters ($4 \times 0.121 = 0.484$ millimoles). The reaction of the dissolved O_2 present in well 504 water reduces the amount of MnO_2 that undergoes reductive dissolution. The fact that 4 protons are consumed rather than 2 for each mole of oxygen consumed appears to be an error in the PHREEQC version 1 code (David Parkhurst, personal communication). Because the MnO_2 mass transfer is 30 times greater than the oxygen mass-transfer, this error does not significantly affect our results.

Out of the 6 minimal models presented in Table 8.5, our favorite models are Models 7, 2, 4, and 3. These models do not involve $AlOHSO_4$ precipitation. As previously mentioned, Glynn and Brown's (1996) view that the evidence for possible Al control by $AlOHSO_4$ precipitation during the evolution from well 402 water to well 503 water was weak is supported by Bigham and Nordstrom's (2000) review of the environmental conditions needed for $AlOHSO_4$ occurrence. Consequently, our preference for amorphous $Al(OH)_3$ as the controlling Al phase is stronger than it was in 1996. Tremolite is present in the alluvial materials and its reaction with acidic water from the Pinal Creek site has been documented by Lind and Stollenwerk (1994). Nevertheless, tremolite dissolution probably contributes much less Ca than dissolution of Ca-rich plagioclase feldspars (such as anorthite and labradorite) during the neutralization of the acidic groundwaters. The accompanying release of Na during feldspar dissolution could be a problem, however, because the PHREEQC models consider Na as a conservative constituent, within the uncertainty of the analytical data. Too much Na dissolution would require a Na sink, which remains elusive. As mentioned earlier, additional cation exchange capacity resulting from the transformation of the feldspars and other aluminosilicates into secondary clay minerals is the only plausible Na sink. Sorption on freshly precipitated iron oxyhydroxides could represent another Na sink, but the affinity of Na for these precipitates is considerably weaker than the affinity of the other cations in the solution.

Two additional PHREEQC inverse modeling simulations were conducted allowing cation exchange of Na for Ca in one simulation and exchange of Na for H in another. The results are discussed in some detail in Glynn and Brown (1996). The cation exchange capacity of the Pinal Creek alluvial sediments is generally low, probably less than 1 meq/100 g, given the coarseness of the sediment and the low organic carbon content (less than 1%). Lacking further information regarding ion-exchange reactions at the Pinal Creek site, Models 7, 4, 2 and 3 from the earlier PHREEQC inverse modeling simulation, which do not include cation exchange (Table 8.5), remain our preferred models. Simulating cation exchange in an inverse geochemical model presumes that the user has knowledge of the thermodynamically preferred directions of exchange. Although we feel that Na replacement for Ca on exchange sites should generally not occur given the preferential dissolution of Ca relative to Na in the mineral dissolution reactions, and the greater charge density (and sorption affinity) of Ca over Na, the direction of exchange for other ion-exchange reactions has much greater uncertainty. Proton release from exchange sites during the neutralization of well 402 water is possible. Our inverse geochemical modeling simulations point out the need for further experiments to determine cation exchange capacities and directions of exchange. The ability of inverse modeling to highlight knowledge gaps is perhaps one of its greatest benefits. As will be demonstrated in the next section, forward geochemical modeling may also provide insight into cation exchange reactions.

Although some mass-transfer processes are likely to occur continuously throughout the flowpath used in inverse modeling, some mass-transfer processes (such as cation exchange reactions) will affect the groundwater chemistry only in narrow portions of the flow system. In the case of continuous processes, an overall rate of reaction (expressed for example in moles per kg of H_2O per travel time or per travelled volume of aquifer) may be provided by the inverse modeling results. In the case of a non-continuous process, however, such rates will have little meaning. Unfortunately, inverse geochemical modeling cannot provide information on the heterogeneous mass-transfer reactions occurring at specific points along a flowpath, but provides, instead, only the *net* amounts of mass transfer between an initial and a final endpoint. Forward geochemical modeling can, however, provide insight on the evolution through time of chemical compositions at specific points along a postulated flowpath. *Field corroboration or confirmation* of forward modeling simulations, however, may require significantly more spatial (and temporal) information than may be available.

8.5 REACTIVE-TRANSPORT MODELING AT THE PINAL CREEK SITE

Inverse modeling is a valuable tool that can be used to gain an improved understanding of the geochemical processes that occur, or have previously occurred, in an aquifer. By itself however, inverse modeling cannot be used to make predictions on the future chemical evolution of a groundwater system, or in the case of a contaminated groundwater, on the movement of contaminants. Forward reactive-transport modeling is needed to make such predictions. Inverse modeling results, nevertheless, can be used to identify potential reactions that should be considered by a reactive transport model.

In metal-contaminated acidic groundwaters, such as those present at the Pinal Creek site, *pH* and *Eh* conditions are the primary chemical variables controlling the transport of metals and determining the quality of the groundwaters. The partially-neutralized, Fe(II)-poor groundwaters (such as the well 503 water used in our inverse modeling exercise) have significantly lower metal concentrations than the more acidic Fe(II)-rich waters (such as the well 402 water). The partially neutralized waters are still contaminated and have high SO_4, Ca, and Mn concentrations that make them unsuitable for most beneficial uses, but nevertheless offer a significant improvement in water quality. Therefore, the ability to predict the movement and reaction of the low-*pH* and Fe(II)-rich groundwater zones is desirable.

Inverse modeling can help identify the possible reactions affecting the neutralization and oxidation of the low-*pH* and Fe(II)-rich groundwaters. Generally however, the movement of the low-*pH* and Fe(II)-rich waters will be controlled by the following factors:

1. the groundwater velocity field,
2. the dilution of the contaminated groundwaters by longitudinal and transverse dispersion,

3. the extents of heterogeneous mass-transfer reactions affecting the *pH* and Fe(II) concentrations in the groundwaters (causing the low-*pH* Fe(II)-rich waters to evolve into higher-*pH* Fe(II)-poor waters), and
4. the initial concentration and composition of mineral and gas phases contacting the groundwaters and responsible for their chemical evolution.

The following sections briefly discuss ongoing research efforts aimed at a better understanding of the movement and chemical evolution of acidic, metal-laden groundwaters at the Pinal Creek site. The research findings from the Pinal Creek site will hopefully provide information, not only on the processes affecting acidic, metal-laden groundwaters in semi-arid alluvial basins, but also on the most efficient techniques to characterize and model the spread of contaminated waters at sites with similarly sparse spatial information.

The groundwater velocity field

Groundwater velocities can be estimated through the construction of a groundwater flow model. Calculations using Darcy's law on observed heads and estimated hydraulic conductivities have provided estimates of groundwater velocities in the Pinal Creek basin (Brown, 1996; Neaville and Brown, 1993). However, USGS efforts (in the mid to late 1990s) to construct a general flow model for both the alluvial and basin-fill aquifers were not successful, in part because of a limited number of piezometric observations, but mainly because the high hydraulic conductivity of the unconsolidated alluvium and its focused distribution along stream channels resulted in wetting and drying oscillations that caused numerical convergence problems. However, geochemical tracers (^{18}O, ^{2}H, ^{13}C, Cl, Ar, N_2) were used successfully by Glynn *et al.* (1999) to understand many characteristics of the origin of groundwaters in the alluvial and basin fill aquifers, including recharge temperature, recharge elevation, the amount of excess air entrained during recharge, and the extent of evaporative losses. Additionally, repeated measurements, from field campaigns in 1991, 1993, 1996, and 1998, of chlorofluorocarbon tracers CFC-11, CFC-12, and CFC-113, and to a lesser extent of SF_6 and $^{3}H/^{3}He$) concentrations, provided estimates of travel times for waters in both the alluvial and basin-fill aquifers. Despite extreme fluctuations in the water table during that period, the independent travel-time estimates obtained from different tracers were generally consistent; and furthermore, the travel times obtained for the deeper waters in both the alluvial and basin-fill aquifers were found to be nearly invariant with time (Glynn *et al.*, 1999). Because of difficulties in the construction of a MODFLOW groundwater-flow model for the Pinal Creek aquifers, a plan to use all the available hydrologic, chemical and isotopic data to calibrate a three-dimensional reactive-transport model for the basin was never achieved. Such a model, and its supporting flow model, would have been useful in refining the understanding the movement of contaminated waters and the impact of anthropogenic or natural remediation processes at the Pinal Creek site, and should generally be considered essential in the investigation of highly dynamic contaminated groundwater systems.

Transport processes and contaminant dilution

The dilution of the acidic metal-contaminated groundwaters is certainly one of the most important processes responsible for the downgradient decrease in dissolved-metal concentrations in the Pinal Creek basin. Although this dilution process was already evident in the dry to normal recharge years (1984–1991), further dilution occurred as a result of the greater than normal recharge events that started in the spring of 1991, continued in 1992, and culminated in a 100-year-magnitude flooding event in spring 1993 (groundwater levels rose as much as 16 meters). Advanced modeling techniques are not needed to demonstrate dilution as a result of these events: plots of metal concentrations as a function of chloride clearly illustrate the process (e.g., Fig. 8.3). However, a 2- or 3-dimensional transient transport model would clearly be useful in determining the relative contributions of transverse dispersion, longitudinal dispersion, flow convergence, and transient high-intensity

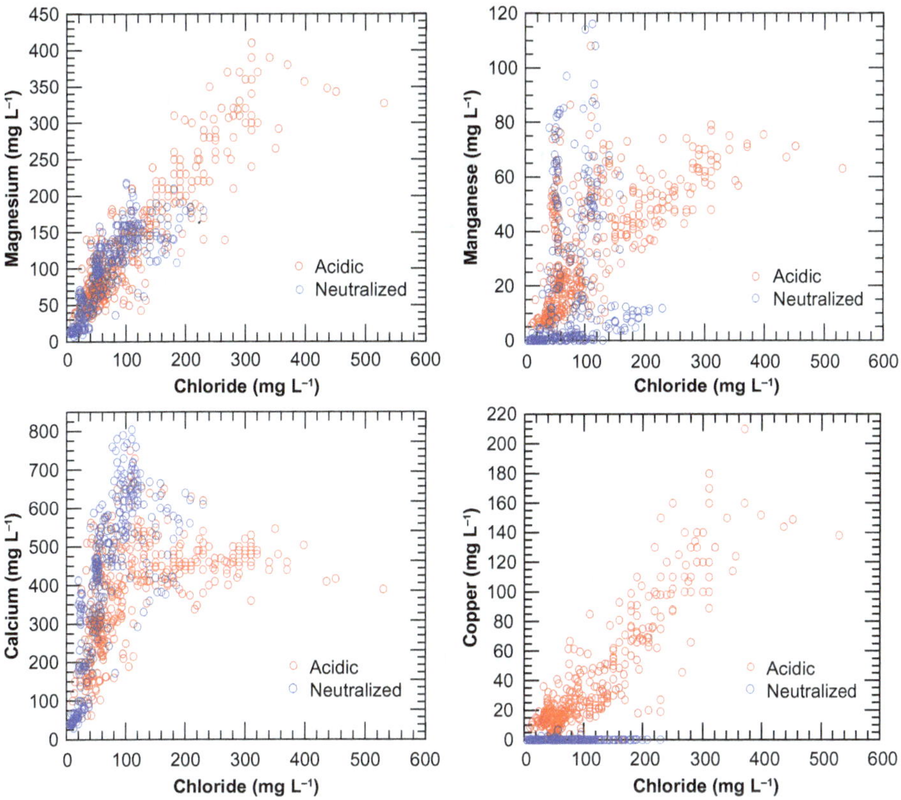

Figure 8.3. Magnesium, manganese, calcium and copper concentrations plotted against chloride concentration in acidic (upgradient) and neutralized (downgradient) groundwaters from the Pinal Creek basin.

recharge in the dilution process. Figure 8.3 also shows the relative importance, for different elements, of dilution and reaction (oxidation, neutralization, precipitation/sorption) processes in controlling groundwater chemical evolution.

8.5.1 *Summary of previous reactive-transport modeling*

Previous reactive-transport modeling conducted for the Pinal Creek site (see, Glynn and Brown, 1996, for more information) is briefly discussed here. Brown (1996) used the PHREEQC (version 1) code to construct a 1-D reactive transport model of the site based on the groundwater chemical data collected by the USGS between 1984 and 1994. The author used the 1984 chloride concentrations to back-fit the observed dilution processes as a function of time and distance along a flowpath that extended from the most acidic waters (well 51) to the most down-gradient waters (well 702). Remarkably, given the dynamic hydrology of the site, only a few adjustments were needed to fit the Cl concentration profiles for subsequent years. Adjustments were made to fit observed concentrations in 1985, 1988 and more importantly in 1993, when the equivalent of a "hundred-year" magnitude flood occurred in the valley. The results of the modeling investigation showed that dilution, rather than sorption or other reaction processes, could account for the decrease of Cu, Zn, Ni, and Co in the acidic ground-water as water flowed from well 51 to well 451. Generally, the simulated Fe(II) and

Mn(II) concentrations did not match the observed concentrations, especially downgradient from well 451. This mismatch was partly caused by a lack of knowledge at the time regarding the reactions controlling Mn and Fe; for example, rhodochrosite precipitation was not considered, and the decrease in Mn(II) concentrations was ascribed to oxidation by in gassing O_2 downgradient from well 503. The mismatch was also partly caused by lack of knowledge on rates of reactions, not only for redox processes, but also for carbonate and silicate mineral reactions and other reactions.

Glynn and Brown (1996) summarized the results of Glynn *et al.* (1991), who simulated the 1-dimensional transport and reactions of a highly acidic water sampled in 1987 from well 51 with a mineral assemblage thought to be reasonable for the Pinal Creek alluvial aquifer. One of the primary purposes of the simulation was to compare the results obtained from two different codes: PHREEQM and MST1D. Despite significant differences in their algorithms, the two codes gave almost identical results. The well 51 water used in the simulations was significantly more contaminated than the well 402 water that was used in the inverse modeling example presented here and in reactive-transport simulations described in this chapter. Relative to the well 51 water, conservative constituents (Cl, SO_4, and Mg) in the well 402 water were diluted by a factor of approximately 2.4; Na decreased only by about 1.7, suggesting the existence of a Na source between wells 51 and 402; Fe and Al decreased significantly more than could be explained by dilution (by factors of 4.7 and 13.5 respectively); SiO_2, *TDIC* and Mn did not change significantly, and Ca and Sr concentrations actually increased by 1.15 and 1.79 times, respectively. The acidity represented by the potential oxidation and precipitation of the dissolved Fe(II), Mn(II) and Al in the water from well 51 was 1.31×10^{-1} moles of protons compared to 2.60×10^{-2} for the well 402 water, or about 5 times greater. Adding the acidity represented by *TDIC* does not significantly change the potential acidity of well 51 water, but does increase the potential acidity of well 402 water to a proton molality of 3.01×10^{-2}, a value still 4.5 times lower than that of well 51 water. [The calculation of potential acidity assumes that (i) each mole of Fe^{2+} oxidation and precipitation (by reductive dissolution of MnO_2, or by reduction of dissolved O_2) produces 2 moles of H^+, (ii) oxidation of dissolved Mn^{2+} and precipitation as MnO_2 (or precipitation of $MnCO_3$) produces 2 moles of H^+, (iii) precipitation of Al^{3+} produces 3 moles of protons, and (iv) each mole of H_2CO_3 produces 1 mole of H^+.] The Glynn *et al.* (1991) simulations for the reactive transport evolution of the well 51 highly acidic water provide an interesting comparison for the reactive-transport simulations presented in the next section.

8.5.2 *A reactive-transport sensitivity analysis on the movement of pH and pe-controlling mineral fronts*

Some of the reactions identified by inverse modeling of the chemical evolution of groundwaters between wells 402 and 503 were used in 1-D PHREEQM and PHREEQC reactive-transport simulations to determine their effect on the movement of the low-*pH* and high-Fe(II) groundwaters in the Pinal Creek basin. Many results of this study will also be applicable to other sites where groundwater is contaminated by sulfuric acid and heavy metals. Before discussing the results of the PHREEQM and PHREEQC transport simulations, however, we will show how the movement of a single mineral dissolution front, critical in controlling the redox state of the Pinal Creek groundwaters, can be modeled without the use of a computer code.

8.5.2.1 *A simple model for advective transport of a reactive front: the MnO_2 dissolution front*

If O_2 ingassing is assumed not to affect the concentrations of Fe(II) at significant depths below the water table (11 m for well 402, 21 m for well 503), the movement of dissolved Fe(II) will probably be determined by the reduction of manganese oxides, as exemplified by the following reaction:

$$2\,Fe^{2+} + MnO_{2,s} + 4H_2O \rightleftarrows 2Fe(OH)_{3,am} + Mn^{2+} + 2H^+ \qquad (8.2)$$

The free energy change associated with this reaction is highly negative, and the reaction can be considered irreversible, regardless of which crystal structure is used for $MnO_{2,s}$ (birnessite, pyrolusite) or for the precipitated iron oxyhydroxide. As a result, if the kinetics of the reductive dissolution reaction are fast relative to the movement of the water (5 m day^{-1}), the velocity of the dissolved Fe(II) front, V_{Fe} can be related to the velocity, V_{H_2O} of the water through an apparent retardation factor R,

$$V_{\mathrm{Fe}} = V_{\mathrm{H_2O}}/R \tag{8.3}$$

where R is related to the amount, M_{ini}, of $MnO_{2,s}$ initially present in the aquifer (expressed in moles kg^{-1} H_2O) and to the amount, ΔM, of $MnO_{2,s}$ dissolved by a unit mass (1 kg of H_2O) of Fe(II)-rich water:

$$R = 1 + \frac{M_{ini}}{\Delta M} \tag{8.4}$$

Equation (8.4) can be related to the more general equation describing the retardation of sharp reaction fronts in systems with advective, but no dispersive, transport (Dria *et al.*, 1987; similar expressions for the "traveling wave" approximation can also be found in Lichtner, 1988, 1985; Ortoleva *et al.*, 1986):

$$R = 1 + \frac{\sum_{k=1}^{K} g_{ik}\left(m_k^D - m_k^U\right)}{\sum_{j=1}^{J} h_{ij}\left(c_j^D - c_j^U\right)} \tag{8.5}$$

where g_{ik} and h_{ij} are the stoichiometric coefficients of element i in mineral k and aqueous species j. m_k and c_j are the mineral and aqueous-species concentrations, respectively. Superscripts D and U indicate downstream and upstream concentrations, respectively.

8.5.2.2 *Determination of the initial $MnO_{2,s}$ and carbonate mineral concentrations*

Although inverse modeling can be used to determine the mass-transfer amounts (ΔM) for various heterogeneous reactions, inverse modeling cannot usually reveal the initial amounts (M_{ini}) of minerals present in an aquifer. Estimates of average initial mineral contents must be made on the basis of: batch or column experiments on unaffected aquifer materials, selective mineral extraction analyses, X-ray evidence, or observation of retardation of reactive fronts in the field.

By determining the net mineral mass transfer (ΔM) experienced by a packet of water between two points along a flowpath, an inverse model may be used to set a lower bound on the initial concentration of that mineral ($M_{ini} \geq \Delta M$), but only if the mineral mass transfer occurs in a unique and localized part of the flowpath, such as a single reaction front. Indeed, although expressed in mol kg^{-1} of H_2O, the mineral mass transfer determined by an inverse model integrates reactions occurring in the volume of aquifer traversed by a packet of water. In the case of a slow mass-transfer reaction (relative to the movement of the water), the value of ΔM determined by an inverse model will change with increasing flowpath length until some equilibrium or steady state is reached. The initial mineral concentration, M_{ini}, in equation (8.4) is also expressed in terms of mol kg^{-1} H_2O, but refers to a static mass of water and therefore a localized aquifer volume. In contrast, the value of ΔM determined by inverse modeling has a Lagrangian frame of reference and its units refer to a dynamic mass of H_2O that has travelled along a specific flowpath length. In the case where the entire mass transfer occurs at a sharp reaction front, the difference in units may be moot, at least if the mineral concentration M_{ini} was initially uniform between the initial and final end points of the flowpath. In any case, equations (8.4) and (8.5) will be of limited use for a slow reaction, one that does not result in the development of a sharp reaction front.[5]

Glynn and Brown (1996) used a value of 2×10^{-2} mol MnO_2 per kg H_2O in their PHREEQC and PHREEQM transport simulations, primarily to stay consistent with the simulations conducted by Glynn *et al.* (1991). A higher value may be more appropriate. Indeed, on the basis of his column elution experiments, Stollenwerk (1994) suggested a value of 7.1 millimoles of MnO_2 per kg of sediment. Depending on values of porosity and bulk density assumed, this solid concentration is equivalent to a value between 3.2×10^{-2} (our estimate) and 4.49×10^{-2} mol kg^{-1} of H_2O (estimate based on Brown, 1996). Based on the selective extraction results of Ficklin *et al.* (1991), Brown (1996) used a value of 7.9×10^{-2} mol kg^{-1} H_2O in his 1-D simulation. The maximum and minimum amounts of MnO_2, dissolved between wells 402 and 503 in inverse modeling simulations were 9.2×10^{-3} (Table 6 in Glynn and Brown, 1996) and 3.9×10^{-3} mol kg^{-1} H_2O (Table 8.5), respectively. These values may be considered lower bounds for the initial amount of MnO_2 present in the aquifer, if it is assumed that the reductive dissolution of MnO_2 is fast relative to the movement of the water; this assumption is used in all the PHREEQM and PHREEQC forward simulations presented in this paper. The lack of accurate groundwater flow and transport models, the large distances between the well sites, and the suggestion that the reductive dissolution of MnO_2 may be slow compared to the groundwater velocities make it difficult to calculate accurately the Fe(II) retardation factor, and therefore the appropriate initial MnO_2 content. In any case, the results of our simulations, and in particular the MnO_2 dissolution front retardation factors that we determine, can easily be extrapolated to other initial MnO_2 concentrations.

Determining appropriate initial carbonate (calcite and dolomite) concentrations for the reactive-transport simulations was also a problem. Brown (1996) used a concentration of 0.18 mol kg^{-1} H_2O in his simulation, a value consistent with the carbonate content determined by Eychaner and Stollenwerk (1985) for a sample of alluvium collected in 1985. Brown (1996) noted that the buffering capacity measurements conducted by Hydro Geo Chem, Inc. (1989) on alluvial samples collected from three different locations in the Pinal Creek basin could be translated into equivalent carbonate concentrations of 0.12 mol L^{-1} for sand and gravel and 0.76 mol L^{-1} for calcareous clay. These measurements, however, did not correct for possible proton adsorption and silicate dissolution reactions. Based on a description of the Pinal Creek site by Eychaner (1989, and pers. comm.), Glynn *et al.* (1991) used an initial carbonate concentration of 0.084 mol kg^{-1} H_2O (0.042 mol kg^{-1} H_2O calcite and 0.021 mol kg^{-1} H_2O dolomite). The resulting *pH* front retardation factor of 5, for the acidic well 51 water used in the simulation, approximately matched the relative rate of advance of the low-*pH* waters at the Pinal Creek site over the last 50 years. The inverse modeling results discussed in the present paper generally show a net dissolution between wells 402 and 503 of 1.5×10^{-3} to 4.6×10^{-3} mol kg^{-1} H_2O of primary carbonate minerals (Table 8.5). If the questionable assumption is made that the carbonate mineral dissolution rates are relatively fast, resulting in localized dissolution, a value of 5×10^{-3} mol kg^{-1} H_2O may be considered a reasonable lower bound on the initial carbonate concentration representative for the Pinal Creek alluvial sediments. In any case, because the initial carbonate mineral concentration chosen for the simulations was the most important adjustable parameter determining movement of the low-*pH* waters in our simulations (and is also the most important factor in the chemical evolution of Pinal Creek groundwaters), a set of 8 different initial carbonate concentrations ranging from 5.25×10^{-3} to 3.32×10^{-1} mol kg^{-1} H_2O were used to test each reaction-model modification.

---

[5]Lichtner (1988) and Ortoleva *et al.* (1986) have shown that reaction kinetics will not affect the rate of front propagation given enough time and distance. The front may not be as sharp but it will still propagate at the same rate as a front resulting from a simulation that uses the Local Equilibrium Assumption.

8.5.2.3 *Setup of the 1-D reactive-transport simulations*

Unless specified otherwise, most of our simulations were conducted with the local equilibrium advective transport code PHREEQC. Dispersion was not usually simulated. The reactive-transport simulations investigated the effect of the following model variations on the retardation of the low-*pH* and high Fe(II) fronts:

1. Changing solid-carbonate concentrations. The initial concentrations chosen were: 5.25×10^{-3}, 1.05×10^{-2}, 2.1×10^{-2}, 3.05×10^{-2}, 4.2×10^{-2}, 8.4×10^{-2}, 1.68×10^{-1}, and 3.32×10^{-1} mol kg^{-1} H_2O.
2. Using a longitudinal dispersivity of 560 m, or no dispersivity. The dispersivity of 560 m represented 10% of the total simulation length, a rule of thumb often applied in groundwater transport modeling. Most simulations, however, used a dispersivity of 0 m, primarily to limit execution times. The PHREEQM and PHREEQC v. 2 (1996 beta version) geochemical transport codes were used for all simulations with a dispersivity of 560 m.
3. Including or excluding dolomite. Most simulations did not include dolomite.
4. Including or excluding rhodochrosite precipitation.
5. Including or excluding $Al(OH)_3$, kaolinite, or $AlOHSO_4$ precipitation. Two different solubility products were used for $AlOHSO_4$ ($10^{-3.2}$ and $10^{-2.2}$). The higher solubility product is the value adopted by Stollenwerk (1994) in fitting the results of laboratory column experiments that reacted Pinal Creek acidic groundwater with Pinal Creek sediments.
6. Including or excluding equilibrium with an infinite reservoir of CO_2 at partial pressures of either $10^{-0.9865}$, a value based on unsaturated zone CO_2 gas measurements at the Pinal Creek site (Glynn and Busenberg, 1994b), or $10^{-1.33}$, the value used in Brown (1996). This latter value was based on the dissolved CO_2 concentration at well 503 (November 1991) reported by Glynn and Busenberg (1994a).
7. Including or excluding cation exchange. Two different cation exchange capacities, 1 meq/100 g and 10 meq/100 g, were tested.
8. Including or excluding diffuse double-layer surface-complexation sorption, which was based on thermodynamic data compiled in Dzombak and Morel (1990) for sorption onto hydrous ferric oxide.
9. Including or excluding the irreversible dissolution of Ca and Mg silicates to match the amounts calculated by two of our inverse modeling simulations (PHREEQC Models 2 and 7 in Table 8.5). These simulations assumed a zero-order kinetic dissolution process for the silicate minerals with an inexhaustible supply of silicate minerals. Two of the simulations were also conducted assuming a zero-order kinetic dissolution process for MnO_2. In all cases, the zero-order kinetic dissolution processes were specified so that the acidic water would receive, during the course of its evolution through the transport column, exactly the silicate mineral mass transfers (and the MnO_2 mass-transfer when applicable) determined by inverse Models 2 and 7 in Table 8.5. The dissolution/precipitation of all other minerals was allowed to proceed to thermodynamic equilibrium at each time step and in each cell.

More than 160 reactive-transport simulations were conducted. All simulations used the water from well 402 (89/1/12) as the infilling solution. The water from well 504 (91/11/22) was used as the background water initially present in the 1-dimensional column. The 5.6 km-long column was subdivided into 10 cells of equal length and with initially homogenous mineral, surface, and aqueous concentrations. (Because mineral concentrations in PHREEQC and PHREEQM are expressed in terms of mol kg^{-1} H_2O, the porosity and bulk density of the sediments are not defined for the programs). A time step of 112 days was used, thereby simulating an average linear groundwater velocity of 5 m day^{-1} (representative of the average groundwater velocity between wells 402 and 503). Up to a maximum of 5000 time steps (1534 years) were simulated.

Numerical dispersion affected the transport of non-conservative constituents in the simulations, and depended on the cell-length discretization, Δx, and on the extent of retardation, R. The corresponding numerical dispersivity, α_{num} can be calculated as follows (Herzer and Kinzelbach, 1989):

$$\alpha_{num} = \left(\frac{\Delta x}{2} - \frac{v\Delta t}{2R}\right) \tag{8.6}$$

or, in the case of our simulations with uniform time and space discretization:

$$\alpha_{num} = \frac{\Delta x}{2}\left(1 - \frac{1}{R}\right) \tag{8.7}$$

The maximum numerical dispersivity is half the cell length or 280 m. Retardation factors of 1, 2, 5 and 10 result in numerical dispersivities of 0, 140, 224 and 252 m, respectively. Ideally, the simulations would have been run with a greater number of cells (and consequently a greater number of time steps and a smaller cell length). Execution times and the number of simulation runs, however, made this impractical. In any case, the effects of numerical dispersion on calculated front retardation factors was small (less than 5%), and decreased as the retardation (and the number of pore volumes needed to observe that retardation) increased: the uncertainty in the exact position of a sharp front within the last cell of a column matters less and less in calculating the retardation factor for a moving front as the number of pore volumes needed to observe the passing of the front increases. The effects of numerical dispersion were also tested in the simulations that assigned a longitudinal dispersivity of 560 m to simulate hydrodynamic dispersion. Hydrodynamic dispersion affects the transport of all constituents, conservative or non-conservative (and therefore retarded). Numerical dispersion has similar characteristics to hydrodynamic dispersion but, in the case of the 1D PHREEQC and PHREEQM transport simulations, it only affects the transport of non-conservative constituents. The results of these simulations are discussed later in this chapter, but support our conclusions regarding the minor effects of numerical dispersion in the Pinal Creek analysis.

Amorphous $Fe(OH)_3$, and one aluminum phase (amorphous $Al(OH)_3$, kaolinite, or $AlOHSO_4$) were allowed to precipitate in all the simulations. Gypsum was allowed to precipitate in all the simulations except when irreversible zero-order kinetic dissolution of Ca and Mg-silicate minerals was included (following Models 2 and 7 in Table 8.5). Rhodochrosite was allowed to precipitate in all sets of simulations, except one. An essentially infinite amount of chalcedony was present in all cells. An initial concentration of 2×10^{-2} mol kg^{-1} H_2O of MnO_2 was specified in all the simulations (except for the simulations that allowed irreversible dissolution of Ca and Mg silicates). Differing initial concentrations of carbonate minerals (calcite and/or dolomite) were specified in all simulations, but dolomite was included together with calcite in only a few of those simulations. Dolomite, however, was the only carbonate mineral present in simulations that attempted to emulate Model 7 in Table 8.5.

A subset of the simulations presented here are referred to as the "Basic Reaction Model" (BRM). The BRM allows one of four possible Al-bearing phases to precipitate, along with secondary gypsum and amorphous $Fe(OH)_3$ precipitation. The BRM specifies an essentially infinite amount of chalcedony, an initial MnO_2 concentration of 2×10^{-2} mol kg^{-1} H_2O, and one of eight fixed initial concentrations of carbonate minerals. Additionally the BRM assumes local equilibrium, no sorption or cation exchange reactions, no CO_2 gas exchange, and zero longitudinal dispersivity.

Graphical presentation of the complete suite of results is limited to a few representative *pH*-breakthrough curves shown in Figures 8.4 and 8.5 for the midpoint of the last cell (cell 10). The "fronts" shown in Figures 8.4 and 8.5 are generally sharp and correspond to spatial transitions between the absence and presence of given minerals. The essential findings of the study are summarized in the sections below.

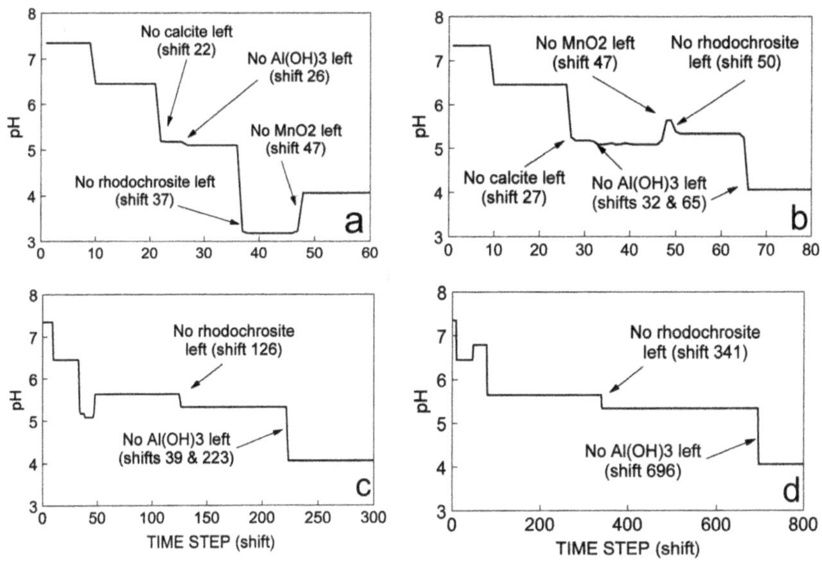

Figure 8.4. *pH* breakthrough curves for four PHREEQC advection-reaction simulations that use the "Basic Reaction Model" (BRM) with amorphous $Al(OH)_3$ as the Al-bearing phase allowed to precipitate. No dolomite was included. Initial calcite concentrations for each of the simulations shown are: a) 2.1×10^{-2} mol kg^{-1} H_2O, b) 3.0×10^{-2} mol kg^{-1} H_2O, c) 4.2×10^{-2} mol kg^{-1} H_2O, d) 8.4×10^{-2} mol kg^{-1} H_2O.

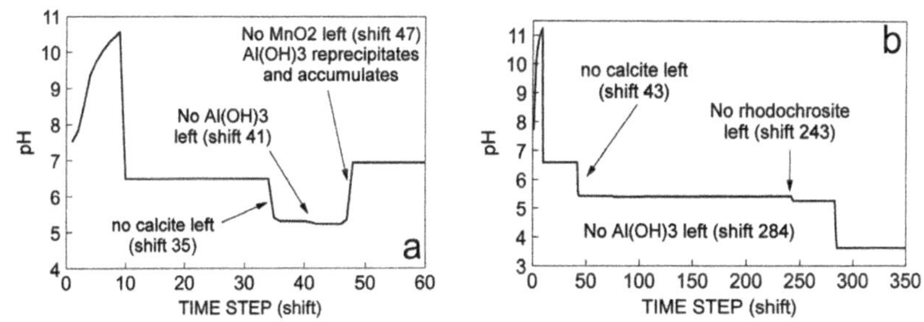

Figure 8.5. *pH* breakthrough curves for two PHREEQC advection-reaction simulations including tremolite and excluding gypsum as determined by inverse Model 2 in Table 8.5. The irreversible dissolution of tremolite (5.00×10^{-5} moles per cell per time step) was simulated as a continuous zero-order reaction process matching over the length of the simulation column the amount of tremolite mass-transfer specified in inverse Model 2. The two simulations shown differed in their consideration of MnO_2: (a) an initial concentration of MnO_2 of 2×10^{-2} mol kg^{-1} H_2O was specified and allowed to react to equilibrium at each time step; (b) MnO_2 was added as a continuous irreversible dissolution process at the rate of 3.903×10^{-4} moles per cell per time step, so as to match over the length of the simulation column the amount of MnO_2 mass transfer specified in inverse Model 2.

8.5.2.4 *Simulation results: movement of the Fe(II)-rich waters and of the MnO_2 dissolution front*

As can be seen from Figures 8.4 and 8.5, the presence, extent, and movement of *pH* fronts in the simulations conducted is dependent on a number of reaction processes, including redox processes and the presence or complete dissolution of primary MnO_2 and secondary rhodochrosite. The movement of the redox front is significantly easier to understand. Indeed, the movement of the low-*pe* Fe(II)-rich groundwater zone was primarily dependent on the initial amount of MnO_2 specified for each simulation and was usually little affected by any other factors. A retardation factor (R) of 4.74 was determined for the BRM simulations containing 2×10^{-2} mol MnO_2 per kg of H_2O. As mentioned earlier, the movement of the MnO_2 dissolution front, and of the attendant Fe(II) and low-*pe* fronts, can easily be calculated using equation (8.4). Indeed, given the Fe(II) concentration of the infilling solution (591 mg L^{-1}, Table 8.1 or 10.58 millimolar) and the stoichiometry (2 moles of Fe(II) to reduce 1 mole of MnO_2), ΔM is 5.29×10^{-3} mol L^{-1} and R is 4.78. The slight difference between this calculated value of R and the value of 4.74 determined from the PHREEQM simulations results from the use of molar concentration units for ΔM in our calculation instead of the molal units used in PHREEQM. Apart from the initial concentration of MnO_2 specified, the retardation of the MnO_2 dissolution front is only affected by processes that affect the concentration of the reductant [dissolved Fe(II)] in the infilling water: namely, dispersion and cation exchange in our simulations. Indeed, the simulations that included dispersion showed a slight decrease in the retardation factor of the MnO_2 dissolution front (an R value of 4.6 for an MnO_2 concentration of 2×10^{-2}); the simulations that included cation exchange, however, resulted in substantially greater MnO_2 retardation factors: 5.2 for the simulations with a cation-exchange capacity (*CEC*) of 1 meq/100 g and 9.2 for the simulations with *CEC* of 10 meq/100 g. (The lower *CEC* is more realistic for the Pinal Creek alluvial sediments). The increase in retardation factor is due to cation-exchange reactions that decrease the dissolved Fe(II) concentrations by, for example, Fe/Ca exchange:

$$Fe^{2+} + CaX_2 \rightleftarrows Ca^{2+} + FeX_2 \tag{8.8}$$

Because Fe(II) surface complexation on hydrous ferric oxide was not simulated, the simulations that included surface-complexation sorption reactions did not result in increased retardation factors for the MnO_2 dissolution front.

8.5.2.5 *Simulation results: evolution of the low-pH waters*

Unlike the movement of the low-*pe* Fe(II)-rich front, the movement of the low-*pH* groundwaters is much more difficult to predict and it is not controlled by a unique mineral dissolution/precipitation reaction. Indeed, the PHREEQM and PHREEQC simulation results show the development of several step changes in *pH* that are controlled by mineral dissolution and precipitation fronts. The consumption of the initial calcite (and dolomite, if present) in the column is typically followed by the complete dissolution of secondary rhodochrosite and, usually much later (if at all) by the dissolution of the secondary Al-bearing phase included in the simulation ($Al(OH)_3$, $AlOHSO_4$ or kaolinite). The complete dissolution of these minerals and the loss of their proton-consuming capacity leads to abrupt *pH* decreases. In contrast, the disappearance of MnO_2 and the cessation of Fe(II) oxidation and $Fe(OH)_3$ precipitation results in a sharp *pH* increase.

The *pH* breakthrough curves shown in Figure 8.4 for the last cell in the column are typical of the *pH* breakthrough curves for simulations without silicate dissolution: *pH* remains high at 7.34, until the first pore volume (at shift 10) flushes through the column, and neutralized infilling water with a *pH* of about 6.5 reaches the last cell. If MnO_2 is the first mineral consumed and $Al(OH)_3$ is allowed to precipitate, the *pH* will increase from 6.5 to 6.8 as the MnO_2 is completely consumed. Later, the complete consumption of calcite will result in a *pH* decrease from about 6.8 to about 5.6, and the ensuing consumption of rhodochrosite will result in another *pH* drop to near 5.3. Finally, the disappearance of the secondary aluminum phase will result in yet another *pH* drop to about 4.06 (given a sufficiently long simulation time), which is close to the initial infilling water *pH* of 4.13. Although the final *pH* values in simulations with silicate mineral dissolution differ somewhat from the *pH* values mentioned above, the general pattern of *pH* evolution in those simulations is also strongly dependent on the disappearance of calcite (or calcite and dolomite), MnO_2, secondary rhodochrosite, and the secondary aluminum phase (Fig. 8.5).

Table 8.6. *pH* values computed with the PHREEQC geochemical code (version 1; Parkhurst, 1995) by equilibrating well 402 water with various mineral phases, including chalcedony, gypsum and amorphous $Fe(OH)_3$ and one of four possible Al bearing phases (cf. table note). The last *pH* column indicates the *pH* value in the absence of any aluminum phase.

Presence of additional minerals?			Computed equilibrium *pH* as a function of the presence of an Al phase				
MnO_2	Calcite	Rhodocrosite	1	2	3	4	no Al phase
Yes	Yes	Yes	6.453	6.442	6.462	6.467	
Yes	No	Yes	5.174	5.071	5.095	5.090	
Yes	No	No	**5.011**	**3.810**	**3.364**	**3.186**	**3.175**
No	Yes	Yes	6.788	6.772	6.748	6.818	
No	No	Yes	5.633	5.576	5.666	5.659	
No	No	No	5.330	4.190	4.554	3.946	4.056

Note: The identity of the Al bearing phase is indexed as follows: (1) $Al(OH)_3$, (2) kaolinite, (3) $AlOHSO_4$ with $\log K_{sp} = -2.2$, (4) $AlOHSO_4$ with $\log K_{sp} = -3.23$).

Many of the *pH* values that are plateaus in Figures 8.4 and 8.5 can be simulated by equilibrating the infilling solution (well 402 water) with various assemblages of coexisting mineral phases found during the transport simulations. *pH* values resulting from PHREEQC equilibrations of well 402 water with several mineral assemblages are given in Table 8.6. The resulting *pH* values usually closely agree with the *pH* values obtained during the transport simulations (and are even within half a *pH* unit of the values reached in the simulations that incorporated CO_2 exsolution and surface-complexation). The PHREEQC equilibration results clearly indicate that, although the loss of calcite results in a *pH* decrease to the range of 5.1 to 5.7, the subsequent loss of rhodochrosite results in a further *pH* decrease to values well below 5 if $Al(OH)_3$ is not present. Table 8.6 indicates that the lowest *pH* value reached during a simulation, after the complete dissolution of rhodochrosite, depends on the aluminum phase included in the simulation. Invariably, simulations that allowed amorphous $Al(OH)_3$ precipitation resulted in higher *pH* values than those simulations that allowed kaolinite, or especially, $AlOHSO_4$ precipitation. In the presence of rhodochrosite (and possibly calcite), however, the choice of aluminum phase allowed to precipitate had little effect on the resulting *pH* values. Finally, Table 8.6 confirms the finding that the lowest *pH* values (near 3.1) were invariably attained when MnO_2 was still present but all other carbonate and Al-phases were absent.

8.5.2.6 *The effect of the initial carbonate to initial MnO_2 ratio on the evolution of the low-pH waters*

The initial carbonate mineral concentration specified in each simulation was the most important control on the movement of the calcite, rhodochrosite, and Al-phase dissolution fronts, and of the associated *pH* fronts (Tables 8.7, 8.8 and 8.9). The initial carbonate to MnO_2 ratio (*CMR*) was the second most important factor affecting the retardation of dissolution fronts for rhodochrosite and the Al-bearing phase (the rhodochrosite dissolution front is important because of the decrease in *pH* associated with the front, particularly in the absence of an $Al(OH)_3$ phase). The set of simulations that allowed irreversible dissolution of Ca- and Mg-silicate minerals gave very different results from the set of simulations that did not allow Ca- and Mg-silicate dissolution. Unless mentioned otherwise, the following discussions apply only to the set of simulations that did not allow silicate mineral dissolution, although many of our conclusions will also be relevant to the simulations with Ca- and Mg-silicate dissolution.

The *CMR* ratio specified in a given simulation determines the timing of the complete dissolution of MnO_2 in relation to the complete dissolution of the carbonate phases. The lowest *pH* values in the transport simulations usually occurred when MnO_2 remained in a

Table 8.7. Rhodochrosite dissolution-front retardation factors for different reaction models and initial carbonate concentrations (in mol kg^{-1} H_2O). All models except one used a dispersivity of 0 meters. All models allowed precipitation of rhodochrosite and of one of four Al-bearing phases. The specified reaction models have an index that refers to the allowed Al-bearing phase. These indices and the meanings of the various reaction models are discussed in the table notes and in the text.

	Initial carbonate mineral concentrations ($\times 10^{-2}$)							
Reaction model	0.525	1.05	2.1	3.0	4.2	8.4	16.8	33.2
Basic Reaction Model, (1)	1.6	2.3	3.7	5.0	12.6	34.1	65.6	128.9
Dispersivity, (1), PHREEQM	1.5	2.3	3.7	4.5	12.7	35.1	67.7	133.0
Dispersivity, (1), PHREEQC	1.5	2.2	3.6	4.5	12.5			
Low *CEC*, (1)	1.6	2.3	3.8	5.1	11.8	33.7	65.4	128.6
High *CEC*, (1)	0.0	3.5	5.1	6.5	8.8	31.3	63.0	126.3
CO_2, $\log p_{CO_2} = -0.9865$, (1)	1.4	2.1	3.5	5.0	12.2	32.5	56.5	104.5
CO_2, $\log p_{CO_2} = -1.33$, (1)	1.5	2.2	3.7	5.9	15.2	37.5	68.8	131.4
Basic Reaction Model, (3)	1.6	2.3	3.6	4.9	14.7	42.3	81.4	159.4
Basic Reaction Model, (4)	1.6	2.3	3.6	4.8	14.5	41.3	80.9	160.2
Low *CEC*, (3)	1.6	2.3	3.7	5.0	13.6	41.8	80.9	158.9
High *CEC*, (3)	0.0	3.5	5.0	6.4	8.6	38.0	77.2	155.4
Surf. Comp., (3), no Cu, Zn, Co, Ni		3.4	4.7	12.3	22.5	47.4	86.8	165.4
Surf. Comp., (3), with Cu, Zn, Co, Ni	2.7	3.4	4.6	7.8	19.8	46.4	79.5	140.8
Tremolite, Model 2, (1)	RA	RA	RA	RA	RA	RA	RA	RA
Tremolite, MnO_2, Model 2, (1)	7.3	10.9	18.2	24.3	32.6	61.6	119.4	235.3
Biotite/K-mont/An, Model 7, (1)	RA	RA	RA	RA	RA	RA	RA	RA
Biotite/K-mont/An, MnO_2, Model 7, (1)	12.1	17.9	29.8	40.1	53.6	101.7	197.5	388.9
Biotite/K-mont/An, Model 7, (3)	RA	RA	RA	RA	RA	RA	RA	RA

Note: The identity of the Al bearing phase is indexed as follows: (1) $Al(OH)_3$, (2) kaolinite, (3) $AlOHSO_4$ with $\log K_{sp} = -2.2$, (4) $AlOHSO_4$ with $\log K_{sp} = -3.23$).

Reaction model abbreviations and meanings:

RA:	rhodochrosite accumulates.
Dispersivity:	basic reaction model with dispersivity of 560 m;
CO_2:	column open to a specified fixed p_{CO_2};
Low *CEC*:	low cation exchange capacity (1 meq/100 g);
High *CEC*:	high cation exchange capacity (10 meq/100 g);
Surf. Comp.:	surface complexation model;
Tremolite, Model 2:	irreversible tremolite dissolution (Model 2 in Table 8.5);
Tremolite, MnO_2, Model 2:	as above but with zero-order MnO_2 dissolution;
Biotite/K-mont/An, Model 7:	irreversible biotite and anorthite dissolution and K-montmorillonite precipitation (Model 7 in Table 8.5);
Biotite/K-mont/An, MnO_2, Model 7:	same as above but with zero-order MnO_2 dissolution.

cell, but all the initial and secondary carbonate (rhodochrosite) phases had completely dissolved. (If additionally the Al-bearing phase had been completely dissolved, the *pH* was even lower). This transient remnant of MnO_2 (third row in Table 8.6) occurred in all simulations with a *CMR* ratio less than 1.5 (5.25×10^{-3}, 1.05×10^{-2} and 2.1×10^{-2} initial carbonate concentrations and 2×10^{-2} initial MnO_2 concentration). Simulations with *CMR* less than 1.5 resulted in lower *pH* values and more rapid dissolution of carbonates, and therefore, in relatively lower retardation factors for the rhodochrosite and Al-phase dissolution fronts (Tables 8.8 and 8.9; Fig. 8.4a).

In simulations with *CMR* ratios near 1.5, the complete dissolution of MnO_2 and rhodochrosite occurred nearly simultaneously, but was preceded by the complete dissolution of calcite (Fig. 8.4b). For most of these simulations, the rhodochrosite dissolution front had a retardation factor (*R*) close to 4.7, the same as the MnO_2 front. Disregarding the Ca and Mg

Table 8.8. Calcite dissolution-front retardation factors. See Table 8.7 caption and notes for explanation of headings and abbreviations. One additional set of simulations reported here used a *CEC* of 1 meq/100 g but no rhodochrosite was allowed to precipitate (low *CEC*, no rhodo).

	Initial carbonate mineral concentrations ($\times 10^{-2}$)							
Reaction model	0.525	1.05	2.1	3.0	4.2	8.4	16.8	33.2
Basic reaction model, (1)	1.2	1.5	2.2	2.7	3.4	8.0	22.7	52.0
Dispersivity, (1), PHREEQM	1.0	1.3	2.0	2.5	3.3	8.0	23.0	53.0
Dispersivity, (1), PHREEQC	1.0	1.3	2.0	2.5	3.3			
Low *CEC*, (1)	1.5	1.9	2.5	3.1	3.8	8.2	22.9	52.3
High *CEC*, (1)	2.5	3.5	4.9	5.9	6.9	10.7	25.4	54.8
CO_2, $\log p_{CO2} = -0.9865$, (1)	1.1	1.4	2.1	2.6	3.3	7.3	18.4	40.8
CO_2, $\log p_{CO2} = -1.33$, (1)	1.2	1.5	2.2	2.8	3.6	8.8	23.0	51.2
Low *CEC*, no rhodo., (1)	1.5	2.1	3.0	3.9	5.0	13.8	32.4	69.5
Basic reaction model, (3)	1.2	1.6	2.2	2.7	3.4	8.7	24.7	56.7
Basic reaction model, (4)	1.3	1.6	2.2	2.7	3.5	8.9	25.6	59.1
Low *CEC*, (3)	1.5	1.9	2.6	3.1	3.8	8.9	25.0	57.0
High *CEC*, (3)	2.5	3.5	4.9	5.9	7.0	11.4	27.4	59.5
Surf. Comp., (3), no Cu, Zn, Co, Ni		2.1	2.7	3.2	3.9	10.2	26.3	58.4
Surf. Comp., (3), with Cu, Zn, Co, Ni	1.7	2.0	2.6	3.1	3.8	8.9	22.2	48.2
Tremolite, Mod. 2, (1)	1.5	1.9	2.8	3.5	4.5	113.7	353.5	>500
Tremolite, MnO_2, Model 2, (1)	1.6	2.2	3.3	4.3	5.5	10.0	18.9	36.8
Biotite/K-mont/An, Model 7, (1)	DA	DA	DA	DA	DA	DA	DA	DA
Biotite/K-mont/An, MnO_2, Model 7, (1)	1.6*	2.2*	3.2*	4.2*	5.4*	9.7*	18.4*	35.7*
Biotite/K-mont/An, Model 7, (3)	DA	DA	DA	DA	DA	DA	DA	DA

*All simulations referring to Model 7 in Table 8.5 have initial dolomite instead of calcite. DA: no calcite specified and dolomite accumulates.

silicate dissolution simulations, the highest rhodochrosite *R* values were obtained for simulations that specified a high ion-exchange capacity (*CEC*), a surface-complexation model, or equilibrium with a low fixed p_{CO_2} ($10^{-1.33}$). In the high *CEC* simulations, however, the retardation factor of the rhodochrosite front was lower than that of the MnO_2 dissolution front ($R = 9.2$), i.e., the rhodochrosite dissolution front preceded the MnO_2 dissolution front.

Finally, in the simulations with a *CMR* ratio greater than 1.5 (4.2×10^{-2}, 8.4×10^{-2}, 1.68×10^{-1} and 3.32×10^{-1} initial carbonate concentrations and 2×10^{-2} initial MnO_2 concentration), complete dissolution of MnO_2 generally occurred before the complete dissolution of secondary rhodochrosite (with the exception of the two high *CEC* simulations). In simulations with an initial carbonate concentration of 8.4×10^{-2} or greater, exhaustion of the initial MnO_2 also occurred before the complete dissolution of the primary calcite (and/or dolomite). The faster movement of the MnO_2 dissolution front relative to the carbonate mineral dissolution fronts resulted in higher *pH* waters contacting the carbonate minerals and Al-phase minerals, and therefore in slower dissolution and higher retardation factors.

8.5.2.7 *Influence of the aluminum mineral allowed to precipitate on the evolution of the low-pH waters*

The choice of Al-bearing phase allowed to precipitate was the third major factor controlling the relative retardation of the rhodochrosite and Al-phase dissolution fronts in the simulations without Ca and Mg silicate dissolution. The simulations that allowed $AlOHSO_4$ to precipitate, instead of amorphous $Al(OH)_3$, exhibited greater *R* values for the rhodochrosite and Al-phase dissolution fronts. Remarkably, the simulations that used the more stable $AlOHSO_4$ (solubility product $10^{-3.23}$) compared to simulations that used the less stable $AlOHSO_4$ phase

Table 8.9. Retardation factors for the final aluminum phase dissolution-front. See Table 8.7 caption and notes for explanation of headings and abbreviations. One additional set of simulations reported here used a *CEC* of 1 meq/100 g but no rhodochrosite was allowed to precipitate (low *CEC*, no rhodo.). In some cases, the aluminum phase was not exhausted during the 5000 time steps allowed for a simulation.

	Initial carbonate mineral concentrations (× 10^{-2})							
Reaction model	0.525	1.05	2.1	3.0	4.2	8.4	16.8	33.2
Basic reaction model, (1)	1.4	1.8	2.6	6.5	22.3	69.6	143.3	290.8
Dispersivity, (1), PHREEQM	1.3	1.8	2.7	8.4	25.3	78.9	162.7	330.3
Dispersivity, (1), PHREEQC	1.2	1.7	2.7	7.6	23.3			
Low *CEC*, (1)	1.7	2.2	3.2	3.9	20.6	68.8	142.6	290.1
High *CEC*, (1)	2.7	3.7	5.3	6.9	9.1	62.6	136.8	284.5
CO_2, $\log p_{CO_2} = -0.9865$, (1)	1.3	1.9	3.2	5.3	20.8	61.9	113.6	217.3
CO_2, $\log p_{CO_2} = -1.33$, (1)	1.5	2.3	3.9	13.9	39.6	106.4	203.9	399.0
Low *CEC*, no rhodo., (1)	1.6	2.2	3.3	4.2	8.6	37.7	91.4	198.6
Basic reaction model, (3)	0.0	2.3	3.6	7.9	30.3	89.1	171.3	335.5
Basic reaction model, (4)	1.8	2.7	4.5	>500	>500	>500	>500	>500
Low *CEC*, (3)	1.6	2.3	3.7	5.0	27.5	87.6	169.9	334.1
High *CEC*, (3)	0.0	0.0	0.0	6.4	8.8	76.0	158.5	323.3
Surf. Comp., (3), no Cu, Zn, Co, Ni		3.5	7.4	24.7	45.9	96.8	176.6	335.8
Surf. Comp., (3), with Cu, Zn, Co, Ni	2.9	3.5	8.8	31.6	48.1	102.8	172.3	300.8
Tremolite, Model 2, (1)	AA	AA	AA	AA	AA	AA	AA	AA
Tremolite, MnO_2, Model 2, (1)	8.2	12.5	21.1	28.4	38.2	72.6	141.1	278.4
Biotite/K-mont/An, Model 7, (1)	AA	AA	AA	AA	AA	AA	AA	AA
Biotite/K-mont/An, MnO_2, Model 7, (1)	AA	AA	AA	AA	AA	AA	AA	AA
Biotite/K-mont/An, Model 7, (3)	9.2	8.0	6.9	6.5	5.0	5.8	6.6	7.3

*All simulations referring to Model 7 in Table 8.5 have initial dolomite instead of calcite. AA: $Al(OH)_3$ accumulates!

(solubility product $10^{-2.2}$) resulted in higher R values for the $AlOHSO_4$ dissolution front, but did not result in significantly different R values for the rhodochrosite dissolution front. More generally, substantial differences in retardation factors (for a given initial carbonate concentration) between the $Al(OH)_3$ and the $AlOHSO_4$ simulations occurred only in simulations with *CMR* ratios above 1.5. The reason for this behavior is not clear, but is probably related to the faster movement of the MnO_2 dissolution front relative to the rhodochrosite dissolution front. In any case, the precipitation of $Al(OH)_3$ generates three times more protons than $AlOHSO_4$, as demonstrated in the following equations:

$$Al^{3+} + SO_4^{2-} + H_2O \rightleftarrows AlOHSO_{4,s} + H^+ \tag{8.9}$$

$$Al^{3+} + 3H_2O \rightleftarrows Al(OH)_{3,am} + 3H^+ \tag{8.10}$$

Precipitation of $Al(OH)_3$ results in substantially lower pH waters and faster carbonate mineral dissolution, which explains the generally smaller retardation factors found in simulations with $Al(OH)_3$.

The precipitation of $Al(OH)_3$ (or $AlOHSO_4$) generally occurred because dissolution of carbonate minerals caused an increase in pH. Reaction with the more acidic infilling waters eventually caused redissolution of the Al-bearing phases. In the case of the simulations with low initial carbonate concentrations and with either $AlOHSO_4$ (solubility product: $10^{-2.2}$) or with $Al(OH)_3$ precipitation allowed, complete consumption of the Al-bearing phase occurred at least twice during each simulation. The retardation factors for the Al-phase dissolution fronts given in Table 8.9 refer to the last dissolution front in each simulation.

8.5.2.8 *Effects of the irreversible dissolution of Ca and Mg silicates on the evolution of low-pH Fe(II)-rich waters*

Several simulations were conducted by using reactions specified in the PHREEQC inverse Models 2 and 7 (Table 8.5). In addition to allowing secondary $Al(OH)_3$, $Fe(OH)_3$, and rhodochrosite to precipitate and specifying initial amounts of calcite (or dolomite) to also react to equilibrium, these simulations forced the dissolution of a fixed number of moles of Ca and Mg-silicate minerals into each cell at each time step, thereby simulating a zero-order kinetic dissolution process that matched (over the 10-cell column) the net amount of Ca- and Mg-silicate minerals dissolved according to PHREEQC inverse Models 2 and 7 in Table 8.5. Irreversible dissolution of tremolite was specified in the simulations based on Model 2 (Fig. 8.5). Irreversible dissolution of biotite and anorthite and removal from solution of K-montmorillonite was specified in the simulations based on Model 7. The silicate dissolution processes would have been more correctly simulated with a *pH*-dependent rate law, but this capability was not present in the PHREEQC code used. An inexhaustible supply of silicate minerals was assumed available to all cells throughout the course of the simulations. Despite the crudeness of these assumptions, it was hoped that these simulations would contribute some insights on the importance of silicate mineral dissolution in the *pH* evolution of acidic contaminated waters.

The Ca- and Mg-silicate dissolution simulations specified an essentially infinite amount of chalcedony that was available to react to equilibrium. Following the results of Models 2 and 7 in Table 8.5, gypsum precipitation was excluded from the simulations. (A few simulations were run allowing gypsum precipitation, but the results were not substantially different and are not presented here.) Although an initial fixed amount of MnO_2 (2×10^{-2} mol kg^{-1} H_2O) was specified in most simulations, two sets of simulations specified instead a zero-order dissolution process for MnO_2, according to the amounts determined by Models 2 and 7 in Table 8.5. Finally, one set of simulations was based on Model 7 (Table 8.5), but allowed $AlOHSO_4$ to precipitate instead of $Al(OH)_3$.

All the simulations that specified a fixed initial MnO_2 concentration resulted in rhodochrosite precipitation and accumulation (Table 8.7, Fig. 8.5a). A rhodochrosite dissolution front never formed; the continuous proton consumption caused by dissolution of the silicate minerals and the high initial Mn and *TDIC* concentrations in the infilling water (were sufficient to ensure that the infilling water never became undersaturated with respect to rhodochrosite. In contrast, the simulations that allowed irreversible zero-order dissolution of MnO_2 produced rhodochrosite dissolution fronts (Fig. 8.5b) because the oxidation of dissolved Fe(II) caused the precipitation of $Fe(OH)_3$ and constantly generated acidity that dissolved rhodochrosite. The *R* values for the rhodochrosite dissolution fronts nearly doubled for each doubling of the initial carbonate concentration if carbonate concentrations were above 4.2×10^{-2} mol kg^{-1} H_2O; at lower initial carbonate concentrations, rhodochrosite *R* values increased only by 50% or less for each doubling in initial carbonate concentration.

The two sets of simulations with irreversible tremolite dissolution (with and without fixed initial MnO_2 concentrations) generated calcite dissolution fronts (Table 8.8, Fig. 8.5). In contrast, of the three sets of simulations based on inverse Model 7 (and irreversible biotite and anorthite dissolution rather than tremolite dissolution), only the simulations that allowed continuous zero-order dissolution of MnO_2 (and consequent production of acidity) generated calcite dissolution fronts. In the simulations with tremolite dissolution and fixed initial MnO_2, the calcite dissolution *R* values jumped by a factor of 25, from 4.5 to 113.7, as the initial calcite concentration used in each simulation increased from 4.2×10^{-2} to 8.4×10^{-2} mol kg^{-1} H_2O (and then tripled with each further doubling of the initial calcite concentrations above that). The jump by a factor of 25 in the *R* values occurred when, as a result of the higher initial calcite concentration, the calcite dissolution front moved more slowly than the MnO_2 dissolution front ($R = 4.7$). In simulations that also used a fixed initial MnO_2 concentration (2×10^{-2} mol kg^{-1} H_2O) but attempted to emulate inverse Model 7, any initial dolomite dissolution was eventually followed by dolomite precipitation and accumulation. Because of the additional contribution of Fe(II)-containing biotite, the retardation factor was 4.5 for the MnO_2 dissolution front in these simulations.

Most of the simulations with Ca- and Mg-silicate dissolution did not produce an Al-phase dissolution front (or produced only a temporary one). The proton consumption caused by the silicate dissolution reactions caused the continuous precipitation of $Al(OH)_3$ from the infilling water (Table 8.9). The set of simulations with irreversible tremolite dissolution and zero-order MnO_2 dissolution did generate an $Al(OH)_3$ dissolution front because of the acidity generated by the continuous addition of MnO_2 by zero-order kinetics, its oxidation of dissolved Fe(II), and the consequent precipitation of $Fe(OH)_3$. The set of simulations emulating Model 7, but allowing $AlOHSO_4$ to precipitate instead of $Al(OH)_3$, produced a $AlOHSO_4$ dissolution front. Surprisingly, the *R* value for that front decreased as the initial carbonate concentration increased to 4.2×10^{-2} mol kg^{-1} H_2O and then continued to increase with higher initial carbonate concentrations.

The *pH* values obtained during, the evolution of the well 402 water during the simulations with irreversible Ca- and Mg-silicate dissolution (Fig. 8.5) were close to the *pH* values predicted for the various mineral assemblages in Table 8.6, but only when carbonate minerals (rhodochrosite or calcite) were still present and MnO_2 was either being added or was still present. The first 10 shifts of the simulations show a large increase in the *pH* of cell 10 because of the constant dissolution of Ca- and Mg-silicates into the background water that initially filled the column; after 10 shifts the water in the column has been replaced by reacted infilling water.

8.5.2.9 *The effect of not allowing rhodochrosite precipitation*

One set of simulations, without Ca- and Mg-silicate dissolution, used a low cation-exchange capacity (1 meq/100 g), but disallowed rhodochrosite precipitation. For those simulations, the calcite dissolution front was the most important *pH* front to consider, because it was either the only, or the slowest (for simulations with initial dolomite and calcite), carbonate mineral dissolution front that needed to be considered. (The rhodochrosite dissolution front was the slowest carbonate mineral dissolution front in all other simulations.) The *R* values for the calcite dissolution front were greater in the simulations without rhodochrosite precipitation than in the simulations that allowed rhodochrosite precipitation. However, the calcite *R* values in the simulations without rhodochrosite precipitation were up to two times smaller than the rhodochrosite *R* values in the runs that allowed rhodochrosite precipitation. The *pH* values after complete dissolution of calcite in the simulations without rhodochrosite precipitation match the low-*pH* values after complete rhodochrosite dissolution in the simulations that allowed precipitation of that mineral (Table 8.6).

In simulations with low initial carbonate (2.1×10^{-2} mol kg^{-1} H_2O and lower), the faster movement of the $Al(OH)_3$ dissolution front relative to the rhodochrosite dissolution front meant that the complete dissolution of rhodochrosite led to a large *pH* decrease (Fig. 8.4a). In simulations with higher initial carbonate concentrations, the second $Al(OH)_3$ dissolution front travelled more slowly than the rhodochrosite front and therefore the disappearance of rhodochrosite resulted in a relatively small decrease in *pH* (Figs. 8.4b, 8.4c, 8.4d). The *R* values for the $Al(OH)_3$ final dissolution-front were smaller (by as much as 1/3 in the runs with high initial carbonate concentrations) in the simulations without rhodochrosite precipitation relative to the $Al(OH)_3$ *R* values obtained in the simulations that allowed rhodochrosite to precipitate. For simulations with the three lowest initial carbonate concentrations (5.25×10^{-3}, 1.05×10^{-2} and 2.1×10^{-2}), however, differences in the *R* values between the simulations that did or did not include rhodochrosite precipitation were relatively small.

8.5.2.10 *The CO_2 open system simulations*

A few reactive-transport simulations allowed the l-dimensional column to equilibrate with an infinite gaseous CO_2 reservoir at a fixed partial pressure of either $10^{-1.33}$ (the calculated p_{CO_2} for water at well 503, Brown, 1996) or $10^{-0.9865}$ (the measured p_{CO_2} of the unsaturated zone at site 500 in November 1991). Opening the system up to CO_2 affected the retardation of the calcite, rhodochrosite and $Al(OH)_3$ dissolution fronts, particularly in the simulations with at least 3×10^{-2} mol kg^{-1} of initial calcite (i.e., with *CMR* ratios of 1.5 and above). The simulations conducted under the higher fixed p_{CO_2} value ($10^{-0.9865}$) generally resulted in lower *R* values for the calcite, rhodochrosite, and $Al(OH)_3$ dissolution fronts, compared to the values determined in the BRM simulations. In contrast, the simulations conducted under the lower fixed p_{CO_2} value ($10^{-1.33}$) resulted in higher *R* values for all three mineral dissolution fronts, compared to the values determined in the BRM simulations. The relevance of these simulations to the migration of the acidic ground waters at Pinal Creek field site is questionable. It is unlikely that groundwaters at the site maintain contact with the unsaturated zone during their evolution. However, the CO_2 equilibration simulations have large effects on the movement of the fronts, and further study is needed to confirm that CO_2 ingassing or outgassing is not significant at the field site.

8.5.2.11 *The effect of longitudinal dispersion*

When the longitudinal dispersivity (α) is increased from 0 to 560 m (which represents 10% of the distance between wells 402 and 503), the retardation of the calcite or rhodochrosite dissolution fronts are not significantly affected, except for the simulations with the lowest initial carbonate concentrations (3×10^{-2} mol kg^{-1} H_2O and lower, corresponding to *CMR* ratios below 1.5). Dispersion typically has little effect on the propagation of sharp fronts caused by simple mineral dissolution reactions, especially if the initial mineral concentrations are sufficient to significantly retard the propagation of the fronts. At the limit, dispersion would have no effect on the retardation of a mineral dissolution front for which an infinite initial mineral concentration had been specified. Of course, using a dispersivity of 560 m has a large effect on the spreading of non-reacting solutes, and is also expected to affect the mineral dissolution fronts that propagate at nearly the same speed as the water. Longitudinal dispersion implies that portions of a tracer or solute of interest move both faster and slower than the average groundwater velocity. The spreading of a conservative solute after a given travel time, or travel distance, can be calculated from the following equation (Appelo and Postma, 1993):

$$\sigma_x = \sqrt{2\alpha vt} = \sqrt{2\alpha x} \tag{8.11}$$

If $\alpha = 560$ m and $x = 5600$ m, the distance between wells 402 and 503, $\sigma_x = 2504$ m. σ_x represents a distance between two specific points in a 1-D homogeneous column. At those two points, a conservative tracer injected continuously at the beginning of the column (i.e., at well 402), with a relative concentration c/c_0 of 1.0, would have achieved relative concentrations of 0.5 and 0.16, respectively. When the relative concentration of a conservative tracer reaches 0.5 at well 503, 5600 m downgradient from well 402, the 0.16 relative concentration level has already moved ahead by 2504 m. In contrast, the 0.84 relative concentration level would be 2504 m upgradient from well 503. The above analysis assumes steady-state flow through a porous medium with homogeneous physical properties and also assumes that there are no "dead-water" zones into which the conservative tracer can diffuse.

The simulations that included longitudinal dispersion were conducted with both the PHREEQM code and with an unpublished version of the PHREEQC v. 2 code obtained in 1996 (David Parkhurst and Tony Appelo, written commun.). Unlike its first 1995 version, PHREEQC v. 2 simulated longitudinal dispersion by using an algorithm similar to the one used in PHREEQM. Efforts were made to ensure that the thermodynamic databases for PHREEQM and PHREEQC were identical, particularly for Al species. Although we are not aware of any differences, some minor differences may have still remained. The database used in the PHREEQC v. 2 runs was identical to that used in all the other PHREEQC runs (the PHREEQC v. 1 runs, without dispersion). Unfortunately, we were unable to make PHREEQC v. 2 converge for the entire set of simulations. Although current versions of

PHREEQC are more stable and have more features than the earlier pre-publication version that was used, modeling comparisons with PHREEQM, and with later versions of PHREEQC v. 2, leave us confident that the results obtained with the earlier version would match those obtained with more recent versions.

8.5.2.12 *The influence of cation exchange and surface-complexation sorption processes*
The reader is referred to Appelo (1996), Appelo and Postma (1993, 2005), Dzombak and Morel (1990) and Davis and Kent (1990) for excellent descriptions of the theories of cation exchange and surface-complexation sorption processes. The cation exchange conventions used in the PHREEQM and PHREEQC codes are described in Appelo and Postma (1993). PHREEQC's simulation of surface complexation sorption processes implements the diffuse double-layer surface-complexation model presented by Dzombak and Morel (1990) and is fully described in Parkhurst (1995) and Parkhurst and Appelo (1999).

Allowing cation exchange with a cation-exchange capacity (*CEC*) of 1 meq/100 g did not substantially affect rhodochrosite and calcite retardation factors (Tables 8.7, 8.8). The ions allowed to exchange were Al^{3+}, Fe^{2+}, Mn^{2+}, Ca^{2+}, Mg^{2+}, Sr^{2+}, Ba^{2+}, Na^+, and K^+. Proton exchange was also simulated. The selectivity coefficients used were the default values present in the PHREEQC thermodynamic database (which are nearly identical to values given in Appelo and Postma, 1993). A *CEC* of 1 meq/100 g [equivalent to 52.2 meq kg^{-1} H_2O appears to be a reasonable estimate for the Pinal Creek sediments, given the low organic carbon content (less than 1%) and relative coarseness of the sediments]. The empirical formula given by Breeuwsma *et al.* (1986) relating the *CEC* to the <2 μm clay fraction and to the organic carbon content is assumed applicable (cited in Appelo and Postma, 1993):

$$CEC\ (\text{meq/100g}) = 0.7\ (\%\ \text{clay}) + 3.5\ (\%\ \text{C}) \qquad (8.12)$$

Direct measurements of the *CEC* of Pinal Creek sediments, and of the exchangeable ion composition of the sediments, would be preferable to an empirical formula. The purpose of our simulations, however, was to determine the effect of ion-exchange processes on the evolution of the low-*pH* and high-Fe(II) contaminated groundwaters at the site. Therefore, simulations were also conducted using an unrealistically high *CEC* of 10 meq/100 g, a *CEC* that would be applicable to sediments with more than 10% clay content (and 1% organic carbon). As discussed earlier, these high *CEC* simulations resulted in a doubling of the retardation of the MnO_2 dissolution front (from 4.7 to 9.2 for an initial MnO_2 concentration of 2×10^{-2} mol kg^{-1} H_2O). The high *CEC* ion-exchange simulations generally increased the retardation factors for the calcite dissolution front but did not result in a uniform increase in the retardation of the rhodochrosite or of the Al-phase dissolution fronts (as a function of initial carbonate). At the lowest initial carbonate concentrations (0.525×10^{-2}), the high *CEC* simulations did not result in rhodochrosite precipitation. At higher initial carbonate concentrations, between 2.1×10^{-2} and 3.0×10^{-2} mol kg^{-1} H_2O, the high *CEC* simulations had higher *R* values for the rhodochrosite dissolution front than in the model runs that did not simulate cation exchange (or those that assumed a low *CEC*). At even higher initial carbonate concentrations (4.2×10^{-2} and above), however, this behavior was reversed: the *R* values for rhodochrosite dissolution for the high *CEC* simulations were lower than the *R* values observed in the simulations with lower or no ion exchange capacity. The retardation factors for the $Al(OH)_3$ dissolution front also exhibited complex behavior with increasing initial carbonate concentration.

The PHREEQC simulations that included a diffuse double-layer surface-complexation-sorption model resulted in retardation factors for the rhodochrosite, calcite and $AlOHSO_4$ dissolution fronts that were generally higher than the *R* values determined in simulations without sorption. Surface protonation and deprotonation reactions were incorporated into the sorption model, along with surface complexation of Mn, Ca, Mg, Cu, Zn, Co, Ni and SO_4 at weak and strong surface sites.

The thermodynamic sorption model was based on the compilation of intrinsic constants for hydrous ferric oxide published by Dzombak and Morel (1990) and used by default in PHREEQC. The number of weak and strong sorption sites used (1.5×10^{-2} and 3.8×10^{-4} respectively) and the amount of surface area per kg of H_2O used (4032 m^2) were identical to the values used by Brown (1996) in his 1-D simulation of reactive transport at the Pinal Creek site, and also in his PHREEQC-simulation fit of the column experiment conducted by Stollenwerk (1994). Using a higher number of sorption sites in our simulations would probably have resulted in more significant effects on the retardation of the mineral dissolution fronts.

Cu (36.2 mg L^{-1}), Zn (4.97 mg L^{-1}), Co (4.14 mg L^{-1}) and Ni (1.57 mg L^{-1}) concentrations measured in the well 402 water were added to one set of simulations. Another set of simulations was conducted without those elements. This second set of simulations resulted in higher *R* values for the rhodochrosite and calcite fronts, but in slightly lower *R* values for the $AlOHSO_4$ front except in the simulations with the two highest initial carbonate concentrations.

8.5.2.13 *Other minor effects on the evolution of the low-pH waters*

Several simulations included an initial concentration of dolomite, in addition to calcite, or allowed kaolinite to precipitate (instead of $Al(OH)_3$ or $AlOHSO_4$). The simulations otherwise used the BRM setup, but all specified the same initial carbonate mineral concentration (8.4×10^{-2} mol kg^{-1} H_2O). The resulting retardation factors for the calcite, dolomite and rhodochrosite dissolution fronts can be found in Glynn and Brown (1996, Table 12). Including or excluding dolomite did not, by itself, affect the retardation of the low-*pH* fronts. The important variable was the total moles of solid carbonate, whether in the form of calcite or dolomite. Allowing kaolinite to precipitate instead of $Al(OH)_3$ resulted in slightly lower retardation factors for the calcite and rhodochrosite dissolution fronts (Table 12 in Glynn and Brown). In contrast, the *R* values obtained for the final kaolinite dissolution front were generally much greater than those obtained for the final $Al(OH)_3$ dissolution front. These effects were due to the greater stability of kaolinite, relative to $Al(OH)_3$, in the presence of acidic water.

8.5.2.14 *Comparison of the reactive transport simulation results with observations at the Pinal Creek site*

Although groundwater samples at the Pinal Creek site have been collected and analysed by the U.S. Geological Survey since 1984, limited resources, and the large size of the basin and of the contaminant plume, prevented the emplacement of a high number of wells. The sparseness of the available spatial information have made it difficult to determine the location of the low-*pH* and high-Fe(II) groundwaters (Fig. 8.6) through time. The longitudinal spreading of the fronts (and especially of the Fe(II) front), which might give valuable information on reaction kinetics, has also been difficult to determine.

Nevertheless some estimates of the velocity of the low-*pH* front can be made, primarily because of the breakthrough of low-*pH* waters (from a *pH* of 4.96 in March 1989 to a *pH* of 4.24 in August 1989) that was observed at well 451 only a few months after its emplacement (December 1988). If the creation of Webster Lake in 1940, approximately 18.5 km upgradient from well 451, provided the principal source of acidic contaminated waters, then an effective velocity of about 1 m day^{-1} can be estimated for the low-*pH* front over that section of the aquifer. This velocity can be compared to an estimated average groundwater velocity of 8.4 m day^{-1} between Webster Lake and well 451, giving an estimated retardation factor of about 8.4 for the movement of the low-*pH* waters. This retardation factor estimate is thought to be a maximum estimate, because the applicable groundwater velocity could be as low as 5 m day^{-1}. Eychaner (1991) and Glynn *et al.* (1991) estimated that a lower *R* value of 5 would be reasonable for the movement of the *pH* front from Webster Lake to well 451.

Similarly, a low-*pH* (4.0 to 4.5) high-Fe(II) water was found to be present during drilling (in February 1995) of a group of wells (LPC wells) emplaced by the Pinal Creek Group (a consortium of copper companies) slightly west of well 503 (on the other side of the creek bed). Given the distance of the LPC wells from well 451 (1.3 km), the minimum velocity of the low-*pH* front is estimated to be greater than 0.65 m day^{-1}. Using this velocity with an estimated average groundwater velocity of about 5 m day^{-1} (between 451 and 503) results in a maximum *R* value of about 7.7 for the low-*pH* front. The acidity of the well 451 water in August 1989 was ~2 times lower than that of the well 402 water used in our simulations, although the potential acidity (cf. discussion of Table 7 in Glynn and Brown, 1996) of the well 451 water did increase with time until a maximum acidity, about 70% of that for well 402, was reached in November 1991.

In comparison to this field evidence, results of a column elution experiment conducted by Stollenwerk (1994), using water from well 51 (of composition similar to that presented in Glynn and Brown, 1996, Table 7) and an 80-cm column containing uncontaminated alluvium from the Pinal Creek site, indicated an *R* value for the low-*pH* front of about 2.5. Many factors can explain the lower retardation found in the column experiment relative to the field values. The low-*pH* front *R* values estimated from the field evidence incorporate significant effects of dilution (Fig. 8.3). The Webster Lake waters were more acidic than the well 51 water used in Stollenwerk's column experiment. The potential acidity of the well 51 water was 4.5 times greater than that of the well 402 water used in the Glynn and Brown (1996) simulations shown here. As a result, the retardation factor of 2.5 determined by Stollenwerk (1994) should translate to a retardation factor of approximately 7.8, $(2.5 - 1) \times 4.5 + 1 = 7.75$, had well 402 water been used in his column experiments. Equation (8.4) is used to normalize these results, assuming that ΔM is proportional to the acidity of the low-*pH* solution and that the field and laboratory experiments both had identical homogeneous mineral concentrations and reactions. Similarly, considering that the water from well 451 in August 1989 was about 2 times less acidic than the water used in the simulations shown here, the *R* value of 7.7 determined between 451 and the LPC site would correspond to an *R* value of 4.35, if normalized to the acidity of the well 402 water (89/1/12). These normalized retardation factors (4.4 and 7.8) can be compared with the *R* values obtained in our numerical simulations of the rhodochrosite dissolution front (the best analog for the low-*pH* front observed in the field). The simulation results in Table 8.7, for runs without irreversible dissolution of Ca and Mg-silicates, indicate that initial carbonate mineral concentrations between 2.1×10^{-2} and 4.2×10^{-2} mol kg^{-1} H_2O and certainly no lower than 1.05×10^{-2}mol kg^{-1} H_2O would give reasonable simulated retardation factors for the low-*pH* waters. This assumes that an MnO_2 concentration of 2×10^{-2} mol kg^{-1} H_2O was also reasonable. If not, two different situations can be considered. A lower initial MnO_2 concentration would result in a higher *CMR* value and therefore (*cf.* section 8.5.2.6) could increase the simulated retardation factors for the low-*pH* front. Consequently, initial carbonate concentrations would have to be adjusted slightly downward in order to match the estimated lab and field retardation factors. Alternatively, a higher initial MnO_2 concentration would probably not have a substantial effect on the simulated low-*pH* front *R* values (for initial carbonate mineral concentrations between 1.05×10^{-2} and 3.0×10^{-2} mol kg^{-1} H_2O).

In the simulations with Ca- and Mg-silicate dissolution, the retardation factors for the rhodochrosite front are so high that an initial carbonate concentration of 5.25×10^{-3} or lower is needed to match the observed retardation of the low-*pH* front in Stollenwerk's experiments and in the field, if we assume that the disappearance of rhodochrosite controls the low-*pH* front. Such a low initial carbonate concentration is not realistic given our knowledge of the alluvial sediments at the site, and we conclude that the simulations with irreversible dissolution of Ca- and Mg-silicate should be conducted by using a finite (as opposed to an inexhaustible) source of these silicates. Further field information on the concentration and rate of dissolution of silicate minerals would however be needed. The calcite *R* values obtained in our forward simulations of inverse Model 2 (Table 8.5) are perhaps reasonable, but only for initial calcite concentrations below 4.2×10^{-2} mol kg^{-1} H_2O. The disappearance of calcite did not by itself, however, result in a sufficient *pH* decrease to match the *pH* values of 4.0 to 4.5 that we associate with the low-*pH* front in the field (Fig. 8.6).

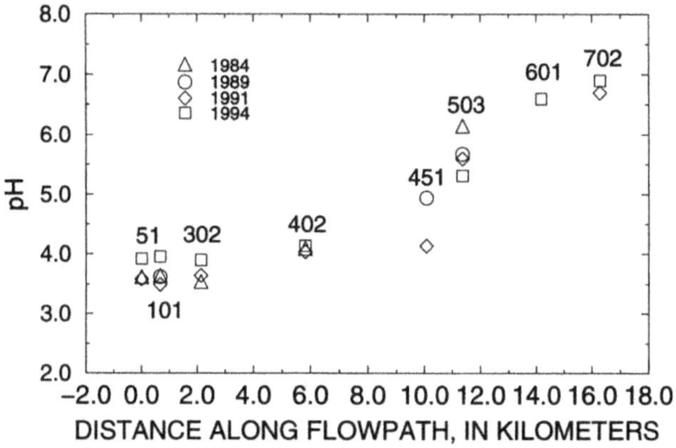

Figure 8.6. *pH* profiles along an aquifer flowpath based on the most contaminated wells in the Pinal Creek alluvial aquifer. Well 451 was drilled in December 1988 and destroyed during the floods of spring 1993. Therefore only January 1989 and November 1991 *pH* data are presented for that well. Water samples taken from 451 show a large *pH* decrease between those dates (most of the decrease occurred between March and August 1989).

The identity of the secondary aluminum phase that precipitates and the rate of redissolution of that phase are also important to consider. Table 8.6 shows that the most acidic waters (below *pH* 5.0) can only be obtained if $Al(OH)_3$ is not present: i.e., it either never precipitates out, or if it does, it has completely redissolved, or its rate of dissolution is so slow that it can be considered “unreactive”. If we assume that secondary $Al(OH)_3$ does precipitate out and that the Local Equilibrium Assumption controls its precipitation and later dissolution, then our simulations of the $Al(OH)_3$ dissolution front show (Table 8.9) that initial carbonate concentrations would have to be lower than 3.0×10^{-2} mol kg^{-1} H_2O to match the retardation of the low-*pH* front observed in the field and in Stollenwerk’s laboratory columns.

As of 1996, there had been no direct observation of the breakthrough of the high-Fe(II) waters at the Pinal Creek site. Waters from wells 451 and the LPC site (near well 503) already had high Fe(II) concentrations during the emplacement of those wells (in November 1991 and February 1995 respectively). The fact that dissolved Fe(II) was present at well 451 before the low-*pH* breakthrough suggests that the Fe(II) front moves faster than the low-*pH* front, or at least precedes the low-*pH* front at that point in time. If we use the creation of Webster Lake as an initial condition, retardation factors between 5 and 8.4 can be estimated for the high Fe(II) front as it moved between Webster Lake and well 451. Similarly, the evidence based on data from well 451 and from the LPC wells suggests a maximum retardation factor of 7.7 over that section of the aquifer. The latter factor is equivalent to a maximum retardation factor of about 3.2 in our simulations, after correcting for the three times higher Fe(II) concentration of well 402 relative to that of well 451 (in August 1989).

In his laboratory column experiments, Stollenwerk (1994) observed retardation factors of approximately 2 for the high-Mn(II) spike, and 2.5 for the high-Fe(II) front. The lag in the Fe(II) front suggests that the rate of $Fe(OH)_3$ precipitation was delayed relative to the rate of MnO_2 dissolution. Normalizing those retardation factors to an infilling solution with 4.7 times less dissolved Fe(II) (for well 402) results in an equivalent retardation factor of between 5.7 and 8.1.

After normalization to the infilling water used in our simulations, the field and laboratory determined retardation factors for the high-Fe(II) front encompass a range from 3.2 to 8.1. This range compares favorably with the retardation factor of 4.7 determined for most of the simulations. Furthermore, according to field and laboratory observations the MnO_2

dissolution front moves slightly faster than the low-*pH* front, but it is not moving much faster, and therefore the *CMR* ratio is probably close to 1.5.

The *pH* values determined for the various *pH*-plateaus in our simulations (Table 8.6, Figs. 8.4 and 8.5) are within the range of the values observed in the field (Fig. 8.6). Therefore the hypothesis that the complete dissolution of secondary rhodochrosite occurs after the dissolution of calcite and is accompanied by a significant decrease in *pH* (in the absence of $Al(OH)_3$) is consistent with the field evidence.

8.6 CONCLUSIONS

The basic theory and assumptions of inverse geochemical modeling were presented. Although much less commonly used than reactive-transport modeling, particularly in investigations of groundwater contamination, inverse geochemical modeling provides a powerful tool to identify the reactions that affect the chemical and isotopic evolution of contaminated groundwaters. The reaction models identified by inverse modeling can be incorporated into a geochemical reactive-transport model to investigate the effects of the various reactions on the transport of reactive contaminants. This approach was demonstrated for the case of acidic groundwater with heavy-metal contamination in the Pinal Creek basin, Arizona.

The interactive inverse geochemical modeling code NETPATH was used to construct a series of inverse models that quantified observed differences in chemical composition between an initial acidic Fe(II)-rich water and an evolutionary, partially-neutralized Fe(II)-poor water, according to a postulated set of reactions. Each inverse model was evaluated according to existing knowledge of the geochemistry of the aquifer and the thermodynamic and kinetic feasibility of the model reactions. Once a plausible set of inverse models was identified, the PHREEQC inverse modeling code was used to further evaluate the set of inverse models. Unlike NETPATH, PHREEQC accounts for uncertainties in the analytical data and, additionally, includes alkalinity-balance, charge-balance and water-balance constraints in solving inverse models.

Inverse modeling with NETPATH and with PHREEQC quantified the reaction processes responsible for the evolution of an acidic Fe(II)-rich groundwater into a partially neutralized Fe(II)-poor water at the Pinal Creek site. The principal reaction processes appear to be the reductive dissolution of solid MnO_2 by aqueous Fe(II), the consequent precipitation of $Fe(OH)_3$, the dissolution of calcite and/or dolomite, the precipitation of an aluminum phase, probably $Al(OH)_3$(rather than $AlOHSO_4$), and, possibly, the precipitation of chalcedony. Results of the inverse modeling simulations also led to the following conclusions:

1. Ca- and Mg-silicate dissolution must be an important process during the neutralization of the low-*pH* waters, given that CO_2 exsolution probably does not occur.
2. Dilution of the acidic groundwaters occurs, and Cl, Na, and SO_4 can be considered conservative constituents that quantify the extent of the dilution. The PHREEQC models revealed that SO_4 could be considered a conservative constituent given its associated analytical uncertainty.
3. Rhodochrosite precipitation (or possibly an MnO_2-$Mn(OH)_3$ electron-transfer or Mn^{2+} sorption mechanism) is responsible for the lower than expected increase in dissolved Mn(II) concentrations caused by the aqueous Fe(II) reduction of MnO_2 solids.

After a brief review of previous reactive-transport modeling at the Pinal Creek site, the results of some new forward simulations were presented. These geochemical transport simulations explored the effects of reactions—identified by inverse modeling and cited by other researchers—on the chemical evolution of an acidic Fe(II) rich water from the Pinal Creek site. The purpose of the PHREEQC and PHREEQM transport simulations was to determine the effect of the reactions on the relative rates of movement of the Fe(II)-rich and low-*pH* groundwaters.

The only factors affecting the retardation of the Fe(II)-rich waters and the propagation of the MnO_2 dissolution front in our simulations were the initial specified concentration of MnO_2 and the concentration of Fe(II) in the inflowing contaminated water. As demonstrated in the text, the rate of movement of the MnO_2 front could be easily calculated by hand, but only in the absence of ion-exchange or other processes substantially affecting the Fe(II) concentration.

The propagation of various *pH* fronts caused by complete dissolution of carbonate and aluminum minerals, however, could not have been easily predicted. The use of a geochemical transport code such as PHREEQM or PHREEQC is necessary to determine the retardation factors applicable to interacting mineral dissolution fronts. The initial amount of carbonate (calcite and/or dolomite) in each simulation was the primary factor determining the movement of the low-*pH* waters. Other important factors were the ratio of initial carbonate to initial MnO_2, the type of Al phase allowed to precipitate, and whether or not rhodochrosite was allowed to precipitate. High initial carbonate to MnO_2 ratios, allowing rhodochrosite precipitation, and allowing $AlOHSO_4$ precipitation (instead of $Al(OH)_3$ precipitation), resulted in higher retardation factors for the movement of the low-*pH* waters. [As mentioned earlier though, Bigham and Nordstrom's (2000) review of environmental conditions required for the occurrence of $AlOHSO_4$ phases reinforces Glynn and Brown's (1996) view that an $Al(OH)_3$ phase was probably a more reasonable phase to consider in modeling the evolution of Pinal Creek groundwaters.] Allowing the irreversible dissolution of Ca- and Mg-silicates, so as to match the mass transfers determined in a few inverse models, resulted in unrealistically high retardation factors for the rhodochrosite and $Al(OH)_3$ dissolution fronts, although the retardation factors for the calcite dissolution front were reasonable. More field and laboratory information is required on the rates and extents of reaction for these silicate minerals to further constrain reactive-transportsimulations.

Inclusion of ion-exchange processes did not substantially affect the retardation of the various *pH* fronts when a reasonable low (1 meq/100 g) *CEC* was used. Because of surface-protonation, allowing surface-complexation reactions resulted in higher retardation factors for the carbonate and $AlOHSO_4$ dissolution fronts. Allowing equilibrium with a p_{CO2} of $10^{-0.9865}$ resulted in lower retardation factors for the carbonate and $Al(OH)_3$ dissolution fronts, particularly at initial carbonate concentrations greater than 3×10^{-2} mol kg^{-1} H_2O. Allowing equilibrium with a lower p_{CO_2} of $10^{-1.33}$ resulted in higher retardation factors for the carbonate and $Al(OH)_3$ dissolution fronts. Simulation of longitudinal dispersion was not an important factor controlling the movement of the calcite and rhodochrosite dissolution fronts except at low initial carbonate concentrations. Longitudinal dispersion also did not affect the simulated rate of movement of the Fe(II)-rich waters. Including dolomite in addition to calcite in the background aquifer and allowing kaolinite to precipitate instead of $Al(OH)_3$ did not significantly affect the propagation of the low-*pH* fronts associated with the dissolution of calcite and rhodochrosite.

A comparison of the retardation factors for the low-*pH* and high-Fe(II) fronts determined in our local equilibrium simulations with retardation factors estimated from field evidence and Stollenwerk's (1994) laboratory column elution tests suggests that an initial carbonate concentration between 2.1×10^{-2} and 4.2×10^{-2} mol kg^{-1} H_2O and the initial MnO_2 concentration of 2×10^{-2} mol kg^{-1} H_2O used in our simulations are reasonable. The *pH* values obtained in our local equilibrium simulations are reasonable given the *pH* values observed in the field.

The retardation factors determined for the mineral dissolution fronts in various simulations are useful not only in estimating the rate of movement of the low-*pH* and high-Fe(II) groundwaters at the Pinal Creek site, but also in identifying the most important chemical parameters controlling the movement of these contaminated waters. Our simulation results may also provide information on the rate of movement of acidic, metal-laden waters at other groundwater contamination sites, and our study illustrates how to adjust for *pH*, dissolved metal (Al, Fe, Mn) concentrations, and mineralogical characteristics of a given site.

This paper presents and demonstrates an approach for the investigation of the evolution and movement of contaminated groundwaters: the use of inverse geochemical modeling to identify important reaction processes, followed by geochemical reactive-transport simulations that incorporate the inverse-model-derived reactions and account for a range of aquifer characteristics. Such an approach results in an improved understanding of the processes that control the past and future evolution of contaminated groundwaters. This information may then lead to better predictions for the transport of reactive contaminants, which, in turn, may result in more effective remediation of groundwater contamination at sites with sparse spatial information.

8.7 THE SENIOR AUTHOR'S FIFTEEN YEAR PERSPECTIVE ON THE GLYNN AND BROWN (1996) PAPER

The geochemical modeling conducted by Glynn and Brown (1996), while confirming and refining knowledge about chemical reactions in the Pinal Creek groundwater system, also highlighted significant knowledge gaps. The Glynn and Brown (1996) study provided a basis for further groundwater geochemical investigations at the Pinal Creek site, including several reactive-transport modeling studies. For example, Brown *et al.* (1998) modified and refined the simulation of groundwater evolution presented in Brown (1996) for 1-D flow (with lateral dilutions as needed) from well 51 to well 701 from 1984 to 1994. Some of the reactions were changed in the modified model as well as the spatial distribution of those reactions. For example, rhodochrosite was allowed to precipitate and CO_2 was allowed to equilibrate with a p_{CO_2} of $10^{-1.33}$ (typical of the unsaturated zone) in the "neutralized" zone downgradient from well 503. Sensitivity analyses examined, for example, surface complexation and sorption of metals on iron oxyhydroxide. Sorption was relatively less important than initially considered by the authors in controlling *pH* and dissolved metal concentrations relative to other processes (such as redox processes, carbonate dissolution, and dilution). Longitudinal hydrodynamic dispersion was found to be relatively unimportant (a beta version of PHREEQC v.2 was used). Generally, good fits were obtained for the simulation of heavy metals such as Cu, Zn, and Ni. However, the simulation of *pH* and dissolved Fe and Mn still proved problematic, and a simulation of the carbonate and redox reaction kinetics was still needed.

Brown *et al.* (2000), again improving on Brown (1996), found that HYDROGEOCHEM and PHREEQC simulations of the Stollenwerk (1994) column experiments provided similar numerical results, but once again, required accounting for reaction kinetics, if the experimental results were to be matched by the numerical simulations. In the absence of adequate knowledge of reaction rates, the observed experimental results could also have been matched by fitting irreversible mass transfers of water and reactants in PHREEQC. A common result of the Brown *et al.* (2000, 1998) studies was that the numerical simulation results were reasonable only if the number of adsorption sites was reduced to 5% of the initial value considered by Stollenwerk (1994). Stollenwerk simulated, through the use of the geochemical code MINTEQA2, the neutralization and later rinse-out of Pinal Creek acidic groundwater that he observed in a column experiment. Because MINTEQA2 was not a reactive transport code, Stollenwerk used a series of "batch" calculations to simulate transport, and explicitly defined the aqueous and solid-phase chemical inputs for each cell at each time step. As mentioned in Brown *et al.* (2000), "Although output from one time step provided much of the input for the next time step, the aqueous and solid-phase concentrations of selected constituents were manually adjusted to fit measured concentrations in column outflow and measured changes in column mineralogy".

To improve the characterisation of calcite, dolomite, and pyrolusite (MnO_2) dissolution rates at the Pinal Creek site, Brown and Glynn (2003) conducted *in-situ* field experiments that suspended well-characterized commercially-obtained mineral samples in acidic waters at the Pinal Creek site (wells 101 and 301) for various lengths of time (from 96 to 595 days) starting in April 1998. *In-situ* dissolution rates for calcite and dolomite were determined, and found to be about three orders of magnitude slower than rates determined by Plummer *et al.* (1978) and Busenberg and Plummer (1982) in laboratory experiments. Interestingly, Brown and Glynn (2003) found that the incorporation of the *in-situ* rates, or alternatively, the faster laboratory-determined rates, in PHREEQC v.2 (Parkhurst and Appelo, 1999) reactive-transport simulations did not result in better fits of Stollenwerk's column experiments. Instead, an equilibrium model adequately simulated carbonate dissolution at the shorter and faster (relative to the field situation) spatial and temporal scales of the column experiment. More surprisingly, Brown and Glynn (2003) observed growth in the mass of the pyrolusite material placed in the *in-situ* experiments, despite thermodynamic conditions that, in theory, favored pyrolusite dissolution. They postulated that this result could be related to the result of Villinski *et al.* (2001), who found that the dissolution of MnO_2 under similar conditions in a laboratory flow-through cell resulted in the precipitation of a mixed oxidation Mn-Fe phase with the structure of jacobsite ($MnFe_2O_4$). Brown and Glynn (2003) then used the rate equations developed by Postma and Appelo (2000) for the reductive dissolution of birnessite (MnO_2) by dissolved Fe(II) to better fit the distribution and evolution of dissolved Mn and Fe concentrations in the transition zone and in the acidic part of the plume (at well 402) from 1984 through 1992. Although the rate constant used in their PHREEQC v.2 simulations was about 3 to 4 orders of magnitude smaller than the range of values used by Postma and Appelo (2000) in their column experiments, Brown and Glynn (2003) did obtain a better fit of the observed Fe, and especially, the observed Mn concentrations in the transition zone. Their simulated *pH* values, however, remained about 1 *pH* unit higher than the observed values.

The field, laboratory and numerical modeling investigations of the Pinal Creek basin provide several lessons that may be useful to other investigations of complex, highly dynamic, systems with a relative paucity of information (and limited funding to obtain information):

1. Constructing, analyzing and interpreting numerical models, regardless of the type of model (hydrologic *vs.* geochemical; inverse vs. forward), forces the modeler(s), and hopefully the user(s) of the models, to reexamine and revise their conceptual model and perceptions of the available information. The modeling process forces the modelers and users to assemble, structure, transform, and assess a wide variety of information.
2. No model is ever final. As numerical modeling leads to greater understanding of a system, it invariably illuminates knowledge gaps and the need for more information. Frequent iteration is needed among field observation, field experiments, laboratory experiments, and the periodic assembly and interpretation of the available information through numerical modeling.
3. The combination of inverse modeling (including inverse geochemical modeling) and forward or "predictive" modeling is particularly powerful in helping assess available information. Inverse modeling forces the modeler to use basic principles (such as conservation of mass and energy) to *interpret* the available observations in terms of properties of the system (e.g., geochemical mass transfers, possible reactions, other intrinsic system properties). Forward, or predictive modeling, forces the modeler to *test* available information by "extrapolating" forward through time, space or other informational dimensions by using the modeler's preconceptions of system behavior. In the case of the Pinal Creek system, inverse modeling helped Glynn and Brown (1996) identify sets of reactions that could explain the observed geochemical evolution of groundwaters, and helped identify field and laboratory observations/experiments that might improve characterization and quantification of active geochemical processes. Sensitivity analyses conducted through "forward" reactive transport modeling identified which minerals and reaction processes, given knowledge of reaction thermodynamics and stoichiometries, provided reasonable rates of migration of chemical fronts and a reasonable distribution of solution properties (such as *pH* and Fe and Mn concentrations) in space and time. More importantly, the geochemical modeling at the Pinal Creek site helped identify which reactions and processes were *unreasonable*.

4. Determining what is *unreasonable* is generally dependent on limited information and observations (through time, space and process domains) and on an *interpretation* of that limited information. In complex, highly dynamic, systems with limited regularity or steadiness of observations, such as the Pinal Creek groundwater system, the problem of eliminating unreasonable processes is exacerbated: what might seem reasonable, or unreasonable, given available information at some point in time and space could change. The studies conducted at the Pinal Creek site illustrate the fact that nature always keeps surprises in reserve for its observers and interpreters. Humility, and frequent testing of assumptions, are needed in modeling nature's systems.
5. Given our often limited knowledge of natural systems, it behooves us to model these systems by considering *general* system behavior before interpreting, matching, and predicting *specific* system behavior. Curve and point-by-point matching of numerical simulations to the observed chemical evolution at the Pinal Creek site, or laboratory columns, was generally difficult, and provided limited, albeit useful, returns on expended effort. Inverse modeling and sensitivity analyses of general system behavior, however, through relatively unconstrained reactive-transport simulations, provided substantial insights on knowledge gaps and information needs that could improve understanding of the complex, highly dynamic processes at the site. These modeling efforts should have been conducted much earlier in the life of the Pinal Creek project investigations.
6. Finally, the Pinal Creek project taught us the value of spatial and temporal observations. Systems that have a complex suite of active processes and that exhibit dynamic changes in observed properties require extensive (and often costly) adaptive monitoring programs that have sufficient resolution in both time and space. Considering a greater variety of geochemical/hydrological information than might initially be considered in setting up a monitoring program is useful and can sometimes help compensate for a lack of spatial/temporal monitoring: different geochemical or hydrological measurements often reflect different degrees of spatial integration and temporal evolution.

Other information on the hydrogeochemistry of acidic metal-contaminated waters, geochemical modeling, and on obtaining U.S. Geological Survey computer codes and the PHREEQM code
Further information on the application of geochemical information and geochemical modeling in understanding and dating groundwater systems can be found in Glynn and Plummer (2005), and in a soon to be published IAEA guidebook on dating of old groundwaters and many chapters therein, including Plummer and Glynn (Radiocarbon dating in groundwater systems; in press) and Plummer *et al.* (Characterization and conceptualization of groundwater-flow systems; in press). An excellent discussion of geochemical modeling tools and philosophy is also provided by Nordstrom (2004; updated in 2007). Finally, the reader is directed to the work of Alpers and Nordstrom (1999) and Nordstrom (2000, 2008). These authors have used geochemical modeling and characterization to inform regulatory decision making in the remediation of acid mine waters.

Readers interested in geochemical and hydrological investigations of the acidic and heavy metal contamination in the Pinal Creek basin should also consider the studies conducted on the role of groundwater-surface water interactions. Narrowing of the basin to the north forces groundwater discharge and maintains perennial flow for 7 kilometers along Pinal Creek. Degassing of carbon dioxide raises the *pH* of the stream from about 6.6 to 7.6 (Choi *et al.*, 1998). The return flow of the streamwater to shallow subsurface flowpaths beneath and to the side of the stream (hyporheic zone) brings this higher *pH* water with high concentrations of dissolved Mn^{2+} back into direct contact with microbes and geochemical coatings in the streambed sediment. Rapid oxidation of Mn^{2+} in the hyporheic zone sediments is catalyzed by Mn oxidizing bacteria on a timescale of hours, and is accompanied by the sorption of Co, Ni, and Zn onto the biogenic manganese oxides (Bargar *et al.*, 2009). Enhanced oxidation of Mn in the hyporheic zone was found to remove 20% of the dissolved load of Mn reaching the stream, before its flow out of the basin (Harvey and Fuller, 1998). Additionally, sorption onto the manganese oxides in the streambed sediments accounted for 52%, 27%, and 24% removal of the dissolved Co, Ni, and Zn loads, respectively (Fuller and Harvey, 2000; Kay *et al.,* 2001).

The latest versions of the USGS PHREEQC and NETPATH geochemical codes, and other related codes, can be downloaded directly from the web site: http://wwwbrr. cr.usgs.gov/projects/GWC_coupled/. Codes available include new versions of NETPATH (NetpathXL; Parkhurst and Charlton, 2008; NETPATH-WIN, El-Kadi *et al.*, 2011) and the 3-dimensional PHREEQC-based reactive transport code PHAST (Parkhurst *et al.*, 2010). USGS codes can also be accessed through the more general USGS National Research Program software web site: http://water.usgs.gov/nrp/models.html. PHREEQM is now largely an obsolete code that has been replaced by PHREEQC version 2 (Parkhurst and Appelo, 1999) and by the Windows interactive code PHREEQCi (Charlton and Parkhurst, 1999). Another interactive code, PHREEQC for Windows, can also be downloaded from Vincent Post's web site: http://pfw.antipodes.nl/index.html. Additional information on the PHREEQC code can also be obtained from Tony Appelo's web site: http://www.xs4all.nl/~appt/, and from the Appelo and Postma (2005) groundwater geochemistry textbook.

ACKNOWLEDGEMENTS

The original version of this manuscript, Glynn and Brown (1996), received substantial help and reviews from David Parkhurst, Joseph Vrabel, Carl Steefel, Eric Oelkers, Don Thorstenson and Paul Ribbe. Additionally, the current manuscript benefitted greatly from reviews provided by Mark Fuhrman and David Parkhurst, and from additional comments, edits, and discussion provided by Leonard Konikow, Kirk Nordstrom and Kevin Breen. The Senior Author retains full responsibility however for any remaining errors of fact, expression or judgment. The work presented here was funded by the Toxics Substance Hydrology and the Hydrologic Research and Development programs of the US Geological Survey. Finally, any use of trade, product, or firm names is for descriptive purposes only and does not imply endorsement by the U.S. Government.

REFERENCES

Alpers, C.N. & Nordstrom, D.K.: Geochemical modelling of water-rock interactions in mining environments. In: G.S. Plumlee & M.J. Logsdon (eds): Reviews in Economic Geology, vol. 6A, *The environmental geochemistry of mineral deposits: Part A. Processes, methods and health issues*, Society of Economic Geologists, Littleton, CO, 1999, pp. 289–324.

Appelo, C.A.J.: Multicomponent ion exchange and chromatography in natural systems. In: C.I. Steefel, P. Lichtner & E. Oelkers (eds): Reactive transport in porous media: general principles and application to geochemical processes. *Reviews in Mineralogy* 34, 1996, pp. 193–227.

Appelo, C.A.J. & Postma, D.: *Geochemistry, groundwater and pollution*. A.A. Balkema, Rotterdam, The Netherlands, 1993.

Appelo, C.A.J. & Postma, D.: *Geochemistry, groundwater and pollution*. 2nd ed., A.A. Balkema, Leiden, The Netherlands, 2005.

Appelo, C.A.J. & Willemsen, A.: Geochemical calculations and observations on salt water intrusions, 1, a combined geochemical/mixing cell model. *J. Hydrology* 94 (1987), pp. 313–330.

Bargar, J.R., Fuller, C.C., Marcus, M.A., Brearley, A.J., Perez De la Rosa, M., Webb, S.M. & Caldwell, W.A.: Structural characterization of terrestrial bacteriogenic Mn oxides from Pinal Creek, AZ. *Geochim. Cosmochim. Acta* 73 (2009), pp. 889–910.

Bigham, J.M. & Nordstrom, D.K.: Iron and aluminum hydroxysulfate minerals. In: C.N. Alpers, J.L. Jambor & D.K. Nordstrom (eds): *Sulfate minerals—crystallography, geochemistry and environmental significance, Reviews in Mineralogy and Geochemistry* 40, 2000, pp. 351–403.

Breeuwsma, A., Wösten, J.H.M., Vleeshouwer, J.J., Van Slobbe, A.M. & Bouma, J.: Derivation of land qualities to assess environmental problems from soil surveys. *Soil Sci. Soc. Amer. J.* 50 (1986), pp. 186–190.

Brown, J.G.: Movement *of metal contaminants in ground water in the Pinal Creek Basin, Arizona: Model assessment and simulation of reactive transport*. MSc Thesis, University of Arizona, Tucson, AZ, USA, 1996.

Brown, J.G. & Glynn, P.D.: Kinetic dissolution of carbonates and Mn oxides in acidic water: measurement of *in situ* rates and reactive transport modeling. *Appl. Geochem.* 18:8 (2003), pp. 1225–1239.

Brown, J.G. & Harvey, J.W.: Hydrologic and geochemical factors affecting metal-contaminant transport in Pinal Creek Basin near Globe, Arizona. In: D.W. Morganwalp & D.A. Aronson (eds): *U.S. Geological Survey Toxic Substances Hydrology Program—Proceedings of the Technical Meeting*, Colorado Springs, Colorado, 20–24 September, 1993: *U.S. Geological Survey Water-Resources Investigations Report* 94–4015, 1994, pp. 1035–1042.

Brown, J.G., Bassett, R.L., Glynn, P.D. & Parkhurst, D.L.: Reactive transport of metal contaminants in ground water in Pinal Creek Basin, Arizona. In: A.M. Geddis (ed.): Water Use in Arizona—Cooperation or Conflict. *Proceedings of the 8th Annual Symposium of the Arizona Hydrological Society*, Tucson, Arizona, 14–15 September, 1995, pp. 104–105.

Brown, J.G., Bassett, R. & Glynn, P.D.: Analysis and simulation of reactive transport of metal contaminants in ground water in Pinal Creek Basin, Arizona. *J. Hydrology* 209 (1998), pp. 225–250.

Brown, J.G., Bassett, R. & Glynn, P.D.: Reactive transport of metal contaminants in alluvium—model comparison and column simulation. *Appl. Geochem.* 15 (2000), pp. 35–49.

Busenberg, E. & Plummer, L.N.: The kinetics of dissolution of dolomite in CO_2-water systems at 1.5 to 65°C and 0.0 to 1.0 atmosphere pCO_2. *Amer. J. Science* 282 (1982), pp. 45–78.

Charlton, S.R., Macklin, C.L. & Parkhurst, D.L.: PhreeqcI—A graphical user interface for the geochemical computer program PHREEQC. *U.S. Geological Survey Water-Resources Investigations Report* 97–4222, 1997.

Charlton, S.R. & Parkhurst, D.L.: PhreeqcI—A graphical user interface to the geochemical model PHREEQC. *U.S. Geological Survey Fact Sheet* FS-031–02, 2002.

Choi, J., Hulseapple, S.M., Conklin, M.H. & Harvey, J.W.: Modeling CO_2 degassing and pH in a stream-aquifer system. *J. Hydrology* 209 (1998), pp. 297–310.

Davis, J.A. & Kent, D.B.: Surface complexation modeling in aqueous geochemistry. In: M.F. Hochella Jr & A.F. White (eds): Mineral-Water Interface Geochemistry. *Reviews in Mineralogy* 23, 1990, pp. 177–260.

Dria, M.A., Bryant, S.L., Schechter, R.S. & Lake, L.W.: Interacting precipitation dissolution waves: the movement of inorganic contaminants in groundwater. *Water Resour. Res.* 23 (1987), pp. 2076–2090.

Dzombak, D.A. & Morel, F.M.M.: *Surface complexation modeling: Hydrous ferric oxide*. Wiley & Sons. New York, 1990.

El-Kadi, A.I., Plummer, L.N. & Aggarwal, P.: NETPATH-WIN: An interactive user version of the mass-balance model, NETPATH. *Ground Water* 49:4 (2011), pp. 593–599.

Engesgaard, P. & Kipp, K.L.: A geochemical transport model for redox-controlled movement of mineral fronts in groundwater flow systems: a case of nitrate removal by oxidation of pyrite. *Water Resour. Res.* 28 (1992), pp. 2829–2843.

Eychaner, J.H.: Movement of inorganic contaminants in acidic water near Globe, Arizona. In: G.E. Mallard & S.E. Ragone (eds):*U.S. Geological Survey Toxic Substances Hydrology Program—Proceedings of the Technical Meeting*, Phoenix, Arizona, 26–30 September, 1989: *U.S. Geological Survey Water-Resources Investigations Report* 89–4220, 1989, pp. 567–575.

Eychaner, J.H.: The Globe, Arizona, research site—Contaminants related to copper mining in a hydrologically integrated environment. In: G.E. Mallard & D.A. Aronson (eds): *U.S. Geological Survey Toxic Substances Hydrology Program—Proceedings of the Technical Meeting*, Monterey, California, 11–15 March, 1991: *U.S. Geological Survey Water-Resources Investigations Report* 91–4034, 1991, pp. 475–480.

Eychaner, J.H. & Stollenwerk, K.G.: Neutralization of acidic ground water near Globe, Arizona. In:K.D. Schmidt (ed):*Groundwater contamination and reclamation.* Proceedings of a Symposium, Tucson, Arizona, 14–15 August, 1985, Bethesda, Maryland. American Water Resources Association, 1985, pp. 141–148.

Ficklin, W.H., Love, A.S. & Papp, C.S.E.: Solid-phase variations in an aquifer as the aqueous solution changes, Globe, Arizona.In: G.E. Mallard & D.A. Aronson (eds):*U.S. Geological Survey Toxic Substances Hydrology Program—Proceedings of the Technical Meeting*, Monterey, California, 11–15 March, 1991: *U.S. Geological Survey Water-Resources Investigations Report* 91–4034, 1991, pp. 475–480.

Fuller, C.C. & Harvey, J.W.: Reactive uptake of trace metals in the hyporheic zone of a mining-contaminated stream, Pinal Creek, Arizona.*Environ. Sci. Technol.* 34 (2000), pp. 1150–1156.

Glynn, P.D.: MBSSAS: A code for the computation of Margules parameters and equilibrium relations in binary solid-solution aqueous-solution systems.*Comput. Geosci.* 17 (1991a), pp. 907–966.

Glynn, P.D.: Effect of impurities in gypsum on contaminant transport at Pinal Creek, Arizona. In: G.E. Mallard & D.A. Aronson (eds):*U.S. Geological Survey Toxic Substances Hydrology Program—Proceedings of the Technical Meeting*, Monterey, California, 11–15 March, 1991: *U.S. Geological Survey Water-Resources Investigations Report* 91–4034, 1991b, pp. 466–474.

Glynn, P.D.: Solid-solution solubilities and thermodynamics: sulfates, carbonate and halides.In: C.N. Alpers, J.L. Jambor & D.K. Nordstrom (eds): Sulfate minerals-crystallography, geochemistry and environmental significance. *Reviews in Mineralogy and Geochemistry* 40, 2000, pp. 481–511.

Glynn, P.D. & Brown, J.G.: Reactive transport modeling of acidic metal-contaminated ground water at a site with sparse spatial information. In: C.I. Steefel, P. Lichtner & E. Oelkers (eds): Reactive transport in porous media: general principles and application to geochemical processes. *Reviews in Mineralogy* 34, 1996, pp. 377–438.

Glynn, P.D. & Busenberg, E.: Dissolved gas and chlorofluorocarbon content of ground waters in the Pinal Creek Basin, Arizona.In: D.W. Morganwalp & D.A. Aronson (eds): *U.S. Geological Survey Toxic Substances Hydrology Program—Proceedings of the Technical Meeting*. Colorado Springs, CO, 20–24 September, 1993: *U.S. Geological Survey Water-Resources Investigations Report* 94–4015, 1994a, pp. 1043–1054.

Glynn, P.D. & Busenberg, E.: Unsaturated zone diffusion of carbon dioxide and oxygen in the Pinal Creek Basin, Arizona. In: D.W. Morganwalp & D.A. Aronson (eds): *U.S. Geological Survey Toxic Substances Hydrology Program—Proceedings of the Technical Meeting*. Colorado Springs, CO, 20–24 September, 1993: *U.S. Geological Survey Water-Resources Investigations Report* 94–4015, 1994b, pp. 1055–1064.

Glynn, P.D. & Plummer, L.N.: Geochemistry and the understanding of ground-water systems. *Hydrogeology J.* 13 (2005), pp. 263–287.

Glynn, P.D., Engesgaard, P. & Kipp, K.L.: Use and limitations of two computer codes for simulating geochemical mass transport at the Pinal Creek toxic-waste site. In: G.E. Mallard & D.A. Aronson (eds). *U.S. Geological Survey Toxic Substances Hydrology Program—Proceedings of the Technical Meeting*, Monterey, California, 11–15 March, 1991: *U.S. Geological Survey Water-Resources Investigations Report* 91–4034, 1991, pp. 454–460.

Glynn, P.D., Busenberg, E. & Brown, J.G.: Use of chlorofluorocarbons, dissolved gases and water isotopes to characterize ground-water recharge in an aquifer contaminated by acidic, metal-laden wastewater. *U.S. Geological Survey Toxic Substances Hydrology Program—Proceedings of the Technical Meeting*, Charleston, South Carolina, 8–12 March, 1999—Volume 1 of 3—Contamination from Hardrock Mining: *U.S. Geological Survey Water-Resources Investigations Report* 99–4018A, 1999, pp. 155–162.

Harvey, J.W. & Fuller, C.C.: Effect of enhanced manganese oxidation in the hyporheic zone on basin-scale geochemical mass balance. *Water Resour. Res.* 34:4 (1998), pp. 623–636.

Herzer, J. & Kinzelbach, W.: Coupling of transport and chemical processes in numerical transport models. *Geoderma* 44 (1989), pp. 115–127.

Hydro Geo Chem Inc.: Investigation of acid water contamination along Miami Wash and Pinal Creek, Gila County, Arizona. Cyprus Miami Mining Corporation, Claypool, AZ, 1989.

Kay, J.T., Conklin, M.H., Fuller, C.C. & O'Day, P.E.: Processes of nickel and cobalt uptake by Mn-oxide forming sediment in Pinal Creek.*Environ. Sci. Technol.* 35: 24 (2001), pp. 4719–4725.

LeBlanc, D.R.: Sewage plume in a sand and gravel aquifer, Cape Cod, Massachusetts. *U.S. Geological Survey Water-Supply Paper* 2218, 1984.

Lichtner, P.C.: Continuum model for simultaneous chemical reactions and mass transport in hydrothermal systems. *Geochim. Cosmochim. Acta* 49 (1985), pp. 779–800.

Lichtner, P.C.: The quasi-stationary state approximation to coupled mass transport and fluid-rock interaction in a porous medium. *Geochim. Cosmochim. Acta* 52 (1988), pp. 143–165.

Lind, C.J. & Stollenwerk, K.G.: Alteration of alluvium of Pinal Creek, Arizona by acidic ground water resulting from copper mining. In: D.W. Morganwalp & D.A. Aronson (eds): *U.S. Geological Survey Toxic Substances Hydrology Program—Proceedings of the Technical Meeting*. Colorado Springs, CO, 20–24 September, 1993: *U.S. Geological Survey Water-Resources Investigations Report* 94-4015, 1994, pp. 1089–1094.

Mackay, O.M., Freyberg D.L., Roberts, P.V. & Cheny, J.A.: A natural gradient experiment on solute transport in a sand aquifer I: Approach and overview of plume movement. *Water Resour. Res.* 22 (1986), pp. 2017–2029.

Neaville, C.C. & Brown, J.G.: Hydrogeology and hydrologic system of Pinal Creek Basin, Gila County, Arizona. *U.S. Geological Survey Water-Resources Investigations Report* 93–4212, 1993.

Nordstrom, D.K.: Advances in the hydrogeochemistry and microbiology of acid mine waters. *Int. Geology Review* 42 (2000), pp. 499–515.

Nordstrom, D.K.: Modeling low-temperature geochemical processes. In: J.I. Drever (ed): Treatise on geochemistry, v. 5, *Surface and ground water, weathering, and soils*. Amsterdam, Elsevier, 2004 (updated in 2007), pp. 37–72.

Nordstrom, D.K.: Questa baseline and pre-mining ground-water quality investigation. 25. Summary of results and baseline and pre-mining ground-water geochemistry, Red River Valley, Taos County, New Mexico, 2001–2005. *U.S. Geological Survey Professional Paper* 1728, 2008.

Nordstrom, D.K., Plummer, L.N., Langmuir, D., Busenberg, E., May, H.M., Jones, B.F. & Parkhurst, D.L.: Revised chemical equilibrium data for major water-mineral reactions and their limitations. In: D.C. Melchior & R.L. Bassett (eds): *Chemical modeling in aqueous system II*, American Chemical Society Symposium Series 416, 1990, pp. 398–413.

Ortoleva, P., Auchmuty, G., Chadam, J., Hettmer, J., Merino, E., Moore, C.H. & Ripley, E.: Redox front propagation and banding molalities. *Physica* 19D (1986), pp. 334–354.

Parkhurst, D.L.: User's guide to PHREEQC--A computer program for speciation, reaction-path, advective-transport, and inverse geochemical calculations. *U.S. Geological Survey Water-Resources Investigations Report* 95–4227, 1995.

Parkhurst, D.L.: Geochemical mole-balance modeling with uncertain data. *Water Resour. Res.* 33:8 (1997), pp. 1957–1970.

Parkhurst, D.L. & Appelo, C.A.J.: User's guide to PHREEQC (Version2)—A computer program forspeciation, batch-reaction, one-dimensional transport, and inverse geochemical calculations. *U.S. Geological Survey Water-Resources Investigations Report* 99-4259, 1999.

Parkhurst, D.L. & Charlton, S.R.: NetpathXL-An Excel® interface to the program NETPATH. *U.S. Geological Survey Techniques and Methods* 6-A26, 2008.

Parkhurst, D.L., Plummer, L.N. & Thorstenson, D.C.: PHREEQE—A computer program for geochemical calculations. *U.S. Geological Survey Water-Resources Investigations Report* 80–96, 1980.

Parkhurst, D.L., Plummer, L.N. & Thorstenson, D.C.: BALANCE-A computer program for geochemical calculations. *U.S. Geological Survey Water-Resources Investigations Report* 82-14, 1982.

Parkhurst, D.L., Kipp, K.L. & Charlton, S.R.: PHAST Version 2—A program for simulating groundwater flow, solute transport, and multicomponent geochemical reactions. *U.S. Geological Survey Techniques and Methods* 6–A35, 2010.

Petersen, N.P.: Geology and ore deposits of the Globe-Miami district, Arizona. *U.S. Geological Survey Professional Paper* 342, 1962.

Plummer, L.N. & Glynn, P.D.: Radiocarbon dating in groundwater systems. *Guidebook on dating of old groundwaters*, Chapter 4. International Atomic Energy Agency, in press.

Plummer, L.N., Wigley, T.M.L. & Parkhurst, D.L.: The kinetics of calcite dissolution in CO_2-water systems at 5 to 60°C and 0.0 to 1.0 atm CO_2. *Amer. J. Sci.* 278 (1978), pp. 179–216.

Plummer, L.N., Parkhurst, D.L. & Thorstenson, D.C.: Development of reaction models for ground-water systems. *Geochim. Cosmochim. Acta* 47 (1983), pp. 665–686.

Plummer, L.N., Prestemon, E.C. & Parkhurst, D.L.: An interactive code (NETPATH) for modeling net geochemical reactions along a flow path. *U.S. Geological Survey Water-Resources Investigations Report*91–4078, 1991.

Plummer, L.N., Prestemon, E.C. & Parkhurst, D.L.: An interactive code (NETPATH) for modeling NET geochemical reactions along a flow PATH. Version 2.0. *U.S. Geological Survey Water-Resources Investigations Report* 94–4169, 1994.

Plummer, L.N., Sanford, W.E. & Glynn, P.D.: Characterization and conceptualization of groundwater-flow systems. *Guidebook on dating of old groundwaters,* Chapter 2. International Atomic Energy Agency, in press.

Postma, D. & Appelo, C.A.J.: Reduction of Mn oxides by ferrous iron in a flow system: column experiment and reactive transport modeling. *Geochim. Cosmochim. Acta* 64 (2000), pp. 1237–1247.

Revesz, K., Coplen, T.B., Baedecker, M.J., Glynn, P.D. & Hult, M.: Methane production and consumption monitored by stable H and C isotope ratios at a crude oil spill site, Bemidji, Minnesota. *Appl. Geochem.* 10 (1995), pp. 505–516.

Robertson, F.N.: Geochemistry of ground water in alluvial basins of Arizona and adjacent parts of Nevada, New Mexico, and California. Regional Aquifer-System Analysis—Southwest Alluvial Basins, Arizona and Adjacent States. *U.S. Geological Survey Professional Paper* 1406-C, 1991.

Steefel, C.I. & MacQuarrie, K.T.B.: Approaches to modeling of reactive transport in porous media. In: C.I. Steefel, P. Lichtner & E. Oelkers (eds): Reactive transport in porous media: General principles and application to geochemical processes. *Reviews in Mineralogy* 34, 1996, pp. 85–129.

Stollenwerk, K.G.: Geochemical interactions between constituents in acidic groundwater and alluvium in an aquifer near Globe, Arizona. *Appl. Geochem.* 9 (1994), pp. 353–369.

Stollenwerk, K.G. & Eychaner, J.H.: Acidic ground water contamination from copper mining near Globe, Arizona. In: B.J. Franks (ed): *U.S. Geological Survey Toxic Substances Hydrology Program—Proceedings of the Technical Meeting,* Pensacola, Florida, 23–27 March, 1987: *U.S. Geological Survey Open-File Report* 87–109, 1987, pp. D19–D24.

Tolle, S. & Arthur, G.V.: Aquifer restoration under the clean water act. In: G.E. Mallard & D.A. Aronson (eds): *U.S. Geological Survey Toxic Substances Hydrology Program—Proceedings of the Technical Meeting*, Monterey, California, 11–15 March, 1991: *U.S. Geological Survey Water-Resources Investigations Report* 91–4034, 1991, pp. 520–523.

Van Cappellen, P. & Gaillard, J.F.: Biogeochemical dynamics in aquatic sediments.In: C.I. Steefel, P. Lichtner & E. Oelkers (eds): Reactive transport in porous media: General principles and application to geochemical processes. *Reviews in Mineralogy* 34, 1996, pp. 335–376.

Villinski, J.E., O'Day, P.A., Corley, T.L. & Conklin, M.H.: *In situ* spectroscopic and solution analyses of the reductive dissolution of MnO_2 by Fe(II). *Environ. Sci. Technol.* 35 (2001), pp. 1157–1163.

Winograd, I.J. & Robertson, F.N.: Deep oxygenated ground water: anomaly or common occurrence? *Science* 216 (1982), pp. 1227–1230.

CHAPTER 9

Models and measurements of porosity and permeability evolution in a sandstone formation

Simon Emmanuel, Jay J. Ague & Olav Walderhaug

"The finest workers in stone are not copper or steel tools, but the gentle touches of air and water at their leisure with a liberal allowance of time."

Henry David Thoreau

9.1 INTRODUCTION

The permeability of porous rock formations is influenced primarily by connected porosity and specific surface area. Elevated porosities typically result in highly permeable rocks, while high specific surface areas lead to low permeability, and this relationship has been described in various forms, such as the well-known Kozeny equation (Kozeny, 1927). Despite the limitations of these relationships (Lukasiewicz and Reed, 1988; Saar and Manga, 1999; Costa, 2006), they often serve as good approximations in sedimentary formations with low clay contents, and their simplicity enables them to be incorporated with ease into numerical models used to describe flow and reaction in porous geological media.

Clearly, to reliably predict permeability in numerical simulations of groundwater systems, accurate estimates are required for porosity and specific surface area. However, while these two parameters can be measured with relative ease, the processes governing textural evolution are poorly understood. As a result, although it is widely recognized that porosity and specific surface area are intimately coupled, conflicting relationships have been proposed: some predict increasing surface areas with increasing porosity, while others predict an opposite trend (Lichtner, 1988; Kieffer *et al.*, 1999; Jové Colon *et al.*, 2004). Consequently, the expression used to describe the surface area-porosity relationship can have a great impact on the evolution of permeability and flow patterns in models that simulate diagenetic processes (e.g., Emmanuel and Berkowitz, 2005).

The primary limitation of simple models linking porosity and specific surface area is that they do not reflect the fact that pore size distribution, rather than total porosity, determines the specific surface area. Models that incorporate pore size distributions, and which are able to predict the way in which these distributions evolve, should therefore be able to more accurately estimate specific surface area and permeability. At present, however, the evolution of pore size distributions is rarely incorporated into numerical models in which the porous matrix is evolving.

One of the primary processes that leads to changes in porosity and pore size distributions in geological media is mineral precipitation and dissolution. Although changes in porosity can be calculated in a relatively straightforward manner from the mass of mineral deposited or dissolved and the mineral's density, predicting the evolution of pore size distributions as a result of mineral reactions remains a non-trivial challenge. While mineralization often leads to an overall reduction in porosity, the formation of new pores between growing crystals can create "honeycomb" cement and bimodal pore size distributions. In other cases, minerals may precipitate as continuous overgrowths on existing grains. Despite the fact that such phenomena are ubiquitous in sedimentary rocks, a quantitative understanding of their impact on pore size distributions and permeability—as well as the rates at which these changes occur—has yet to be fully achieved. Crucially, the combination of field observations and appropriate theoretical models can provide critical insight into system evolution during mineral precipitation and dissolution in porous rocks.

9.2 POROSITY MEASUREMENTS IN MINERALIZED ROCK

In porous and fractured rock formations, mineral precipitation usually involves two primary steps: transport of a dissolved mineral phase through the matrix, followed by the precipitation of the mineral in the pore space. When mineralization proceeds in such a manner, gradients in porosity can develop, and systems that have evolved under such conditions provide useful data concerning the effect of mineral reactions on pore size distributions. Furthermore, systems possessing such porosity gradients can be used to test reactive transport models that predict pore size evolution.

However, relating such models to field data typically represents an extremely challenging task: flow patterns are often poorly constrained and the source of the precipitating mineral phase is usually unknown, both of which complicate efforts to construct realistic models. Thus, selection of a suitable diagenetic scenario is key to successfully modeling the impact of mineral precipitation on pore size.

One diagenetic scenario that has an important impact on the evolution of the porous matrix and also lends itself to modeling is stylolitization. Stylolites are irregular suture-like or tooth-like interlocking surfaces most commonly observed in carbonate and quartz-rich sedimentary rocks. These surfaces develop where mineral grains—often quartz or calcite—are dissolved at the contacts with clay-rich or micaceous laminae. Although the precise origin of stylolites is hotly debated, most models agree that they represent sharp interfaces at which rock is being dissolved; once in solution, the mineral phase typically diffuses into the porous matrix where it reprecipitates, reducing the matrix porosity (Fig. 9.1). As the interface between the stylolite and the porous rock represents a suitably defined source for the precipitating phase, a well constrained model can be constructed to simulate the evolution of porosity and permeability.

Although stylolites are common features in sedimentary rocks, they are usually too closely spaced to allow the detection of porosity gradients in the matrix. However, in the Jurassic Stø Formation, located in the Barents Sea, well developed porosity gradients adjacent to stylolites have been observed (Walderhaug and Bjørkum, 2003). The studied parts of the Stø Formation comprise fine-grained to medium-grained shallow marine quartz sandstones exhibiting individual stylolites that are spaced at intervals of around 50 cm to 3 m. Quartz mineralization and reduced porosity are observed in the matrix near the stylolites and an inspection of cores reveals that the regions directly adjacent to the stylolites are slightly

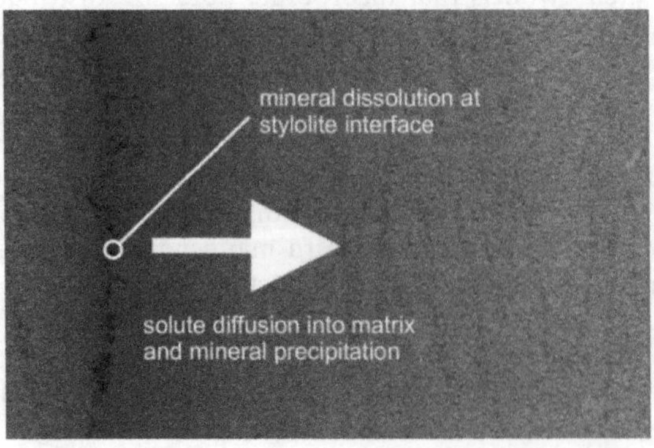

Figure 9.1. Schematic representation of mineral precipitation in the matrix adjacent to stylolites. Dissolution dominates at the stylolite interface (the jagged vertical dark line to the left ofthe image); the dissolved mineral phase diffuses into the porous matrix where it precipitates, leading to a reduction in porosity.

darker and better cemented than the surrounding matrix. A detailed description of the geology of the formation and the core is given by Walderhaug and Bjørkum (2003).

To determine the influence of mineral precipitation on the pore size distributions, Emmanuel *et al.* (2010) sampled a 25 cm section of the core at different distances from a representative stylolite in the formation. Total porosity, pore size distributions, and surface areas were measured by mercury injection porosimetry, a technique used to characterize reservoir rock, such as sandstone (Uchida, 1987; Giesche, 2006). In addition, high-resolution electron imaging was used to obtain images of both polished and unpolished samples.

Consistent with previous studies, Emmanuel *et al.* (2010) found that the total porosity in the Stø Formation changed sharply in the region closest to the stylolite, increasing from 12.1% at a distance of 0–0.5 cm to 20.9% at 3–4 cm (Fig. 9.2); at greater distances from the stylolite, the porosity was found to maintain an average value of approximately 20%. Such patterns of porosity reduction are consistent with a scenario of diffusion of dissolved silica away from the stylolite and precipitation of quartz in the matrix.

Backscattered electron (BSE) images of polished samples also show a much lower proportion of voids in the zone nearest the stylolite interface, with most of the pores falling in the range 10–100 μm in diameter (Fig. 9.3). Moreover, the presence of euhedral and subhedral grains strongly suggests that quartz precipitates as overgrowths on existing quartz grains, consistent with cathode luminescence imaging carried out in a previous study of the Stø Formation (Walderhaug and Bjørkum, 2003). Critically, however, the quartz overgrowths seem to be size dependent: although the pore in the center of Figure 9.3d (approximately 30–40 μm in diameter) exhibits euhedral overgrowths, euhedral crystal faces do not appear in the smaller scale intergranular porosity (<5 μm) found in the same image.

Importantly, the pore size distribution data also suggest that mineral precipitation occurs primarily in large pores. For all the samples, pores <10 μm in size represent a cumulative porosity of approximately 2.5–3%. By contrast, in larger pores (>10 μm) cumulative porosity varies from around 9% in the samples exhibiting the highest level of mineralization (Fig. 9.4a–c) to approximately 18% in samples least affected by quartz precipitation (Fig. 9.4d–f). Such a result indicates that quartz precipitation takes place in pores >10 μm in size but is inhibited in smaller pores (i.e., <10 μm). Similar pore size distribution patterns were observed in halite cemented sandstone studied by Putnis and Mauthe (2001), suggesting that the inhibition of

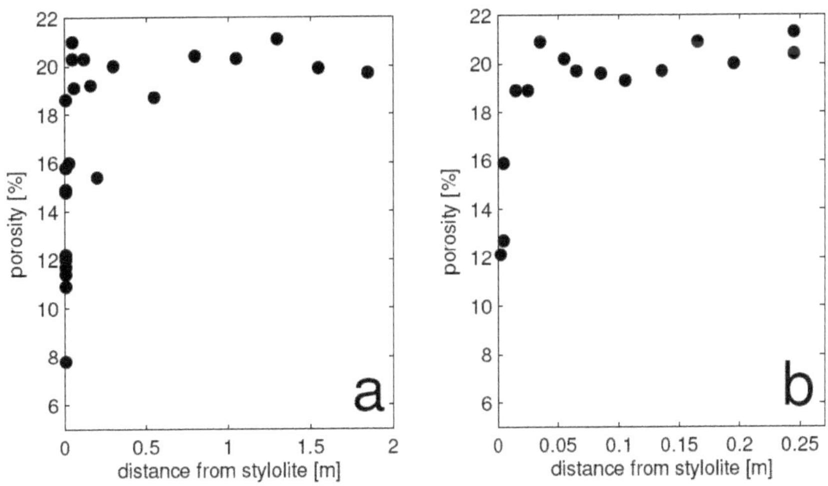

Figure 9.2. Total porosity as a function of distance from stylolite interface in the Stø Formation sandstone. (a) Data compiled by Walderhaug and Bjørkum, 2003 from measurements near different stylolites and (b) a detailed profile, reported by Emmanuel *et al.* (2010), adjacent to a single stylolite.

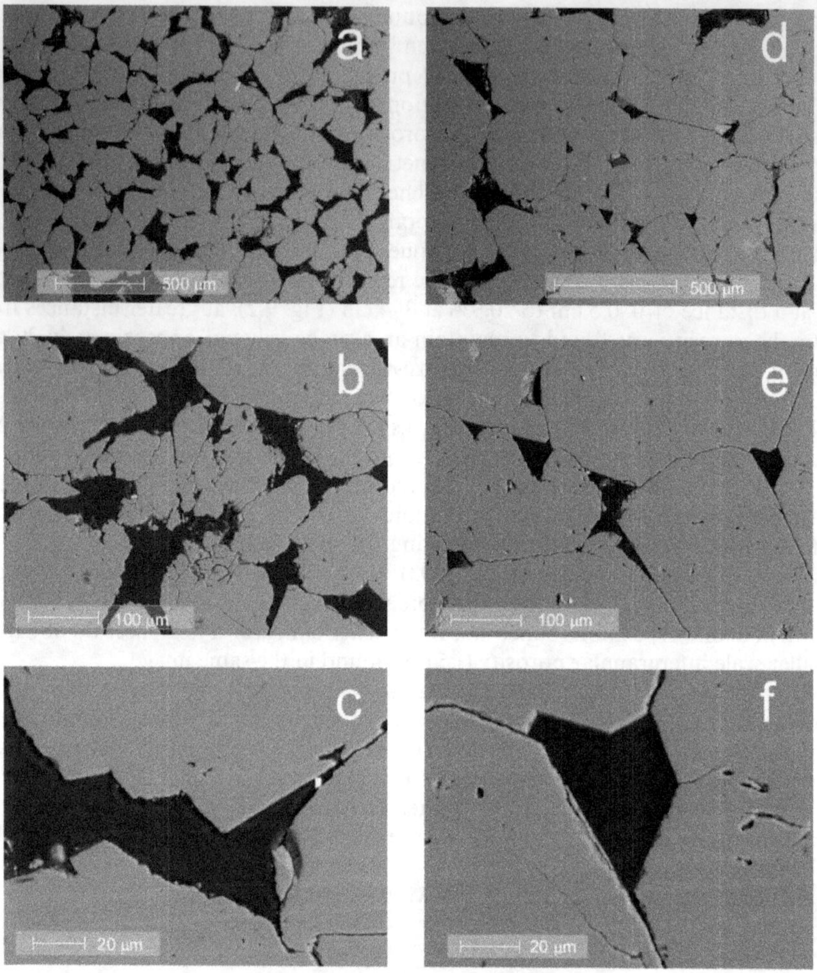

Figure 9.3. Backstattered electron (BSE) images of polished sandstone samples from the Stø Formation from (a)–(c) 24–25 cm and (d)–(f) 0–1 cm from the stylolite. The images reveal that most of the pore space comprises voids 10–100 µm in diameter. The medium gray regions are quartz while the dark gray zones are epoxy filling in the pores. Note the euhedral to subhedral quartz overgrowths in the intergranular voids in (c) and (f). From Emmanuel *et al.* (2010).

mineral precipitation in tiny pores may be a common feature in porous rocks. Clearly, any model that aims to capture the impact of mineral precipitation on the porous medium should be able to account for such pore size distribution patterns.

9.3 THEORY AND NUMERICAL MODELING OF POROSITY EVOLUTION

9.3.1 *Conceptual model of the porous medium*

To simulate the patterns of mineral precipitation in the porous rock, Emmanuel *et al.* (2010) constructed a numerical model which described the evolution of pore size distributions during mineral precipitation in a 1 D porous domain. A brief description of the governing equations is given in this section, and more detailed discussions of the model are provided by Emmanuel and Ague (2009) and Emmanuel *et al.* (2010).

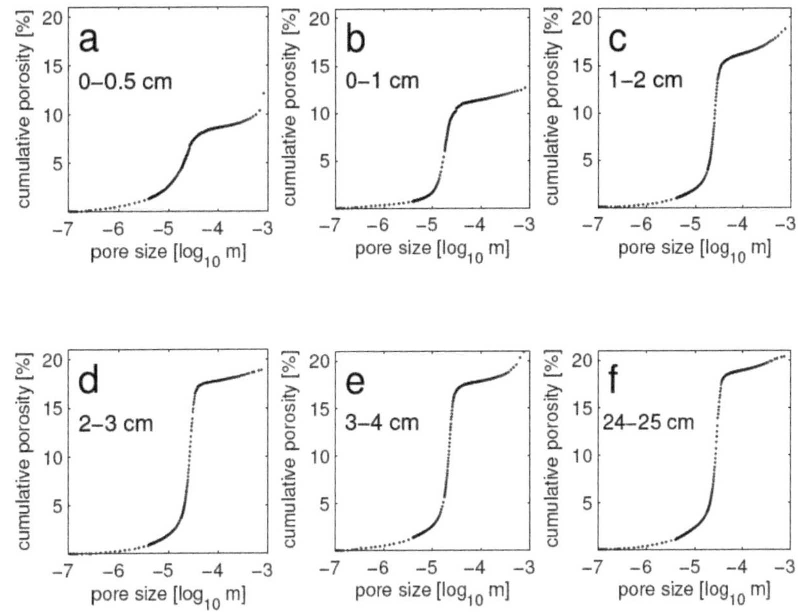

Figure 9.4. Measured cumulative porosity curves at six distances from the stylolite: (a) 0–0.5 cm; (b) 0–1 cm; (c) 1–2 cm; (d) 2–3 cm; (e) 3–4 cm; and (f) 24–25 cm. Note that the curves do not differ significantly in the range 10^{-7}–10^{-5} m, indicating that quartz precipitation has not occurred in the smallest pores. Adapted from Emmanuel *et al.* (2010).

In the construction of a mathematical model to simulate the evolution of the porous medium, one of the fundamental stages is to develop a conceptual model that can be compared with the pore size distributions measured using mercury porosimetry. One of the simplest and most effective ways to model the porous medium is to assume that the porosity comprises interconnected voids with a continuous range of sizes. Each pore size d is assigned a partial porosity, $\varnothing_d$, which is related to the total porosity, $\varnothing$, by the expression:

$$\varnothing = \int_0^{\infty} \varnothing_d \, \mathrm{d}(d) \tag{9.1}$$

In such a framework, the partial porosity associated with each pore size can be defined as:

$$\varnothing_d = X_d L_d \tag{9.2}$$

where X_d is the cross sectional area of a pore of size d and L_d is the total length of pores per unit volume of pores of size d, and for pores with square cuboid-shaped cross sections (Fig. 9.5)—similar to the pore geometry observed in the sandstone formation—the cross sectional area of the individual pores is given as:

$$X_d = d^2\left(1 - \frac{\omega - \sin\omega\,\cos\omega}{\sin^2\omega}\right) \tag{9.3}$$

where $\omega = (\pi/2 - \theta)/2$ and θ is the dihedral angle (Fig. 9.5). In addition, the partial specific surface area associated with each pore size is:

$$A_d = \frac{4L_d\omega d}{\sin\omega} \tag{9.4}$$

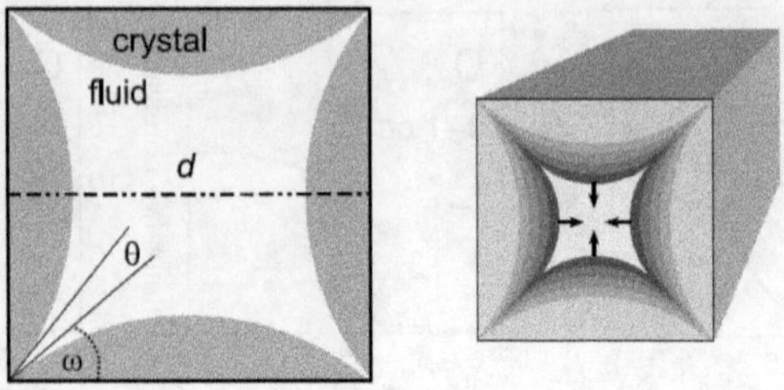

Figure 9.5. Schematic cross section of the pore geometry assumed in the simulations of Emmanuel *et al.* (2010). Crystal growth is perpendicular to the walls of a square cuboid-shaped pore created as a result of contact between four quartz grains of equal size. As the pore is filled during mineralization, the pore size decreases, leading to an increase in crystal curvature. The angle θ corresponds to the dihedral angle, while the dimension *d* represents the characteristic size of the pore. Arrows indicate the direction of crystal growth. From Emmanuel *et al.* (2010).

such that the total specific surface area in a given population of pore sizes is given by:

$$A = \int_0^\infty A_d \mathrm{d}(d) \tag{9.5}$$

In a system in which a mineral is precipitating at a bulk rate of Q, the rate of change of porosity is given by:

$$\frac{\partial \varnothing}{\partial t} = -v_m Q \tag{9.6}$$

where v_m is the molar volume. While this equation determines the overall rate of change of porosity, in a porous system with a polymodal pore size distribution the key to determining the evolution of pore size distributions is to think of the rate of change of total porosity as the sum of the rates of change of the partial porosities such that:

$$\frac{\partial \phi}{\partial t} = \int_0^\infty \frac{\partial \phi_d}{\partial t} \mathrm{d}(d) \tag{9.7}$$

where

$$\frac{\partial \phi_d}{\partial t} = -v_m Q_d \tag{9.8}$$

and Q_d is the partial rate of precipitation associated with pores of size d. Thus, the reaction rates associated with the different pore sizes will determine the rates at which those pores are filled by mineral precipitation.

9.3.2 *Reaction kinetics*

Quartz dissolution and reprecipitation is the dominant mineral reaction in the sandstone rocks of the Stø Formation and can be summarized as:

$$SiO_2(\text{quartz}) + 2H_2O \rightleftarrows H_4SiO_4 \tag{9.9}$$

from which the bulk solubility, S_0 of quartz is defined as:

$$S_0 = \frac{a_{\mathrm{H_4SiO_4}}^{eq}}{(a_w)^2} \tag{9.10}$$

where $a_{\mathrm{H_4SiO_4}}^{eq}$ and a_w indicate the equilibrium activities of orthosilicic acid and water respectively.

In order to describe the rate at which quartz precipitates within the pores of the Stø Formation sandstone, it is crucial to consider the effect of pore size on reaction kinetics. However, accurately representing the influence of pore scale processes in kinetic models represents a non-trivial challenge. In standard geochemical kinetic theory, the main influence of pore size is to determine the specific surface area of the medium (A), which is usually related to the bulk precipitation rate by an equation of the form (Berner, 1980; Lasaga, 1998):

$$Q = KA\left(\frac{S}{S_0} - 1\right)^{\beta} \tag{9.11}$$

where Q is the reaction rate (in units of mol m^{-3} s^{-1}), K is a kinetic rate coefficient, S is the ion activity product (defined for quartz $S = \frac{a_{\mathrm{H_4SiO_4}}}{(a_w)^2}$), and β is the rate order. In general, a porous medium with small pores will have a much higher specific surface area—and accordingly a much higher reaction rate—than a porous medium with large pores.

However, pore size could play an additional role in determining reaction rates in porous geological media. In small nanometer to micron-scale pores, surface free energy usually increases the solubility of crystals, and it has recently been suggested that such standard kinetic models can be modified by replacing the constant solubility by one that is pore size controlled (Emmanuel and Berkowitz, 2007; Emmanuel and Ague, 2009; Emmanuel *et al.*, 2010). If the porous medium comprises pores of uniform size, the effective reaction rate can be defined as:

$$Q = KA\left(\frac{S}{S_d} - 1\right)^{\beta} \tag{9.12}$$

where S_d is the solubility in the pores determined by interfacial energy effects.

Clearly though, rocks and soils possess a wide range of pore sizes, and the effective bulk rate can be given as the sum of all the rates associated with the different pore sizes such that (Emmanuel *et al.*, 2010):

$$Q = K\int_0^{\infty} Q_d \mathrm{d}(d) \tag{9.13}$$

where the partial reaction rate, Q_d, is given by:

$$Q_d = KA\left(\frac{S}{S_d} - 1\right)^{\beta} \tag{9.14}$$

In order to evaluate these rate expressions, the impact of pore size on solubility needs to be determined. This can be achieved by considering the impact of interfacial energy during crystal growth, which yields an expression for the effective solubility, S_d (Adamson, 1990; Scherer, 2004):

$$S_d = S_0 \exp\left(\frac{\upsilon_m \gamma \zeta}{RT}\right) \tag{9.15}$$

where γ is the interfacial energy, R is the gas constant, and T is temperature. In addition, ζ is the crystal curvature, defined as the rate of change of surface area with respect to volume, which for a spherical crystal possessing a radius of curvature r_c, is simply given by $\zeta = 2/r_c$. Thus, the expression for effective solubility becomes:

$$S_d = S_0 \exp\left(\frac{2\upsilon_m \gamma}{RTr_c}\right) \tag{9.16}$$

A similar approach can also be adopted to describe the solubility of minerals growing within the pores of a porous medium. However, for crystals confined within a rigid porous medium the maximum radius of curvature becomes a function of the pore size and is no longer equal to the radius of the crystal. For a crystal growing in a pore with the geometry displayed in Figure 9.5, the radius of curvature is related to the dimension of the pore, d, by $r_c = d/2\cos\theta$, where θ is again the dihedral angle. Substituting this relationship into equation (9.16) yields (Scherer, 2004):

$$S_d = S_0 \exp\left(\frac{4\upsilon_m \gamma \cos\theta}{RTd}\right) \tag{9.17}$$

Substituting typical parameter values for mineral systems into equation (9.17) reveals that the ratio S_d/S_0 can be significantly larger than unity in microscopic crystals (Fig. 9.6). It is important to differentiate between two measures of supersaturation: (i) the bulk supersaturation index, S/S_0 and (ii) the effective supersaturation index S/S_d; in both these expressions S refers to the ion activity product. The bulk supersaturation index is the standard measure for supersaturation and is independent of pore size; by contrast, the effective supersaturation index can vary from pore to pore depending on the interfacial energy effect. Although the bulk solubility may also be affected by fluid chemistry, it is important to note that it will not be modified by pore size effects.

Substitution of equation (9.17) into equation (9.13) yields an expression that describes the impact of interfacial energy on bulk reaction rates:

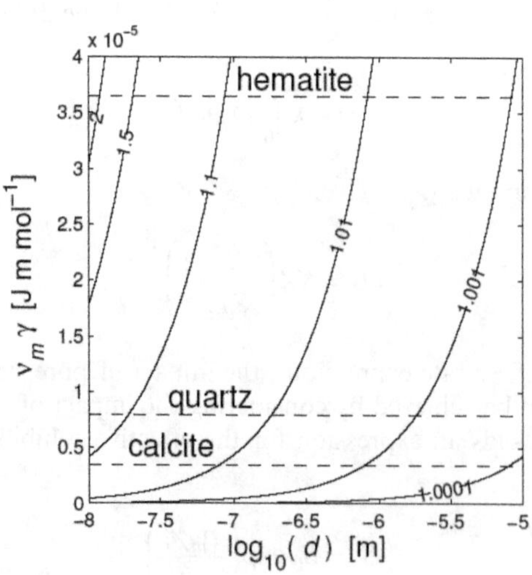

Figure 9.6. Ratio of effective solubility to bulk solubility (S/S_0) as a function of interfacial energy and pore size. The y axis is given by the product of the molar volume and interfacial energy ($v_m\gamma$) as both are required to define a specific mineral system in equation (9.17). The $v_m\gamma$ values are indicated for three common minerals (calcite, quartz, and hematite).

$$Q = K\int_0^{\infty} A_d \left(\frac{S}{S_0 \exp(\frac{4\upsilon_m \gamma \cos\theta}{RTd})} - 1 \right)^{\beta} \mathrm{d}(d) \tag{9.18}$$

Importantly, equation (9.18) (interfacial energy model) reduces to equation (9.11) (standard rate model) when interfacial energy effects are negligible.

It is worth emphasizing that pore size is often assumed to have only a minimal effect on reaction rates, and the inclusion of interfacial energy effects in small pores remains a non-standard approach. However, when a system possessing a high proportion of micron-nanometer scale pores is very close to chemical equilibrium, models predict that crystal growth might be inhibited in small pores at the same time that crystals continue to grow in large pores (Emmanuel and Berkowitz, 2007; Emmanuel and Ague, 2009). As a result the total rate of mineral precipitation in the porous medium should be much lower than expected due to the loss of reactive surface area associated with the small pores and this aspect of the model is discussed in greater detail later on.

9.3.3 *Reactive transport equations*

To model the temporal and spatial evolution of the pore size distributions, Emmanuel and Ague (2009) used a standard diffusion-reaction equation for porous media; however, in addition to calculating solute concentrations and reaction rates, the model was capable of simultaneously tracking the evolution of different pore sizes. The model assumes that at any given distance from the stylolite, solute concentrations are identical in all pores, irrespective of size, allowing a continuum approach to be applied; as a result, the bulk supersaturation remains independent of pore size, although the *effective* supersaturation will vary from pore to pore. Such an assumption should hold for porous systems in which the reaction rate is slow compared with diffusion (Kang *et al.*, 2006), as is the case here for quartz precipitation.

The two main processes that determine the transfer of solute through the sandstone matrix are diffusion and precipitation, and a non-dimensional reactive transport equation describing these processes can be given by:

$$\frac{\partial(\phi\bar{C})}{\partial\bar{t}} = \frac{1}{Pe}\frac{\partial\left(\phi\frac{\partial\bar{C}}{\partial\bar{x}}\right)}{\partial\bar{x}} - \bar{Q} \tag{9.19}$$

where $\bar{C}$ is the dimensionless H_4SiO_4 concentration (defined as $\bar{C} = \frac{C}{S_0}$, where C is the dimensional concentration), $\bar{t}$ is dimensionless time ($\bar{t} = \frac{tv}{H}$, where t is dimensional time, v is the velocity at which the stylolite boundary recedes due to dissolution, and H is the initial domain size), and $\bar{x}$ is the dimensionless distance (related to the dimensional distance, x by $\bar{x} = \frac{x}{H}$). In addition, Pe is the Peclet number (given by $Pe = vH/D_e$), and $\bar{Q}$ is the dimensionless reaction rate expression. In the definition of the Peclet number, the effective diffusion coefficient D_e is defined as $D_e = D_w/\tau$, where D_w is the free water diffusion coefficient and τ is tortuosity. Advection is not considered here as large fluid fluxes are unlikely to be sustained in a marine sedimentary environment relatively far from the continental landmass.

As described in section 9.3.2, the reaction term can be defined either by the standard rate equation (eq. 9.11) or by the interfacial energy model (eq. 9.18). For the standard kinetic expression, the non-dimensional rate becomes:

$$\bar{Q} = \kappa\bar{A}\left(\frac{\bar{C}}{(a_{H_2O})^2} - 1\right)^{\beta} \tag{9.20}$$

where κ is a non-dimensional rate constant ($\kappa = \frac{KH}{vS_0}$) and $\bar{A}$ is the non-dimensional specific surface area defined as $\bar{A} = AH$. Similarly, non-dimensionalization of the interfacial energy model yields:

$$\bar{Q} = K\int_0^{\infty} \bar{A}d\left(\frac{\bar{C}}{(a_{\mathrm{H_2O}})^2 \exp\left(\frac{4v_m\gamma\cos\theta}{RTd}\right)} - 1\right)^{\beta} d(d) \tag{9.21}$$

where $\bar{A}_d = A_dH$. Furthermore, in non-dimensional terms, the evolution of total porosity is described by:

$$\frac{\partial\phi}{\partial\bar{t}} = -S_0v_m\bar{Q} \tag{9.22}$$

Due to the dissolution of the domain at the stylolite interface, non-standard boundary conditions apply: the size of the domain is given by $\bar{h} = 1 - \bar{t}$, yielding:

$$\frac{\partial\bar{C}}{\partial\bar{x}}\Big|_{\bar{x}=1-\bar{t}} = 0 \tag{9.23}$$

and

$$\bar{J}\,|_{\bar{x}=0} = \frac{1-\phi}{S_0v_m} \tag{9.24}$$

where $\bar{J}$ is the dimensionless solute flux at the stylolite boundary. Details of the method used to solve the moving boundary problem can be found in Emmanuel and Ague (2009).

In the approach described here, the two definitions for the reaction rate yield two possible models: a standard rate model and an interfacial energy model. Although these models behave differently, the evolution of porosity in both is controlled primarily by two non-dimensional parameters: (i) the Peclet number (*Pe*), which determines both the rate at which the stylolite interface moves and the flux of dissolved quartz into the porous domain and (ii) the non-dimensional rate coefficient (κ), which controls the rate at which quartz precipitates. During diagenesis, *Pe* is estimated to vary from 10^{-9}–10^{-2}, while κ is expected to be in the range 10^0–10^9 (Emmanuel and Ague, 2009).

9.3.4 *Numerical solution and model optimization*

Ultimately, the objective of the model is to simulate the evolution of pore size distributions at different distances from the stylolite. Moreover, it should be possible to compare the solutions with measured pore size distributions to identify optimal values for model parameters. However, due to the complexity of the model, the series of the equations cannot be solved analytically, necessitating the use of numerical methods.

Emmanuel *et al.* (2010) obtained numerical solutions to the equations describing solute transport and porosity evolution using the COMSOL Multiphysics software package based on finite elements. The numerical model determines the evolution of porosity in a porous domain with an initially uniform porosity and polymodal pore size distribution. The total porosity is discretized into sixteen different pore sizes in the range 10^{-7}–10^{-4} m, yielding an initial cumulative porosity curve ($\psi_d = \int_0^d \phi_d \,\mathrm{d}(d)$) that matches the measured curve in the sample farthest from the stylolite interface (24–25 cm); as this sample is least affected by

Table 9.1. List of symbols and parameter values.

Symbol	Definition	SI Units	Value
H	initial domain size	m	0.25
S_0	quartz solubility product	mol m^{-3}	0.1995[a]
T	temperature	K	357[b]
a_w	seawater activity		0.98[a]
β	empirical rate order		1
γ	interfacial free energy	J m^{-2}	0.35[a,c]
θ	dihedral angle		$\pi/4$[d]
v_m	molar volume	m^3 mol^{-1}	2.269×10^{-5}[e]

[a]Stumm and Morgan (1996)
[b]Current temperature in the Stø Formation, although quartz may have precipitated at higher temperatures (Walderhaug and Bjørkum, 2003)
[c]Parks (1984)
[d]Emmanuel and Ague (2009)
[e]Lide (1996)

quartz precipitation, it is assumed to represent the state of the porous matrix prior to stylolitization. Additional parameter values used in the model are given in Table 9.1.

Matching the numerical model to the measured pore size distribution curves was carried out by applying an error minimization technique that compared model solutions with the measured data. The minimization method—based on the Nelder-Mead method—was applied to an objective function, defined as the root mean square of the error between the measured and predicted values. This technique was applied both to the standard kinetic model and the interfacial energy model, and only Pe, κ, and duration of the simulation were varied to find the best fit; all other values in the models—including the interfacial energy parameter γ—were estimated from literature values (Table 9.1) and kept constant throughout the simulations. The optimization method yielded values of $Pe = 1.93 \times 10^{-8}$, $\kappa = 3.31 \times 10^{6}$, and $\bar{t} = 0.0061$ for the standard kinetic model, and $Pe = 2.31 \times 10^{-8}$, $\kappa = 3.14 \times 10^{6}$, and $\bar{t} = 0.0064$ for the interfacial energy model.

9.4 COMPARISON BETWEEN NUMERICAL MODELS AND MEASUREMENTS

The curves produced by the two models describing cumulative porosity as a function of pore size reveal important differences that can be used to assess the role played by interfacial energy during mineralization. At high degrees of quartz cementation, (Fig. 9.7d–f), the standard kinetic model predicts that voids smaller than ~10 μm should be completely filled in the region closest to the stylolite interface, a result that is clearly at odds with the pore size measurements in the Stø Formation sandstone. Importantly, no combination of parameters (Pe, κ, and duration of the simulation) was able to inhibit precipitation in the smallest pores in the standard rate model. By contrast, the interfacial energy model was able to preserve micron-scale porosity, producing an excellent match with the measured curves (Fig. 9.7).

The agreement between the interfacial energy model and the measured pore size distributions suggests that quartz precipitation in small pores can be inhibited by interfacial energy. Alternative mechanisms, however, could produce similar results. Previous studies have suggested that mineral precipitation may be inhibited by clusters of clay minerals and clay coatings on quartz grains (Heald and Larese, 1974; McBride, 1989; Walderhaug *et al.*, 2006; Aharonov and Katsman, 2009). While this mechanism may play an important role in rocks with high clay content, the amount of clay in the Stø Formation is extremely low, comprising

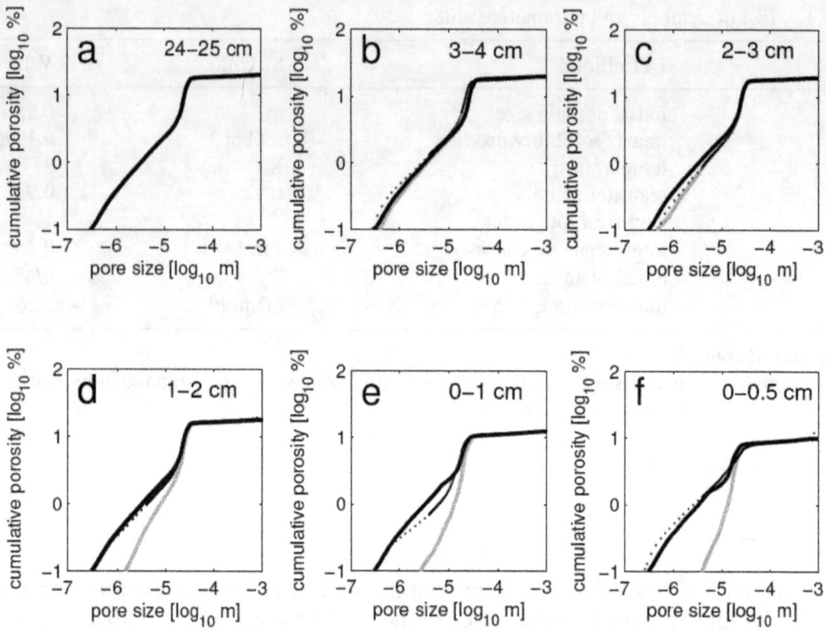

Figure 9.7. Comparison of simulated cumulative porosity curves at 6 distances from the stylolite: (a) 24–25 cm; (b) 3–4 cm; (c) 2–3 cm; (d) 1–2 cm; (e) 0–1 cm; and (f) 0–0.5 cm. Measured porosity (black dots) and model curves (black—pore size dependent solubility model, gray—constant solubility model) are given as a function of pore diameter. Note that very near the stylolite (d–f), the pore size dependent solubility model agrees well with the measured data, while the constant solubility model overestimates the degree of mineralization in <10 μm pores. Adapted from Emmanuel *et al.* (2010).

only ~0.2% by volume (Walderhaug and Bjørkum, 2003). In addition, BSE images show that clay minerals are absent from most of the small pores between the quartz grains (Fig. 9.3).

Mineral precipitation in micron-scale pores might also be limited if supersaturated fluids were unable to penetrate tiny voids, which could occur if pores are physically isolated or unwetted by fluid. However, the fact that submicron scale pores are measured by the porosimetry method indicates that they are indeed connected, and at the elevated fluid pressures and high concentrations of dissolved salts in the Stø Formation, even submicron scale pores are expected to have contained fluid (Zhang and Austad, 2006; Yu *et al.*, 2009).

Reduced solute flux through small pores represents an additional mechanism capable of producing the pore size distributions observed in the sandstone formation. Such a mechanism is expected to have a significant impact on pore size distribution patterns when solute transport is dominated by advection and when reaction rates are rapid enough to be limited by the rate of mass transfer. However, in the Stø sandstone, diffusion is likely to have been the primary mode of mass transfer during mineralization, and quartz precipitation is typically thought to be a very slow process, taking place over thousands to millions of years. In any case, the impact of advection on reaction rates in small pores cannot simply be modeled by introducing an advective term into equation (9.19), and the good fit with the data suggests that diffusion need not be supplemented by advection to adequately describe solute transport in the Stø Formation.

Preservation of micron-scale porosity might also be facilitated by the creation of small pores during mineral precipitation. Highly porous intergranular cement is sometimes observed in rocks, and some SEM images do in fact show that micropores form between the crystals as quartz overgrowths intrude into intergranular voids. However, in order to reproduce the measured pore size distributions, newly formed micron-scale porosity would have to balance, at every stage, the amount of primary micron-scale porosity filled, which would seem unlikely. Furthermore, inspection of the electron miscroscope images indicates that porous cement makes only a minor contribution to the overall porosity in the Stø Formation sandstone.

Yet another mechanism that might influence mineralization is quartz dissolution along individual grain contacts, a process that is often associated with compressive stress. Such dissolution could contribute dissolved quartz to the fluid phase, thereby elevating levels of supersaturation and accelerating the bulk rate of mineral precipitation and porosity reduction. Although not taken into account in the models presented here, the mechanism cannot on its own account for the preservation of micron-scale pores. In any case, while some degree of dissolution may have taken place, petrographic studies have concluded that intergranular dissolution is unlikely to have represented an important source of dissolved silica in the Stø Formation (Walderhaug and Bjørkum, 2003).

Regardless of the mechanism responsible, the preservation of micron scale porosity in sedimentary rocks is crucial for mass transfer in many geological systems. During diagenesis and metamorphism, mineral precipitation is often believed to clog porosity, ultimately producing a barrier to further diffusion and advection of reactants. The measurements and model, however, indicate that a significant proportion of small pores can remain open and connected, helping to account for the pervasive mineralization often observed in geological systems.

9.5 IMPLICATIONS FOR BULK REACTION RATES

In addition to influencing mass transfer in hydrogeological settings, the preservation of micron scale porosity has important implications for mineral reaction rates. While small pores in the Stø Formation comprise only 15–25% of the total porosity, they represent at least 70% of the surface area (Emmanuel *et al.*, 2010). As these small pores do not take part in the reaction, the available surface area for quartz precipitation in the sandstone is reduced by an equivalent proportion, which acts to reduce the bulk reaction rate. Critically for geochemical reaction rates, the proportion of reactive surface in finer grained sedimentary rocks could be much lower, suggesting that reaction rates may be orders of magnitude lower than those predicted from total surface areas.

Such an effect on the overall impact on kinetics can be calculated by defining a threshold pore size, d_{crit}. This threshold determines the size below which mineral precipitation ceases at a given level of S/S_0 as a result of interfacial energy effects. Rearranging equation (9.17), d_{crit} for any system can be defined as:

$$d_{crit} = \frac{1}{\ln\left(\dfrac{S}{S_0}\right)} \frac{4\upsilon_m \gamma \cos\theta}{RT} \tag{9.25}$$

and an approximation of the effect on reaction rate can be obtained by evaluating

$$F = \frac{1}{A}\int_0^{d_{crit}} A_d \mathrm{d}(d) \tag{9.26}$$

where F is approximately equal to the ratio between the reaction rate in the interfacial model and that in the standard kinetic model. When the interfacial energy model and the standard model are equivalent, $F = 1$; when interfacial energy effects are prominent $F << 1$. Thus, for a rock with a given pore size distribution, reaction kinetics will be determined by two parameters: (i) the bulk supersaturation index, S/S_0; and (ii) the parameter Λ (defined as $\Lambda = \frac{4v_m \gamma \cos\theta}{RT}$), which serves as a measure of the interfacial energy effect. In many diagenetic scenarios, bulk supersaturation is expected to be low (<1.01; Aase *et al.*, 1996; Oelkers *et al.*, 1996); furthermore, assuming temperatures >273 and <500 K during diagenesis and γ in the range 0–1.6 J m^{-2} (Stumm and Morgan, 1996), produces values of Λ between 0 and 2×10^{-8} m. Accordingly, d_{crit} in many geological systems should be greater than 1 μm, and in common sedimentary rocks, such as sandstone, limestone, chalk, and shale, it is not uncommon for more than 99% of the total surface area to be associated with submicron scale porosity (Kate Gokahle, 2008). As a result, in such systems $F < 0.01$ and bulk reaction rates could be 2 or more orders of magnitude lower than those predicted by standard rate laws.

Strikingly, such a result is entirely compatible with the observation by numerous studies that the rate coefficients of mineral reactions measured using field-based methods are often far lower than those found in laboratory (Blum and Stillings, 1995; Baxter and DePaolo, 2000; Yokoyama and Banfield, 2002; Baxter, 2003; Yo and Mudd, 2008). Although a range of mechanisms—including changes in surface area during reaction, depletion of reactive surfaces, and the buildup of leached layers and secondary precipitates (White and Brantely, 2003; Maher *et al.*, 2006)—have been proposed to account for this discrepancy, the calculations here demonstrate that interfacial energy effects could also play a significant role. In order to reduce experimental time scales, laboratory experiments are often conducted at levels of supersaturation that are much greater than those typically found in diagenetic systems; at high S, mineral precipitation should occur on most mineral surfaces so that $F \rightarrow 1$ and standard rate equations apply. However, in natural systems lower levels of supersaturation may cause mineral growth to be restricted to the largest pores, in which case $F \ll 1$ and standard rate laws are no longer valid. While it has been suggested that reduced reactive surface area could help account for the disagreement between laboratory and field reaction rates, the interfacial energy model offers a strong theoretical basis for the phenomenon.

9.6 IMPLICATIONS FOR PERMEABILITY EVOLUTION IN AQUIFERS

The difference in pore size distributions produced by the two models suggest that interfacial energy effects could have important implications not only for the rates of mineral precipitation but also for the evolution of permeability in the porous medium. Although, the models do not predict permeability directly, the parameter can be approximated using the Kozeny relationship (Kozeny, 1927):

$$\frac{k = c_0 \varnothing^3}{A^2} \tag{9.27}$$

where c_0 is a constant, usually assumed to have a value of 0.2 (Bear, 1972); importantly, the choice of a different value will not affect the analysis described below. Despite its simplicity, the Kozeny relationship is often effective at predicting permeability in sedimentary rocks and is widely used in numerical models.

A comparison of measured values of porosity and specific surface area in the sandstone matrix in the Stø Formation with the standard rate and interfacial energy models yields significant differences. Although total porosity is predicted well by both models (Fig. 9.8a), the standard kinetic model overestimates the reduction in specific surface area in mineralized samples (Fig.9.8b). As a result, when porosity reduction is greatest, the permeability predicted by the standard rate model is more than an order of magnitude greater than that predicted by the interfacial energy model (Fig. 9.8c). Moreover, while the interfacial energy model produces an overall trend that is in agreement with the permeabilities based on the measured values, the standard rate model predicts an unusual porosity-permeability relationship: as porosity decreases due to mineralization, permeability initially increases, reaching a peak at around 17% porosity, after which permeability declines. Thus, in addition to strongly

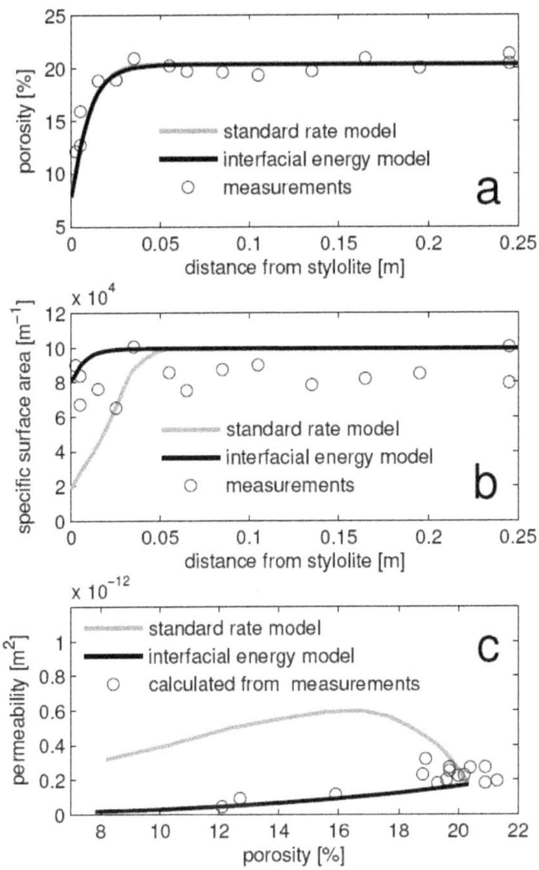

Figure 9.8. Measurements and model predictions of porosity, specific surface area, and permeability in the sandstone matrix adjacent to the stylolite. (a) Porosity as a function of distance; (b) specific surface area profile; (c) permeability calculated using the Kozeny relationship. In contrast to the standard rate model, the interfacial energy model produces a good agreement with permeabilities calculated from the porosity and surface area measurements.

impacting reaction rates, the choice of model will have a major influence on permeability predictions. This coupling of chemical and physical effects will have important implications for the modeling of fluid flow and geochemical reactions in sedimentary systems in which significant micron and nanometer scale porosity is present.

9.7 CONCLUDING REMARKS

The analysis discussed here challenges the standard models used to describe mineral reaction rates in porous media. Measurements demonstrate that mineral precipitation is inhibited in micron and submicron scale pores, an observation that can be explained by the effects of interfacial energy on the solubility of crystals in tiny micron scale pores. In addition to offering insight into the evolution of porosity evolution in porous systems, the model helps us understand how diffusive and advective transport may persist in geological media even when mineralization is pervasive.

To show the potential impact that the mechanism can have on mineral precipitation and dissolution patterns, one can consider a solute being transported through a fluid-saturated monomineralic porous rock in which the pore space is dominated by tiny pores so that mineral solubility is determined by pore size. Given enough time, the solute concentration will eventually approach an equilibrium value for the ion activity product that exceeds bulk solubility product. If the fluid is then transported into a fracture or region containing larger pores, the fluid phase will effectively become oversaturated with respect to the dissolved mineral phase so that precipitation will take place. This scenario is similar in many respects to the mechanism first proposed by Yardley (1975) for vein formation in metamorphic systems. Although a clear oversimplification of the reactive transport in fractured systems, such a mechanism could be crucial in explaining the way in which supersaturated fluids move through porous media without precipitating minerals along the way, while concentrating mineral deposits in fractures. Although the concentration of mineralization in fractures is usually explained by elevated solute fluxes, such an explanation does not account for the apparent inhibition of mineral precipitation in the porous matrix.

Whether or not interfacial energy ultimately proves to be the primary factor responsible for the absence of mineral precipitation in small pores, the implications for reaction rates and permeability evolution are significant. In many models, minerals are assumed to precipitate evenly on all the surfaces in the porous medium, an assumption that is clearly not the case in the Stø Formation; thus, standard models are likely to overestimate reaction rates and permeability in porous rocks. To determine whether this phenomenon is widespread in geological systems, however, additional field studies of well constrained systems are required; just as importantly, experimental systems that can observe directly the effects of interfacial energy on reaction rates will further help the evaluation of the importance of interfacial energy in geological systems.

ACKNOWLEDGEMENTS

The authors thank an anonymous reviewer for thoughtful comments and suggestions. In addition, J.O. Eckert is thanked for assistance with the BSE imaging. This material is based upon work supported by the National Science Foundation under Grant No. 0744154. Acknowledgment is also made to the Donors of the American Chemical Society Petroleum Research Fund for support of this research. We would also like to express our appreciation to the editors for their suggestions and their patience.

REFERENCES

Aase, N.A., Bjørkum, P.A. & Nadeau, P.H.: The effect of grain-coating microquartz on preservation of reservoir porosity. *Amer. Assoc. Petrol. Geol. Bull.* 80 (1996), pp. 1654–1673.

Adamson, A.W.: *Physical chemistry of surfaces.* 5th ed., Wiley, Hoboken, NJ, 1990.

Aharonov, E. & Katsman, R.: Interaction between pressure solution and clays in stylolite development: Insights from modeling. *Am. J. Sci.* 309 (2009), pp. 607–632.

Baxter, E.F.: *Geochronology— linking the isotopic record with petrology and textures.* Volume 220, Chapter Natural Constraints on Metamorphic Reaction Rates, Geological Society of London, Special Publication, 2003, pp. 183–202.

Baxter, E.F. & DePaolo, D.J.: Field measurement of slow metamorphic reaction rates at temperatures of 500–600°C. *Science* 288 (2000), pp. 1411–1414.

Bear, J.: *Dynamics of fluids in porous media.* American Elsevier Publishing Company, New York, NY, 1972.

Berner, R.A.: *Early diagenesis: a theoretical approach.* Princeton University Press, Princeton, NJ, 1980.

Blum, A. & Stillings, L.: *Chemical weathering rates of silicate minerals.* Volume 31, Chapter Feldspar dissolution kinetics. Mineralogical Society of America, Washington, DC, 1995, pp. 291–351.

Colon, C.F.J., Oelkers, E.H. & Schott, J.: Experimental investigation of the effect of dissolution on sandstone permeability, porosity, and reactive surface area. *Geochim. Cosmochim. Acta* 68 (2004), pp. 805–817.

Costa, A.: Permeability-porosity relationship: A reexamination of the Kozeny-Carman equation based on a fractal pore-spaces geometry assumption.*Geophys. Res. Lett.* 33 (2006), L02318.

Emmanuel, S. & Ague, J.J.: Modeling the impact of nano-pores on mineralization in sedimentary rocks. *Water Resour. Res.* 45 (2009), W04406.

Emmanuel, S. & Berkowitz, B.: Mixing-induced precipitation and porosity evolution in porous media. *Adv. Water Resour.* 28 (2005), pp. 337–344.

Emmanuel, S. & Berkowitz, B.: Effects of pore-size controlled solubility on reactive transport in heterogeneous rock. *Geophys. Res. Lett.* 34 (2007), L06404.

Emmanuel, S., Ague, J.J. & Walderhaug, O.: Interfacial energy effects and the evolution of pore-size distributions during quartz precipitation in sandstone. *Geochim.Cosmochim. Acta* 74 (2010), pp. 3539–3552.

Giesche, H.: Mercury porosimetry: a general (practical) overview. *Part. Part. Syst. Charact.* 23 (2006), pp. 9–19.

Heald, M.T. & Larese, R.E.: Influence of coatings on quartz cementation. *J. Sediment. Res.* 44 (1974), pp. 1269–1274.

Kang, Q., Lichtner, P.C. & Zhang, D.: Lattice Boltzmann pore-scale model for multicomponent reactive transport in porous media. *J. Geophys. Res.* 111 (2006), B05203.

Kate, J.M. & Gokhale, C.S.: A simple method to estimate complete pore size distribution of rocks. *Eng. Geol.* 84 (2008), pp. 48–69.

Kieffer, B., Colon, S.F.J., Oelkers, E.K. & Schott, J.: An experimental study of the reactive surface area of the Fontainebleau sandstone as a function of porosity, permeability, and fluid flow rate. *Geochim. Cosmochim. Acta* 63 (1999), pp. 3525–3534.

Kozeny, J.: ÜberkapillareLeitung der WassersimBoden (Aufstieg, Versickerung und Andwendungdie Bewässerung). Sitzungsber.Akad.Wiss. Wien, Abt. IIa 136, 1927, pp. 271–306.

Lasaga, A.C.: *Kinetictheory in the earth sciences*. Princeton University Press, Princeton, NJ, 1998.

Lichtner, P.C.: The quasi stationary state approximation to coupled mass transport and fluid-rock interaction in porous media. *Geochim. Cosmochim. Acta* 52 (1988), pp. 143–165.

Lide, D.R.: *CRC Handbook of chemistry and physics.* 77th edition, CRC Press, Boca Raton, FL, 1998.

Lukasiewicz, S. & Reed, J.: Specific permeability of porous compacts as described by a capillary model. *J. Am. Ceram. Soc.* 71 (1988), pp. 1008–1014.

Maher, K., Steefel, C.I., DePaolo, D.J. & Viani, B.E.: The mineral dissolution rate conundrum: Insights from reactive transport modeling of U isotopes and pore fluid chemistry in marine sediments. *Geochim. Cosmochim. Acta* 70 (2006), pp. 337–636.

McBride, E.F.: Quartz cement in sandstones: a review. *Earth-Sci. Rev.* 26 (1989), pp. 69–112.

Oelkers, E.H., Bjørkum, P.A. & Murphy, W.M.: A petrographic and computational investigation of quartz cementation and porosity reduction in North Sea sandstones. *Am. J. Sci.* 296 (1996), pp. 420–452.

Oelkers, E.H., Bjørkum, P.A., Walderhaug, O., Nadeau, P.H. & Murphy, W.M.: Making diagenesis obey thermodynamics and kinetics: the case of quartz cementation in sandstones from offshore mid-Norway. *Appl. Geochem.* 15 (2000), pp. 295–309.

Parks, G.A.: Surface and interfacial free energies of quartz. *J. Geophys. Res.* 89 (1984), pp. 3997–4008.

Putnis, A. & Mauthe, G.: The effect of pore size on cementation in porous rocks. *Geofluids* 1 (2001), pp. 37–41.

Saar, M. & Manga, M.: Permeability-porosity relationship in vesicular basalts. *Geophys. Res. Lett.* 26 (1999), pp. 111–114.

Scherer, G.W.: Stress from crystallization of salt. *Cement Concrete Res.* 34 (2004), pp. 1613–1624.

Stumm, W. & Morgan, J.J.: *Aquatic chemistry.* 3rd ed., Wiley Interscience, Hoboken, NJ, 1996.

Uchida, T. Pore-size distributions and the evaluation of permeability in reservoir rocks a proposal of empirical expressions with regard to petrological properties of pores. *J. Jpn. Assoc. Pet. Technol.* 52 (1997), pp. 1–11.

Walderhaug, O. & Bjørkum, P.A.: The effect of stylolite spacing on quartz cementation in the Lower Jurassic Stø Formation, Southern Barents Sea. *J. Sediment. Res.* 73 (2003), pp. 146–156.

Walderhaug, O., Bjørkum, P.A. & Aase, N.E.: Kaolin-coating of stylolites, effect onquartz cementation and general implications for dissolution at mineral interfaces. *J. Sediment. Res.* 76 (2006), pp. 234–243.

White, A.F. & Brantley, S.L.: The effect of time on the weathering of silicate minerals: Why do weathering rates differ in the laboratory and field? *Chem. Geol.* 202 (2003), pp. 479–506.

Yardley, B.: On some quartz-plagioclase veins in the Connemara schists, Ireland. *Geol. Mag.* 112 (1975), pp. 183–190.
Yokoyama, T. & Banfield, J.F.: Direct determinations of the rates of rhyolite dissolution and clay formation over 52,000 years and comparison with laboratory measurements. *Geochim. Cosmochim. Acta* 66 (2002), pp. 2665–2681.
Yoo, K. & Mudd, S.M.: Discrepancy between mineral residence time and soil age: Implications for the interpretation of chemical weathering rates. *Geology* 36 (2008), pp. 35–38.
Yu, L., Evje, S., Kleppe, S., Karstad, T., Fjelde, I. & Skjaeveland, S.M.: Spontaneous imbibitions of seawater into preferentially oil-wet chalk cores—experiments and simulations. *J. Petrol. Sci. Eng.* 66 (2009), pp. 171–179.
Zhang, P. & Austad, T.: Wettability and oil recovery from carbonates: effects of temperature and potential determining ions. *Colloids Surf., A Physicochem. Eng. Asp.* 279 (2006), pp. 179–187.

CHAPTER 10

Geochemical modeling of water chemistry evolution in the Guarani Aquifer System in São Paulo, Brazil

Ondra Sracek & Ricardo Hirata

"The purpose of science is not to analyse or describe but to make useful models of the world. A model is useful if it allows us to get use out of it."

Edward de Bono

The Guarani Aquifer System (the SAG based on Portuguese abbreviation) is one of the largest aquifer systems in the world (second after the Nubian Sandstone Aquifer). It is a transboundary aquifer, since it extends to four countries including Brazil, Argentina, Uruguay, and Paraguay (Araujo *et al.*, 1999). In São Paulo state (Brazil), the Guarani system comprises the Botucatu and Piramboia aquifers (Fig. 10.1), which are confined by the underlying Passa Dois and Rio do Rastro aquitards and the overlying Serra Geral Formation. The aquifer is dipping towards the west, e.g., towards the Paraná River (Fig. 10.2). The Piramboia Formation consists of silty-clay sandstones of aeolian and fluvial origin; the Botucato Formation is composed of well-sorted sandstones of aeolian origin with dominating quartz plus feldspar and micas (Hirata *et al.*, 2011). Cation exchange capacity is around 1 meq/100 g. Average value of hydraulic conductivity of the most productive Botucato Aquifer is 1.5×10^{-4} m s^{-1} with resulting average transmissivity of 3.9×10^{-3} m^2 s^{-1}. Calculated flow velocities are low due to low hydraulic gradients, in the order of tens of meters per year (Sracek and Hirata, 2002). Flow in São Paulo state is also towards the Paraná River (Fig. 10.1). Discharge zone of the SAG is not well known, but is located out of the study area in São Paulo state and probably out of Brazil in Uruguay and Argentina (Araujo *et al.*, 1999).

There is an evolution from Ca-HCO_3 groundwater type with low mineralization and neutral *pH* in recharge zone towards Na-HCO_3 groundwater type with high mineralization and high *pH* values (above 9.0) in the deep confined zone (Fig. 10.3) (Silva, 1983; Meng and Maynard, 2001; Gastmans *et al.*, 2010). Simultaneously, concentration of F^- increases up to 13 mg L^{-1} and temperature reaches 63°C in Presidente Prudente well close to the Paraná River. Redox conditions are oxic everywhere in the study area in São Paulo state and redox processes do not play any role in groundwater chemistry evolution (Sracek and Hirata, 2002). This is a consequence of the SAG sedimentary deposition in an aeolian environment without organic matter (Hirata *et al.*, 2011).

Mass balance calculations aiming to explain the origin of Na are in Table 10.1. It is evident that the increase of Na concentrations observed downgradient cannot be explained only by dissolution of minerals such as halite (NaCl), mirabilite ($Na_2SO_4 \cdot 10H_2O$) or silicates such as albite ($NaAlSi_3O_8$). Amounts of Na contributed by the dissolution of halite is calculated on the basis of the change of Cl^- concentration, the same calculation for mirabilite is based on SO_4^{2-}, and for albite is based on dissolved Si.

On the basis of ionic ratios and mass balance calculations presented in Table 10.1, it has been concluded that the increase of Na concentration cannot be caused only by the dissolution of evaporites like halite (NaCl) and thenardite ($NaSO_4$), not by dissolution of plagioclases such as albite, and an alternative source of Na has to be present. It has been suggested by Sracek and Hirata (2002) that cation exchange combined with dissolution of carbonates may explain the formation of Na-HCO_3 groundwater type in confined zone. The conceptual model of groundwater chemistry evolution can be summarized as:

$$CaCO_3 + CO_2 + H_2O + Na_2\text{–}X \rightleftarrows 2Na^+ + 2HCO_3^- + Ca\text{–}X \tag{10.1}$$

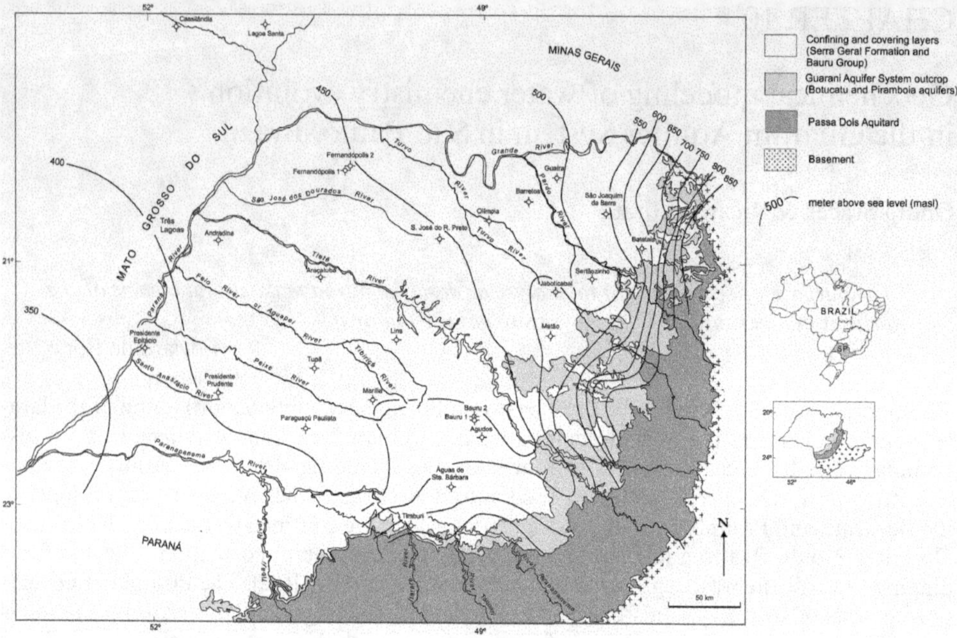

Figure 10.1. Guarani Aquifer System in São Paulo state, Brazil (after Sracek and Hirata, 2002) with potentiometric lines. The profile B'–B is shown in Figure 10.2.

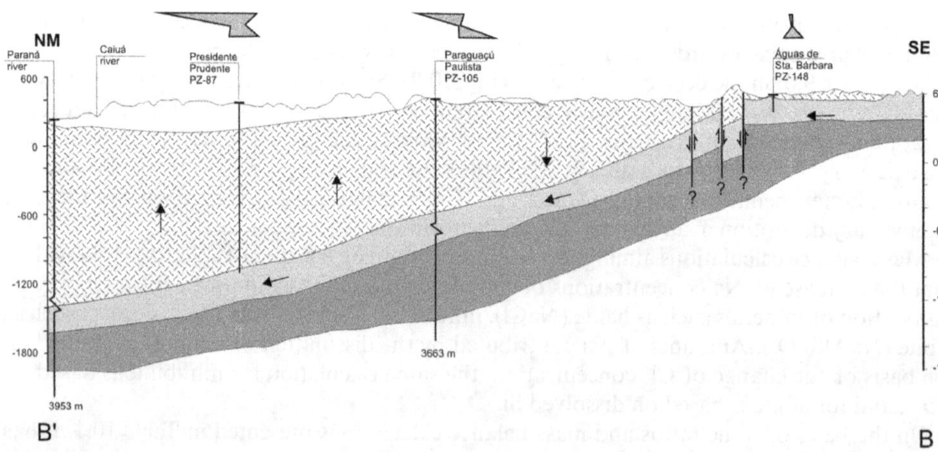

Figure 10.2. Cross section of the Guarani Aquifer System (in light gray shading) in São Paulo state, Brazil (after Sracek and Hirata, 2002). Groundwater chemistry in the SAG is indicated by the Stiff diagrams above the profile (upper left corner indicates Na^+ concentration, lower right corner indicates HCO_3^- concentrations).

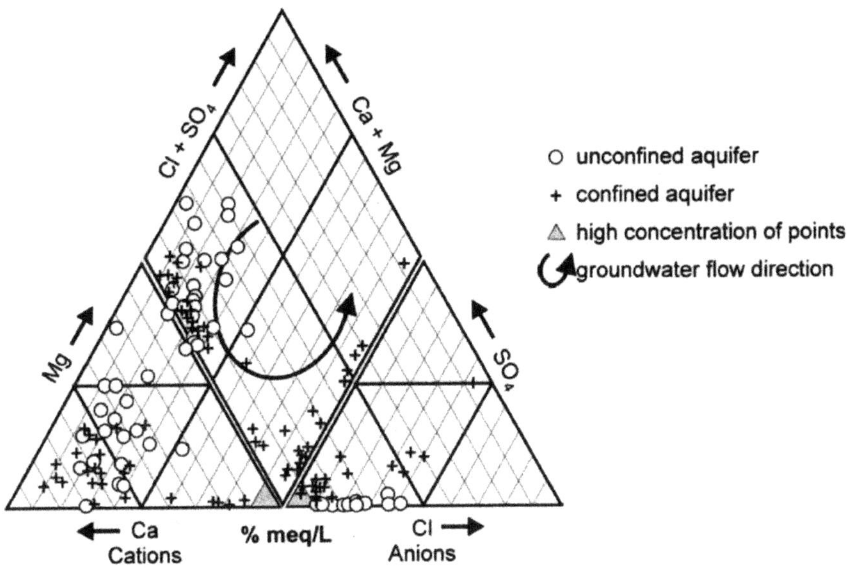

Figure 10.3. The Piper diagram for groundwater samples from the Guarani Aquifer in São Paulo, Brazil (after Sracek and Hirata, 2002). There is an evident trend for groundwater evolution from Ca-HCO_3 type in unconfined zone towards Na-HCO_3 groundwater type in confined zone.

Table 10.1. Manual mass balance calculation for Na (based on Sracek and Hirata, 2002). Units are mmol L^{-1}.

Na source/profile	+ΔNa total	+ΔNa halite	+ΔNa mirabilite	+ΔNa albite	Unaccounted Na
PZ 34-17	3.79	0.367	0.270	0.082	3.07
PZ 148-105	5.16	0.446	0.104	0.117	4.49
PZ 105-87	4.21	3.289	0.854	–	0.07

where X denotes exchange sites. This reaction removes from groundwater Ca^{2+} produced by dissolution of calcite and increases pH and dissolved inorganic carbon (DIC) concentrations. The fluorite (CaF_2) may also dissolve due to removal of Ca^{2+} from groundwater,

$$CaF_2 \rightleftarrows Ca^{2+} + 2F^- \tag{10.2}$$

The conceptual model has been tested using speciation modeling followed by inverse geochemical modeling. A similar approach has been used, for example, by Weaver and Bahr (1991), Gerla (1993), and more recently by Güler and Thyne (2004). Mineralogical information about the solid phase was limited, but basic solid phase characteristics were known.

Selected results of speciation calculations for groundwater samples performed by the code PHREEQC (Parkhurst and Appelo, 1999) are shown in Table 10.2.

There is an increase of DIC concentration between wells PZ-148 and PZ-105. This behavior indicates calcite dissolution, which is also supported by negative $SI_{calcite}$ values. Another piece of evidence for calcite dissolution are $\delta^{13}C(DIC)$ values increasing towards the Paraná River (Gastmans *et al.*, 2010). Values of $\delta^{13}C$ in calcite are close to 0‰ and dissolution of calcite results in enrichment of dissolved carbon in heavy isotope ^{13}C and also in decreasing activity of radioactive ^{14}C. On the other hand, the situation changes in well PZ-87 (Presidente Prudente).

Table 10.2. Selected results of speciation calculation for wells located on the profile B'–B in the Guarani Aquifer System.

Parameter/well	$SI_{calcite}$	SI_{gypsum}	$SI_{fluorite}$	p_{CO_2} [atm]	*DIC* [mmol L^{-1}]
PZ-148	−1.21	n.a.	−3.91	−3.22	1.45
PZ-105	−0.70	−4.67	−2.83	−2.75	4.68
PZ-87	0.63	−2.93	−0.48	−3.47	3.13

n.a.—not available

There, saturation with respect to calcite is reached and calculated p_{CO_2} becomes very low. This is linked to the increase of groundwater temperature reaching 63°C (Table 10.1). Based on the SAG conceptual model (Sracek and Hirata, 2002; Gastmans *et al.*, 2010), the well is already located behind cation exchange/calcite dissolution front. This is also consistent with calcitic cement found in the SAG material in this deep confined zone, which is missing in the recharge zone (Hirata *et al.*, 2011).

Next step included inverse geochemical modeling based on the presented conceptual model and speciation calculations. This type of modeling requires initial and final wells water chemistry and composition of solid phase between both wells (Zhu and Anderson, 2002). Furthermore, these wells have to be hydraulically connected and the system must be in geochemical steady-state, i.e., groundwater chemistry does not change in time, only in space.

The inverse geochemical modeling performed by the NETPATH code (Plummer *et al.*, 1994) included Ca, Na, C, S, and Cl as elements (constraints in NETPATH terminology) and calcite, halite, gypsum, CO_2(g), and Ca/Na exchange as phases. The values of $\delta^{13}C$ were not used directly in the modeling, but only for evaluation of the conceptual model. Results are presented in the Table 10.3. They indicate that dissolution of calcite and Na-input from exchange sites may explain the changes of groundwater chemistry on the first profile between well PZ-148 in recharge zone and well PZ-105 in deep confined zone.

The situation becomes different on the second profile between wells PZ-105 and PZ-87, where the Na^+ concentration still increases and so does Cl^- concentration, although HCO_3^- and, consequently, *DIC* decrease. The value of log p_{CO_2} also decreases, reaching almost atmospheric value of −3.47. This is consistent with positive value of saturation index for calcite in well PZ-87 (Presidente Prudente) (Table 10.2). This profile is already located behind cation exchange front, which moves slowly downgradient and presently is located in the proximity of well PZ-105. Thus, most of Na^+ increase on the profile can be accounted for only by the dissolution of halite (Table 10.1).

The results of inverse geochemical modeling for second profile located already in deep confined zone close to the Paraná River are also in Table 10.3. There is a significant input of Cl^-, presumably from dissolution of halite in underlying formations. Gypsum also dissolves and accounts for increasing sulfate concentration. Cation exchange is still in operation, but it is much more limited compared to the upgradient zone between wells PZ-148 and PZ-105. On the other hand, calcite precipitates and there is some degassing of CO_2. These processes may be caused by high groundwater temperature, which reaches 63°C in flowing artesian well PZ-87 (Presidente Prudente).

This conceptual model was confirmed in recent study with expanded data from Gastmans *et al.* (2010), where the authors evaluated a larger area (including both Mato Grosso do Sul west of the Paraná River and São Paulo states). They concluded that cation exchange/calcite dissolution occurs at a relatively narrow front located in the transition zone between unconfined zone and deep confined zone. The petrological study of the SAG sandstones confirmed the presence of carbonate cement in deep confined zone. However, in recharge zone sandstone is carbonate-free as a consequence of calcite dissolution and, in contrast, only dissolution of silicates, formation of kaolinite and precipitation of amorphous silica occur.

Table 10.3. Output of inverse geochemical modeling for zone between wells PZ-148 and PZ-105, and wells PZ-105 and PZ-87, sign minus indicates removal from ground water. Mass transfer units are mmol L^{-1}.

Phase	PZ-148–PZ-105 Mass transfer	PZ-105–PZ-87 Mass transfer
Halite	0.45	3.31
Calcite	2.25	−0.49
Gypsum	0.10	0.853
CO_2(g)	0.97	−0.96
Na_2-*X*	2.36	0.528
Ca-*X*	−2.36	−0.264

The SAG in São Paulo state (and also in Mato Grosso do Sul state west of the Paraná River) can be divided into three zones (Gastmans *et al.*, 2010):

- Zone 1: This is the recharge zone where the aquifer is unconfined. The main processes are feldspar and remaining carbonate cement dissolution. Groundwater is of Ca-HCO_3 or Ca-Mg-HCO_3 type and the sandstones contain chalcedony cement.
- Zone 2: This represents intermediate zone between recharge zone and deep confined zone, where the aquifer is confined by the Serra Geral basalt. The main processes are cation exchange and dissolution of carbonate cement and groundwater becomes of Na-HCO_3 type. The sandstones contain significant amounts of carbonate cement. This zone corresponds to the first profile PZ-148–PZ-105 in inverse geochemical modeling.
- Zone 3: This is a deep confined zone where the Serra Geral basalts are more than 500 m thick. Groundwater is enriched in Cl and SO_4, probably as a consequence of their input from the underlying formations. Groundwater is supersaturated with respect to calcite and dissolution of carbonate cement does not occur (Hirata *et al.*, 2011). This zone corresponds to the second profile PZ-105–PZ-87 in inverse geochemical modeling. Increasing temperature results in mobilization of fluorine. However, it is unknown if F^- source is directly in the SAG or in the overlying Serra Geral Formation.

The study demonstrates how investigations of groundwater chemistry and solid phase composition combined with speciation and inverse geochemical modeling can be used to verify a conceptual model of processes in a large aquifer system. Even when solid phase characterization is limited, concentration changes along a flow path combined with values of saturation indices and, in an ideal case, with isotopic data, may provide an input for inverse geochemical modeling. Especially valuable for evaluation of water/rock interactions are $\delta^{13}C$ values for dissolved inorganic carbon. The SAG system is oxic everywhere in São Paulo state and carbonates in solid phase are the only source of inorganic carbon in deep confined zone. However, when conditions are reducing and processes such as methanogenesis have to be taken into account, situation becomes more complicated. Multiple sources of organic matter have to be considered and the modeling becomes more complex (Aravena *et al.*, 1995; Noseck *et al.*, 2009).

A common problem of inverse geochemical modeling is its non-uniqueness, i.e., several models may explain observed water chemistry evolution. However, application of isotopes generally reduces greatly the number of possible models.

REFERENCES

Araujo, L.M., Franca, A.B. & Potter, P.E. Hydrogeology of the Mercosul aquifer system in the Parana and Chaco-Parana Basins, South America, and comparison with Navajo-Nugget aquifer system, USA. *Hydrogeol. J.* 7 (1999), pp. 317–336.

Aravena, R., Wassenaar, L.I. & Plummer, L.N.: Estimating ^{14}C groundwater ages in a methanogenic aquifer. *Water Resour. Res.* 31 (1995), pp. 2307–2317.

Gastmans, D., Chang, H.K. & Hutcheon, I.: Groundwater geochemical evolution in the northern portion of the Guarani Aquifer System (Brazil) and its relationship to the diagenetic features. *Appl. Geochem.* 25 (2010), pp. 16–33.

Gerla, P.J.: Pathline and geochemical evolution of groundwater in a regional discharge area, Red River Valley, North Dakota. *Groundwater* 30 (1992), pp. 743–754.

Güler, C. & Thyne, G.D.: Hydrologic and geologic factors controlling surface and groundwater chemistry in Indian Wells-Owens Valley area, southeastern California, USA. *J. Hydrol.* 285 (2004), pp. 177–198.

Hirata, R., Gesicki, A., Sracek, O., Bertolo, R., Giannini, P.C. & Aravena, R.: Relation between sedimentary framework and hydrogeology in the Guarani Aquifer System in São Paulo state, Brazil. *J. South Amer. Earth Sci.* 31 (2011), pp. 444–456.

Meng, S.X. & Meynard, J.B.: Use of statistical analysis to formulate conceptual models of geochemical behavior: water chemical data from the Botucato aquifer in Sao Paulo state, Brazil. *J. Hydrology* 250 (2001), pp. 78–97.

Noseck, U., Rozanski, K., Dulinski, M., Havlová, V., Sracek, O., Brasser, T., Hercik, M. & Buckau, G.: Carbon chemistry and groundwater dynamics at natural analogue site Ruprechtov, Czech Republic: Insights from environmental isotopes. *Appl. Geochem.* 24 (2009), pp. 1765–1776.

Parkhurst, D.L. & Appelo, C.A.J.: User's guide to PHREEQC: A computer program for speciation, reaction-path, 1-D transport, and inverse geochemical calculations. *U.S. Geol. Surv. Water Res. Invest. Rep.* 99-4259, 1999.

Plummer, R.N., Prestemon, E.C. & Parkhurst, D.L.: An interactive code (NETPATH) for modeling of NET geochemical reactions along a flow PATH, Version 2.0. *U.S. Geol. Surv. Water Res. Invest. Rep.* 94–4169, 1994.

Silva, R.B.G.: *Hydrogeochemical and isotopic study of groundwater of the Botucato Aquifer in São Paulo state* (in Portuguese). PhD thesis, University of São Paulo, Brazil, 1983.

Sracek, O. & Hirata, R.: Geochemical and stable isotopic evolution of the Guarani Aquifer System in the state of São Paulo, Brazil. *Hydrogeology J.* 10 (2002), pp. 643–655.

Weaver, T.R. & Bahr, J.M.: Geochemical evolution in the Cambrian-Ordovician Sandstone Aquifer, Eastern Wisconsin, 2. Correlation between flow paths and groundwater chemistry. *Groundwater* 29 (1991), pp. 510–515.

Zhu, C. & Anderson, G.: *Environmental applications of geochemical modelling*. Cambridge University Press, 2002.

CHAPTER 11

Modeling of reactive transport at a site contaminated by petroleum hydrocarbons at Hnevice, Czech Republic

Ondra Sracek & ZbynekVencelides

> *"Modeling has become so entrenched in our professional culture that it has become common for reviewers of papers presenting observational studies to call for "validation" of the observational results through the running of a model!! This is a grotesque inversion of the scientific method; we should not ask that observations fit some model but, rather, that the model fit the observations. My experiences along these lines obviously are a partial source of the frustration that has prompted this diatribe."*
>
> Chuck Doswell

11.1 SITE CHARACTERIZATION AND CONCEPTUAL MODEL

Natural attenuation (NA) of petroleum hydrocarbons (PHC) has become increasingly popular as a substitute for classical remediation approaches such as pump-and-treat. At PHC-contaminated sites, the NA processes are in operation when a sufficient supply of electron acceptors such as oxygen and nitrate is available (Wiedemeier *et al.*, 1999). However, other processes than the PHC degradation can consume electron acceptors (Chapelle *et al.*, 2002; Schreiber and Bahr, 1999). Ferric iron may play a key role in electron mass balance (Christensen *et al.*, 2000), especially at sites like Bemidji in Minnesota, USA, where concentrations of other electron acceptors are limited (Cozzarelli *et al.*, 2001).

There is a contamination by petroleum hydrocarbons at Hnevice site close to the river Labe (Elbe), north of Prague in the Czech Republic (Fig. 11.1). Several PHC releases occurred since 1940s. In late 1980's, the free phase plume shrunk as a consequence of dissolution and remediation pumping. Groundwater flow in alluvial sediments composed of gravel and coarse sand is roughly northward i.e., parallel to the river Labe. Velocity of flow is high, up to 1.0 m day^{-1} and seasonal water level changes in Labe are causing transient flow direction changes in the aquifer. However, these changes do not impact the flow path evolution of geochemical processes in a very wide plume. The study site is located upgradient of the free phase plume, where groundwater with electron acceptors enter the zone previously contaminated by free phase and the recent free phase plume (Vencelides *et al.*, 2007).

Several multilevel piezometric nests were installed in the background zone, in the transition zone, which was previously contaminated, but is without free phase now, and in the recent free phase plume (Fig. 11.1). Depth-resolution of sampling close to water table was 10 cm and increased to 1 m towards the base of the aquifer at a depth of about 10 m bgs. Profiles of selected species concentrations are in Figure 11.2. In piezometric nest HJ506 located at the limit of free phase plume in the past, the concentration of total PHC is close to zero, just like at the piezometric nest HJ507. Then total PHC concentration increases at piezometric nest HJ508 and reaches maximum values at piezometric nest PJ519 and HJ520 located in recent plume. Concentrations of dissolved oxygen (*DO*) are depressed in the recent plume, but they are at maximum values at intermediate depth even at uncontaminated piezometric nest HJ506, probably reflecting natural stratification. Maximum concentrations of

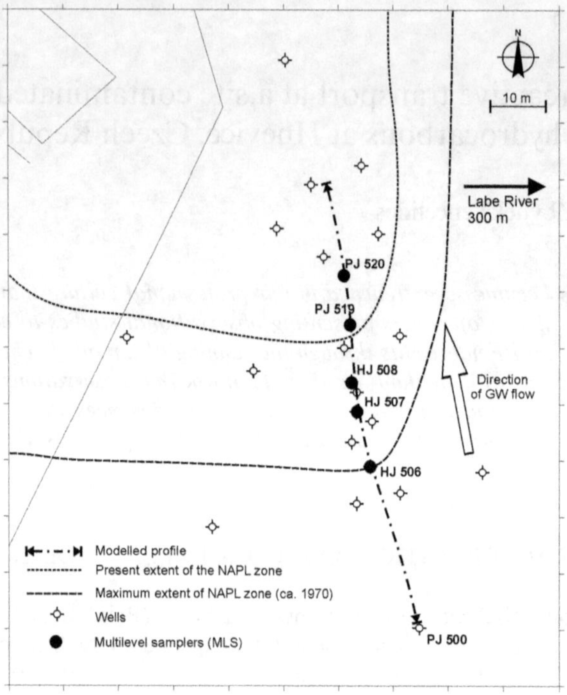

Figure 11.1. Location of the Hnevice site, previous and recent free phase plumes boundaries and piezometric nests (after Vencelides *et al.*, 2007).

Fe^{2+}, Mn^{2+} and HCO_3^- are at HJ519 and HJ520 close to water table, as a consequence of total PHC degradation. A notable feature is the overlap between several redox zones, for example at PJ520 consumption of SO_4^{2-} starts when *DO* and NO_3^- are still present and dissolution of Fe(III) minerals takes place. This may be caused by redox zonality at a finer scale than that of sampling. However, the overlap has been observed at several sites like Bemidji (Cozzarelli *et al.* (2001) and may also be caused by other factors like limited reactivities of Fe(III) minerals (Jakobson and Postma, 1999).

Significant concentrations of methane are present only at HJ519 and HJ520 in recent plume close to water table. Relatively high background concentrations of NO_3^- (>100 mg L^{-1}) and SO_4^{2-} (>300 mg L^{-1}) are caused by their upgradient sources (hop orchards for nitrate and fly ash lagoons for sulfate, respectively).

Currently there are 4 different geochemical zones: background zone never invaded by PHC free phase (Zone I), former PHC free phase zone, now re-oxidation zone (Zone IV); recent PHC free phase plume (Zone III), and fringe zone between recent free phase plume and deep uncontaminated zone, Zone II (Fig. 11.3). The detailed study site was located at the rear of the plume, where background ground water with high concentrations of electron acceptors mixes with contaminated ground water. Conceptual model of this site includes (i) initial dissolution of free phase, dissolution of Fe(III) minerals such as ferrihydrite, precipitation of mixed Fe(II)-carbonates and possibly Fe-sulfides, degassing of CO_2 and CH_4, and (ii) later flux of electron acceptors into the former free phase zone and oxidation of Fe(II) minerals.

The aim of modeling presented here was the verification of conceptual model of geochemical processes at the Hnevice site and the assessment of the role of secondary ferrous minerals formed in the zone contaminated by PHC in the past, but free of PHC recently.

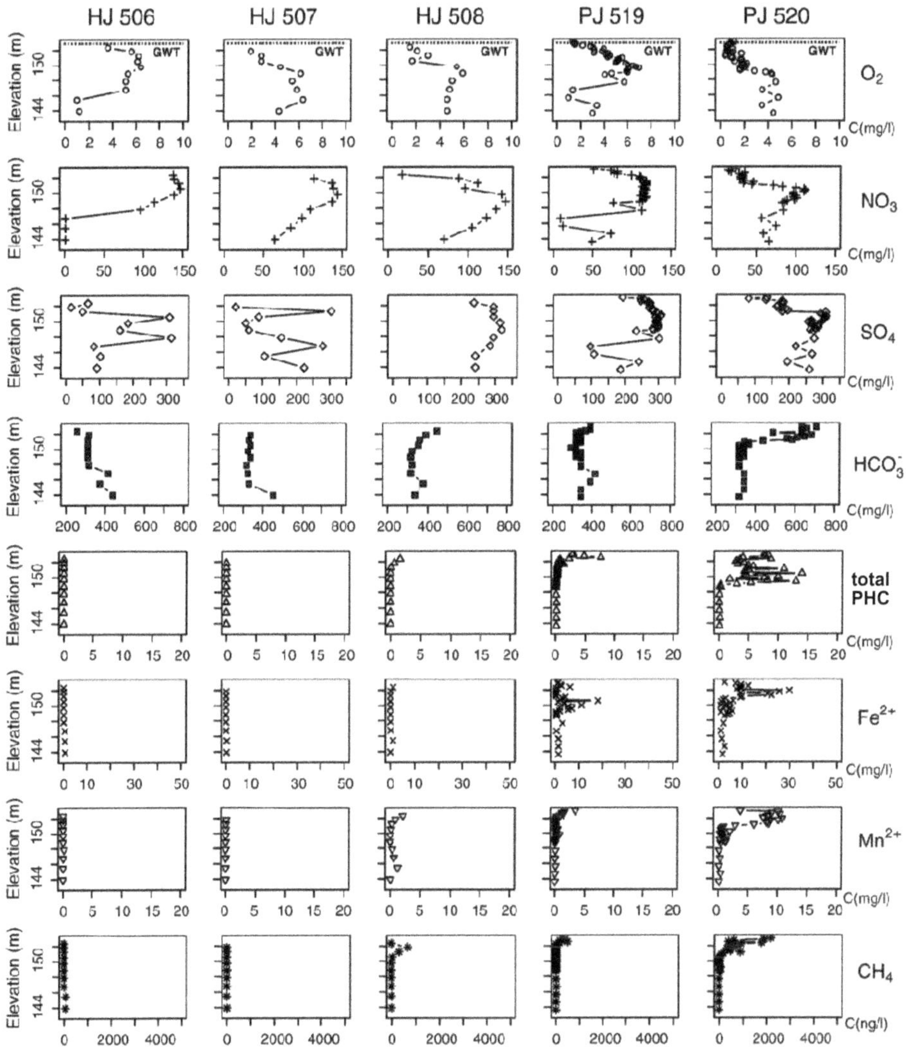

Figure 11.2. Vertical concentration profiles for selected species sampled in piezometric nests (after Vencelides *et al.*, 2007).

11.2 SPECIATION AND INVERSE GEOCHEMICAL MODELING

Speciation and inverse geochemical modeling was performed by the program PHREEQC (Parkhurst and Appelo, 1999). Results of speciation calculations for selected sampling points located close to water table are in Table 11.1.

At site HJ507, which is currently without PHC contamination, groundwater is supersaturated with respect to calcite, but undersaturated with respect to other carbonate minerals. Values of log p_{CO_2} and *DIC* are already relatively high. Site HJ519 represents the margin of the recent plume where saturation is also reached with respect to rhodochrosite and siderite and values of log p_{CO_2} and *DIC* increase compared to the upgradient uncontaminated site. Site HJ520 is located in the middle of the recent plume and saturation is reached with respect to all carbonate minerals except dolomite. Extremely high values of p_{CO_2} reaching

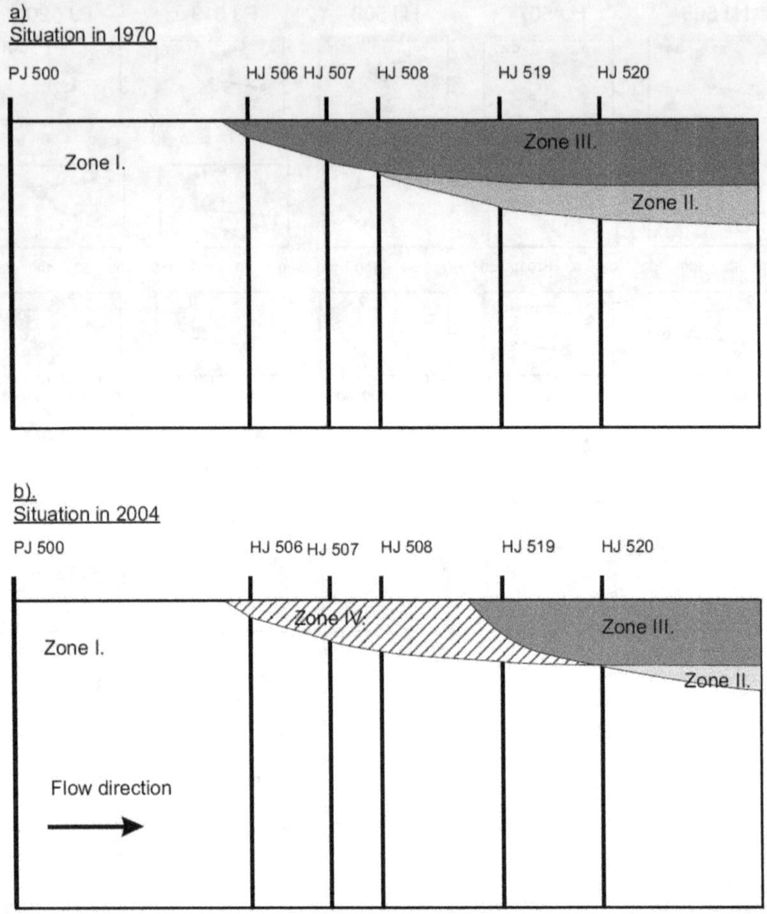

Figure 11.3. Conceptual model of geochemical zones at Hnevice site (after Vencelides *et al.*, 2007). Cross section follows profile at Figure 11.1.

Table 11.1. Selected speciation results.

Sample	$SI_{calcite}$	$SI_{dolomite}$	$SI_{rhodochrosite}$	$SI_{siderite}$	$\log P_{CO_2}$ [atm]	*DIC* [mmol L^{-1}]
HJ507	0.21	–0.47	–0.92	–0.14	–1.80	6.48
HJ519	0.15	–0.41	0.40	0.37	–1.65	7.64
HJ520	0.08	–0.41	1.23	0.78	–1.15	15.66

–1.15 and *DIC* reaching 15.66 mmol L^{-1} are a consequence of intense degradation of PHC, producing inorganic carbon. The results of a mineralogical investigation of the solid phase were consistent with calculated saturation indices, i.e., the mixed Fe(II), Mn(II)-carbonates plus FeS mineral phase found in the present PHC plume (Vencelides *et al.*, 2007).

Inverse modeling was performed for the segment between sampling points PJ519 and PJ520 (Fig. 11.2). This segment is located in the zone between the margin and central zone of the recent contaminant plume. Selected phases were: goethite, manganite, calcium-rich siderite

Table 11.2. Selected output of inverse geochemical modeling at Hnevice site.

Phase	Run 1: Mass transfer [mmol L^{-1}]	Run 2: Mass transfer [mmol L^{-1}]
Toluene C_7H_8	9.13	8.84
$Fe_{0.8}Ca_{0.2}CO_3$	–3.01	–3.01
Ferrihydrite, $Fe(OH)_3$(a)	3.57	2.46
Manganite MnOOH	0.14	0.14
Mackinawite FeS	–1.11	–
N_2(g)	–0.46	–0.46
CH_4(g)	–3.73	–3.68
CO_2(g)	–2.07	–1.83
H_2S(g)	–	–1.11

$Fe_{0.8}Ca_{0.2}CO_3$, organic matter with composition C_7H_8 (toluene), mackinawite, and degassing of CO_2(g), N_2(g), and CH_4(g). The composition of Fe(II) carbonate mineral composition is consistent with data from the Bemidji site (Tuccillo *et al.*, 1999) and gave the best results. Any other Ca/Fe ratios generally caused convergence problems in the inverse modeling calculations. The organic matter composition roughly corresponds to toluene. Data on $\delta^{13}C(DIC)$ were used for calibration of inverse geochemical modeling. The δ^{13} C values for the mineral phases used in the model were (in per mil.): Fe(II) carbonate –1.0, C_7H_8 –27.0, CO_2(g) –15.0, and CH_4(g) –60.0. The ranges of uncertainties were ±2 per mil for all C phases except for methane, were the value was ±5 per mil. The uncertainty for dissolved concentrations was 0.165 (e.g., 16.5% range around entered concentration values was allowed).

First, SOLUTION 1 (here PJ519) and SOLUTION 2 (here PJ520), i.e., initial solution and final solution in the terminology of inverse geochemical modeling, are defined. Samples are located on the same flowline. Values of $\delta^{13}C$ isotope with range of possible values are entered for relevant phases (carbonate minerals, toluene, and methane). In Run 1 precipitation of mackinawite, FeS, is allowed, in Run 2, reduced sulfur is degassed as H_2S.

Selected output of inverse modeling is in Table 11.2. First column shows mass transfers in mol L^{-1}. Positive sign indicates dissolution or degradation (for example, $Fe(OH)_3$(a) and toluene), negative sign indicates precipitation (for example, mackinawite). Negative signs for gases (here N_2, CO_2 and CH_4) indicate degassing. In Run 1 more ferrihydrite is dissolved because dissolved iron is incorporated into precipitated mackinawite. In Run 2, less ferrihydrite is dissolved because no sulfide minerals precipitate. In both runs degassing of CO_2(g) is necessary to account for relatively constant *pH* along flowpath. Degassing of N_2(g) account for nitrate reduction in the plume. Transfer coefficients for other phases are similar or the same (e.g., manganite). The modeling results are also constrained by $\delta^{13}C$ values.

11.3 MODELING OF REACTIVE TRANSPORT

Results of inverse geochemical modeling were considered in the choice of suitable phases for modeling of reactive transport. The reactive transport modeling was performed by the program PHT3D (Prommer *et al.*, 2003). The mixture of hydrocarbons was modeled as toluene plus residual fraction of low solubility. Initial fraction of toluene in the mixture was 0.5. The dissolution of the 2-component mixture was calculated by Raoult's Law. The degradation of toluene was modeled by 1st order kinetics, where toluene was first transferred to the pool available for degradation and then degraded. The dissolution/precipitation of ferrihydrite, siderite and mackinawite were modeled as kinetic reactions depending on saturation index SI. Degassing of CO_2, N_2 and CH_4 was allowed when their partial pressures exceeded hydrostatic pressure.

A 2-D cross-sectional area 80 m and 2 m thick oriented in the direction of groundwater flow was used for modeling. Horizontal discretisation was 1.0 m and vertical discretisation was 0.1 m. The finer vertical discretisation was necessary to account for steep vertical concentration gradients at plume fringe. Hydraulic conductivity of 1.9×10^{-3} m s^{-1} and effective porosity of 0.25 were constant over the domain and constant head boundary conditions were used at the inflow and outflow boundaries. Running time was 10 years.

Selected results of reactive transport modeling in 2-D domain are shown in Figure 11.4. In the base case, intermediate dispersivities $\alpha_L = 0.1$ m and $\alpha_V = 0.001$ m based on performed tracer test results were used. In the high dispersivity case, values of $\alpha_L = 0.5$ m and $\alpha_V = 0.005$ m were used; and in the low dispersivity case values of $\alpha_L = 0.05$ m and $\alpha_V = 0.0005$ m were used. The highest concentrations of toluene are found for low dispersivities, while the lowest concentrations were found for the highest dispersivities. In the high dispersivity case the end of dissolved plume is found about 42 m from upstream end of the domain. The depletion of ferrihydrite starts at about 31 m. Upstream of this location presence of dissolved oxygen and nitrate prevent consumption of ferrihydrite as an electron acceptor. An enrichment in ferrihydrite which is secondary with respect to the contamination event starts at lower fringe of the plume at elevation of about 151.5 m. There is re-oxidation of ferrous iron transported out of the plume and precipitation of ferrihydrite,

$$Fe^{2+} + 0.25O_2 + 2.5H_2O \rightleftarrows Fe(OH)_3 + 2H^+$$

The thickness of the secondary ferrihydrite precipitation zone is lowest in the case of the smallest value of vertical dispersivity because less Fe(II) is transported out of the plume. Re-oxidation of reduced species (i.e., secondary redox reactions (SRR)) at a plume fringe was found to be important also in other modeling studies (e.g., Hunter *et al.*, 1998; Van Breukelen and Griffioen, 2004).

The values of dispersivities have a strong impact on the amount of precipitated Fe(II) minerals. The amount of precipitated siderite is highest for lowest dispersivity values and lowest for higher dispersivity values. High dispersivity decreases dissolved Fe(II) concentrations, thus decreasing the amount of precipitated ferrous minerals such as siderite and *vice versa*.

The same trends as those for siderite are observed for FeS, but no precipitation is observed for highest dispersivity values because dispersion caused low dissolved iron concentrations. Furthermore, the FeS precipitation zone is located closer to the plume core than the siderite precipitation zone and, thus, FeS precipitation occurs in the more reducing environment. This is consistent with the onset of sulfate reduction after the ferrihydrite reduction.

Other important factors determining the amounts of precipitated siderite and FeS are kinetic rate constants. An extreme case are equilibrium reactions with maximum precipitated amounts, but with low Fe(II) concentrations in groundwater. However, based on high observed dissolved iron concentrations both phases precipitate and dissolve kinetically. For this reason, kinetic rate constants were adjusted to fit dissolved iron concentrations.

The modeling was able to reproduce partial overlap of several electron acceptors consuming zone indicated by both field observations and previous reactive transport modeling studies (Cozzarelli *et al.*, 2001; Prommer *et al.*, 1999; Schreiber *et al.*, 2004). Possible explanations were adsorption of Fe(II) and heterogeneity of Fe(III) availability in the solid phase.

The modeling also confirmed an important role of the re-oxidation zone (Zone IV in Fig. 11.3.) in mass balance of electron acceptors. After retreat of free phase plume large amounts of minerals such as siderite and FeS are left behind and their oxidation consumes a significant portion of dissolved oxygen and nitrate, which are not available for attenuation of recent free phase plume (Vencelides *et al.*, 2007). For example, when kinetic constants based on fits of the observed dissolved iron were used, only about 90% of electron acceptors reached the recent plume during the re-oxidation period. This means that the presence of a zone with reduced minerals increases the time necessary for natural attenuation of the remaining contamination.

Figure 11.4. Results of reactive transport modeling for different values of dispersivities: left-base case, middle-high dispersivities, right-low dispersivities (after Vencelides *et al.*, 2007).

In summary, reactions at the plume fringe and in the re-oxidation zone have a strong impact on natural attenuation because they influence the availability of electron acceptors within a plume. The modeling presented indicates a significant role of Fe(II) re-oxidation at the lower edge of the plume. However, this remains a controversial issue (Van Breukelen and Griffoen, 2004) and supporting field data are limited. Some modeling studies included cation exchange of Fe(II), which eliminated re-precipitation of iron (Brun *et al.*, 2002).

This study shows an important role of modeling in the identification of processes in contaminant plumes and prediction of their behavior in future. The approach following the steps: collection and analysis of water from multilevel piezometers, calculation of saturation indices combined with mineralogical investigation of solid phase, inverse geochemical modeling supported by isotopic data and finally reactive transport modeling can be used at other natural attenuation sites.

REFERENCES

Brun, A., Engesgaard, P., Christensen, T.H. & Rosbjerg, D.: Modeling of transport and biogeochemical processes in pollution plumes: Vejen landfill, Denmark. *J. Hydrol.* 256:3–4 (2002), pp. 228–247.

Chapelle, F.H., Bradley, P.M., Lovley, D.R., O'Neill, K. & Landmeyer, J.E.: Rapid evolution of redox processes in a petroleum hydrocarbon-contaminated aquifer. *Ground Water* 40:4 (2002), pp. 353–360.

Christensen, T.H., Bjerg, P.L., Banwart, S.A., Jakobsen, R., Heron, G. & Albrechtsen, H.-J.: Characterization of redox conditions in groundwater contaminant plumes. *J. Contam. Hydrol.* 45 (2000), pp. 165–241.

Cozzarelli, I.M., Bekins, B., Baedecker, M.J., Aiken, G.R., Eganhouse, R.P. & Tuccillo, M.E.: Progression of natural attenuation processes at a crude oil spill site: I. Geochemical evolution of the plume. *J. Contam. Hydrol.* 52 (2001), pp. 369–385.

Hunter, K.S., Wang, Y. & Van Capellen, P.: Kinetic modeling of microbially-driven redox chemistry of subsurface environments: coupling transport, microbial metabolism and geochemistry. *J. Hydrol.* 209 (1998), pp. 56–80.

Jakobsen, R. & Postma, D.: Redox zoning, rates of sulfate reduction and interaction with Fe-reduction and methanogenesis in a shallow sandy aquifer, Romø, Denmark. *Geochim. Cosmochim. Acta* 63:1 (1999), pp. 137–151.

Parkhurst, D.L. & Appelo, C.A.J.: User's guide to PHREEQC: A computer program for speciation, reaction-path, 1-D transport, and inverse geochemical calculations. U.S. Geol. Surv. Water Res. Invest. Rep. 99-4259, 1999.

Prommer, H., Barry, D.A. & Zheng, C.: PHT3D-A MODFLOW/MT3DMS based reactive multi-component transport model. *Ground Water* 42:2 (2003), pp. 247–257.

Prommer, H., Davis, G.B. & Barry, D.A.: Geochemical changes during degradation of petroleum hydrocarbons: field investigations and biogeochemical modelling. *Org. Geochem.* 30 (1999), pp. 423–435.

Schreiber, M.E. & Bahr, J.M.: Spatial electron acceptor variability: implications for assessing bioremediation potential. *Bioremed. J.* 4 (1999), pp. 363–378.

Schreiber, M.E., Carey, G.R., Feinstein, D.T. & Bahr, J.M.: Mechanism of electron acceptor utilization: implications for stimulating anaerobic biodegradation. *J. Contam. Hydrol.* 73 (2004), pp. 99–127.

Tuccillo, M.E., Cozzarelli, I.M. & Herman, J.S.: Iron reduction in the sediments of a hydrocarbon-contaminated aquifer. *Appl. Geochem.* 14 (1999), pp. 655–667.

Van Breukelen, B.M. & Griffioen, J.: Biogeochemical processes at the fringe of landfill leachate plume: potential for dissolved organic carbon, Fe(II), Mn(II), NH_4, and CH_4 oxidation. *J. Contam. Hydrol.* 65 (2004), pp. 245–268.

Vencelides, Z., Sracek, O. & Prommer, H.: Modelling of iron cycling and its impact on the electron balance at a petroleum hydrocarbon contaminated site in Hnevice, Czech Republic. *J. Contam. Hydrol.* 89 (2007), pp. 270–294.

Wiedemeier, T.H., Rifai, H.S., Newell, C.J. & Wilson, J.T.: *Natural Attenuation of Fuels and Chlorinated Solvents in the Subsurface*. John Wiley and Sons, New York, NY, 1999.

CHAPTER 12

Numerical modeling for preliminary assessment of natural remediation of phosphorus in variably saturated soil in a peri-urban settlement in Kampala, Uganda

Robinah N. Kulabako, Roger Thunvik, Maimuna Nalubega & Leigh A. Soutter

"As far as the laws of mathematics refer to reality, they are not certain, and as far as they are certain, they do not refer to reality".

Albert Einstein

12.1 INTRODUCTION

The limited financial and technological resources available to clean polluted aquifers particularly in the developing countries such as in Sub-Saharan Africa where degradation of groundwater is recognized as one of the most serious water resources problem implies that contamination of groundwater resources remains a growing public health hazard (Xu and Usher, 2006). In this case, natural attenuation (or monitored natural attenuation—MNA) appears to be a possible contaminated land management option necessitating investigation. Attenuation is generally most effective in the unsaturated (vadose) zone.

The American Society for Testing and Materials (ASTM) (1998) defines natural attenuation as the reduction in mass or concentration of a compound in groundwater over time or distance from the source of constituents of concern due to naturally occurring physical, chemical, and biological processes, such as biodegradation, dispersion, dilution, adsorption, and volatilization. A particular natural attenuation site may include any or all of these processes which, act without human intervention. Cleanup strategies of contaminated soils and groundwater based on natural attenuation are the predominant choices being pursued today because of the high costs and limited successes of traditional engineered groundwater cleanup methods (U.S NRC, 2000).

However, before using monitored natural attenuation, it must be clearly demonstrated that natural attenuation processes protect the environment from harmful impacts efficiently and persistently, which may be the case under favorable conditions mainly depending on the contaminant properties and the specific hydrogeological settings (Rügner *et al.*, 2006; Nichols, 2003). A previous study undertaken in a peri-urban (informal) settlement in Kampala, Uganda (Bwaise III Parish) that included monitoring of groundwater quality demonstrated widespread shallow groundwater contamination (microbial and organic contamination with relatively high phosphorus and nitrate levels) (Kulabako *et al.*, 2007 and 2008). With the groundwater being an important source of water supply in this and similar peri-urban settlements in Kampala (Kulabako *et al.*, 2010a), application of a risk-based approach for contaminated land management means that in this case, groundwater is a principal resource necessitating protection.

12.2 SETTING

Bwaise III Parish (32°34′E, 0°21′N) is found in the northern part of Kampala, in Kawempe Division, approximately 4 km from the city centre. The parish has an area of 57 ha and is divided into six local administrative zones namely; Kalimali, Bokasa, Bugalani, St. Francis,

Katoogo and Kawaala. The area though mainly residential has some economic activity for the low-income residents. The area, a typical urban poor settlement in the city, is largely unplanned with lack of basic services, poor road access and deplorable housing. It has one of the highest population growth rates in Kampala District with an annual average rate of 9.6% more than twice the average city's growth rate (3.7%) and a population density of about 27,000 persons/km^2 (UBOS, 2005). More than 40% of the 57 ha parish is a low-lying flood plain that was initially a wetland (Kulabako *et al.*, 2007).

Waste management in Bwaise III is poor and therefore surface contaminants remain within the area, degrading the surficial environment and causing widespread contamination of the shallow groundwater. Sources of contaminants are scattered almost uniformly between the dense settlements. The identified major anthropogenic sources of shallow groundwater contamination in the area are pit latrines, solid waste dumps and the grey discharge to unlined

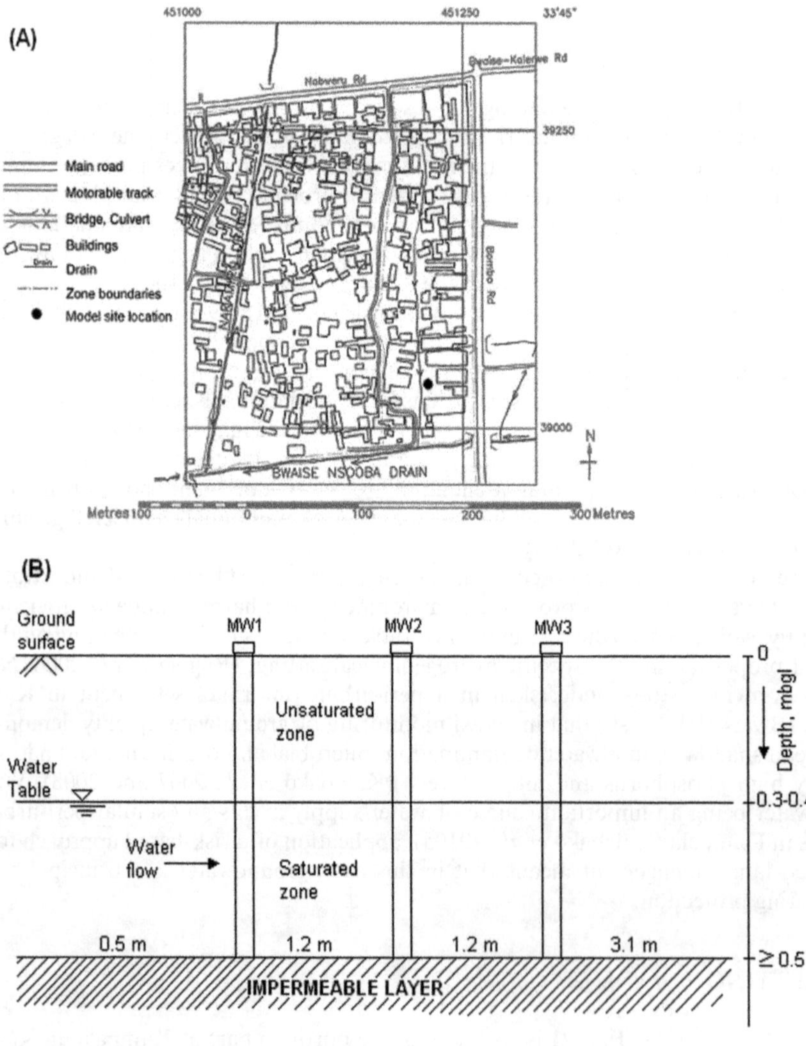

Figure 12.1. Model site location in Kalimali zone (A) and cross section (not to scale) of the model site with installed monitoring wells (B).

drains (Kulabako *et al.*, 2004). These sources together with animal farming contribute to the occurrence of phosphorus in the shallow groundwater underlying the area (Kulabako *et al.*, 2008).

Rainfall is the major force driving transport of the surface contaminants to shallow groundwater in the area (Kulabako *et al.*, 2007). However, wastewater is also disposed of via on-site sanitation systems, open drains and spaces and it has now become apparent that this mode of managing this wastewater generates rather high rates of infiltration, often referred to as "modern recharge." The increased recharge due to urbanization is somewhat contradictory, since urbanization implicates the construction of roofs and paved surfaces which reduce recharge (Nyenje *et al.*, 2010).

In Bwaise III, the impervious surfaces do indeed reduce recharge from precipitation at the upgradient part of the area and on many occasions, cause fast surface runoff responses leading to flooding of the lower lying parts (Gard and Torstensson, 2006).

Given the location of our model site (within the low lying part of Bwaise III, Fig. 12.1A), here the infiltration rates are considered much higher than the rainfall intensity due to some overland flow adding to the net infiltration. This assumption appears to agree with a study undertaken by Kelbe *et al.* (1991) in which he observed a reduction of peak discharge from catchments inhabited with informal settlements as compared to similar pristine catchments due to land practices there.

Groundwater data (water table depth, electrical conductivity-EC and total phosphorus) previously obtained by the author over a period of 17 months for the three monitoring wells, MW1-MW3 installed at the model site (Fig. 12.1B) indicated that the groundwater flow was from well 1 to well 3 (Kulabako *et al.*, 2008). Hence to assess the potential for natural remediation at the model site, the data for wells 1 and 3 are used in this chapter. The two wells are about 2.4 m apart.

12.3 NUMERICAL MODEL

The mathematical model built using COMSOL Multiphysics (Li *et al.*, 2009) consists of two components: a flow model and a solute transport model and was based on the conceptual model developed for a cross section within the study area (Fig. 12.2). The conceptual model has three geological layers; the bottom layer consists of stiff clay silts (layer 1), the middle layer consists of silty sands (layer 2), while the top layer (layer 3) is largely sand. Both the top and middle layer also contain various types of fill material ranging from plastics, broken glass, rugs and decaying organic material. The layer thicknesses for the cross section are 0.45, 0.55 and 0.5 m for top, middle and bottom respectively. The horizontal length of the studied cross section was 12 m with the first 3 m containing monitoring wells installed (with a separation distance of 1.2 m) along a presumed groundwater flow line (Fig. 12.1B).

12.3.1 *Flow model*

To study how long it would take to contaminate the subsurface with the discharge from the modeled cross section having phosphorus concentrations exceeding the national effluent discharge standard (0.005 kg m^{-3}), the flow is modeled using a generalization of Darcy's law used within Richards' equation. This equation (eq. 12.1) (COMSOL, 2008) simulates soil-water movement in a vertical profile under rainfall infiltration as follows:

$$[C+S_eS]\frac{\partial p}{\partial t}+\nabla\cdot\left[-\frac{k_s}{\mu}k_r\nabla(p+\rho_f gD)\right]=Q_S \qquad (12.1)$$

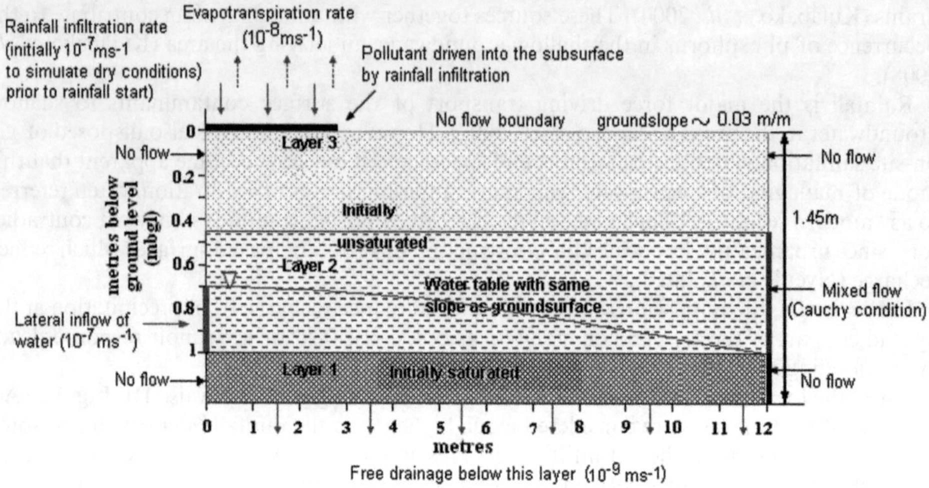

Figure 12.2. Conceptual model of cross section with Bwaise III Parish with initial boundary conditions.

The specific or Darcy flux q [m s^{-1}]

$$q = -\frac{k_s}{\mu} k_r \nabla (p + \rho_f g D) \tag{12.2}$$

appears inside the divergence operator in equation (12.1), which utilizes a constitutive relation for the storage coefficient S (m^{-1}), defined as $S = (\chi_p(1-\theta) + \chi_f\theta_s) \cdot \chi_f$ and χ_p are the fluid and matrix compressibility, respectively [m s^2 kg^{-1} = Pa^{-1}].

The dependent variable of the system (eqs. 12.1 and 12.2) is the term for the fluid pressure p[kg m^{-1} s^{-2}]. Other terms include μ the dynamic viscosity [kg m^{-1} s^{-1}], C the specific moisture capacity [m^{-1}], Se the effective saturation (dimensionless), k_s the saturated permeability [m^2], k_r the relative permeability (dimensionless), ρ_f the fluid density [kg m^{-3}], g the acceleration of gravity [m s^{-2}], D the vertical coordinate [m], t time [s], and Q_s the volumetric flow rate per unit volume of reservoir for a fluid source [s^{-1}]. Additionally S, (eq. 12.2), describes the change to fluid stored within the medium as a function of time using the liquid volume fraction at saturation θ_s (dimensionless) with the particle and fluid compressibilities, χ_p and χ_f [m s kg^{-1}], respectively.

The flow model (eqs. 12.1 and 12.2) uses expressions defined in terms of moisture retention and permeability for each layer in the unsaturated zone (COMSOL, 2008), and estimated through the relations defined by Brooks and Corey (1964):

$$\theta = \theta_r + S_e(\theta_s - \theta_r) \tag{12.3}$$

where θ_s and θ_r denote the volume fraction of the fluid at saturation and after drainage or residual respectively, and S_e is the effective saturation. COMSOL uses pressure head H_p [m] in equation (12.4) to define the dimensionless parameters, effective saturation S_e (eq. 12.5), specific moisture capacity C (eq. 12.6) and the relative permeability (eq. 12.7). The system is as follows:

$$H_p = \frac{p}{\rho_f g} \tag{12.4}$$

$$S_e = \frac{1}{|\alpha H_p|^n} \tag{12.5}$$

$$C = \frac{-n}{H_p}(\theta_s - \theta_r)\frac{1}{|\alpha H_p|^n} \tag{12.6}$$

$$k_r = S_e^{\frac{2}{n}+l+2} \tag{12.7}$$

Equations (12.5) to (12.7) use constants α, n and l that specify a particular type of medium:

$$\alpha = \frac{1}{\psi_a};\, n = \lambda$$

where α is an empirical parameter [m^{-1}] that equals the inverse of ψ_a, the air-entry value or bubbling pressure head [m], and λ denotes the pore size distribution index, which affects the slope of the retention function. The parameter l is related to the connectivity and tortuosity of the pore space and is considered to be about 0.5 for many soils (Looney and Falta, 2000).

The flow model was set up using (i) material and fluid parameters in Tables 12.1 and (ii) boundary and initial conditions that represent the conceptual model of the cross section (Fig. 12.2). The inward flux (eq. 12.8a) was set equal to rainfall infiltration (five times the rainfall intensity-section 12.2) over the interval $0 < x < 3$ m of the upper boundary (Fig. 12.2). The rest of the upper boundary had zero flux (eq. 12.8b) as the dense settlement compacts the ground surface making it mostly impervious (Gard and Tortensson, 2006).

The left and right boundaries of the topsoil layer were assumed to have negligible flow (hence a zero prescribed flow) considering the hydraulic conductivity contrasts and barring preferential flow paths leading to smaller lateral flow in the unsaturated zone than in the saturated zone (Kulabako *et al.*, 2010b). It should also be noted that the water table at the model site has never been observed to occur beyond the middle layer. If needed, it would be easy to apply the Cauchy boundary condition and the procedure developed for dealing with it as described below and in the Appendix.

Table 12.1. Hydraulic parameters for the soils in the different layers.

Layer variable	Top (3)	Middle (2)	Bottom (1)	Data origin
K_s [m s^{-1}]	10^{-3}	10^{-4}	10^{-7}	Double infiltration rate measurements for the top layers and standard falling head permeability test for the bottom layer
θ_s [m^3 m^{-3}]	0.29	0.41	0.56	Porosity measurements using the water volume displacement method
θ_r [m^3 m^{-3}]	7.056	5.050	11.448	Soil water retention measurements
α [m^{-1}]	38.46	38.46	9.009	Soil water retention measurements
n [-]	0.341	0.180	0.078	Soil water retention measurements
l [-]	0.5	0.5	0.5	Literature
X_f [Pa^{-1}]	$4 \cdot 10^{-10}$	$4 \cdot 10^{-10}$	$4 \cdot 10^{-10}$	Literature
X_p [Pa^{-1}]	10^{-5}	10^{-5}	10^{-5}	Assumption
ρ_f [kg m^{-3}]	1000	1000	1000	Literature
g [m s^{-2}]	9.81	9.81	9.81	Literature

Table 12.2. Data for the solute transport model.

Variable layer	ρ_b	K_{plin}	K_L	$\overline{s}$	α_1	α_2	D_{mL}
Top (3)	1211	0.00695			0.1	0.03	$1.752 \cdot 10^{-9}$
Middle (2)	1211		283.2	$3.20 \cdot 10^{-4}$	0.1	0.03	$1.752 \cdot 10^{-9}$
Bottom (1)	1205		25.5	$3.90 \cdot 10^{-4}$	0.1	0.03	$1.752 \cdot 10^{-9}$

Where ρ_b is the soil dry bulk density [kg m^{-3}], K_{plin} is the partition coefficient for the linear sorption isotherm [m^3 kg^{-1}], K_L are Langmuir constant [m^3 kg^{-1}] and sorption maximum [kg kg^{-1}], respectively. α_1 and α_2 are the longitudinal and transverse dispersivities [m] respectively estimated from bromide tracer test data for the model site while D_{mL} is the coefficient of molecular diffusion [m^2 s^{-1}] calculated for the site using the modified version of Wilke-Chang (Tecle, 2008).

The bottom layer is relatively impervious and hence considered to have an insignificant lateral influx and outflow on its left and right boundaries respectively. A zero flux was therefore prescribed for these boundaries:

Flux

$$\mathbf{n} \cdot \frac{k_s}{\mu} k_r (\nabla p + p_f g \nabla D) = N_0 \tag{12.8a}$$

Zero flux

$$\mathbf{n} \cdot \frac{k_s}{\mu} k_r (\nabla p + p_f g \nabla D) = 0 \tag{12.8b}$$

where **n** represents the normal vector to the boundary and N_0 is a prescribed flux.

The groundwater pressure gradient in the middle layer (H_p, eq. 12.4) is about the same as the slope of the ground surface, i.e., about 0.01–0.05 mm^{-1}. The flux in the middle layer therefore equals the Darcy flux with respect to the physical properties of this layer. To model the flow through, a lateral inward flux (eq. 12.8a) was imposed on the left hand side boundary to the middle layer for depths less than 0.55 m below ground level (Fig. 12.3), which achieves a water table slope that mimics the topographic slope (0.31 m/12 m $\approx$ 0.03, Fig. 12.1). The imposed lateral flux (10^{-7} ms^{-1}) into the domain was based on findings by Herzog (2007) following application of a transient groundwater flow model using Groundwater Modeling System (GMS) for the larger area (Fig. 12.1A) that had 16 monitoring wells.

A Cauchy boundary condition (eq. 12.9) was used at the right boundary of middle layer (Fig. 12.1), which allows one to simultaneously specify the flux through the boundary for the head specified just beyond the boundary:

Cauchy (mixed)

$$\mathrm{n} \cdot \frac{k_s}{\mu} k_r (\nabla p + \rho_f g \nabla D) = N_0 + R_b [p_b - p + \rho_f g (D_b - D)] \tag{12.9}$$

where R_b represents the conductance to flow in a hypothetical layer adjacent to the boundary, and p_b is the pressure at the edge of the resistive layer. In our case, the term R_b was modified with logical operators so as to create a flow condition below the water table ($R_b > 0$) and a no flow condition above it ($R_b \leq 0$). The mixed boundary condition is used to simulate the seepage face boundary condition that occurs at the outlet to a ditch. The seepage face

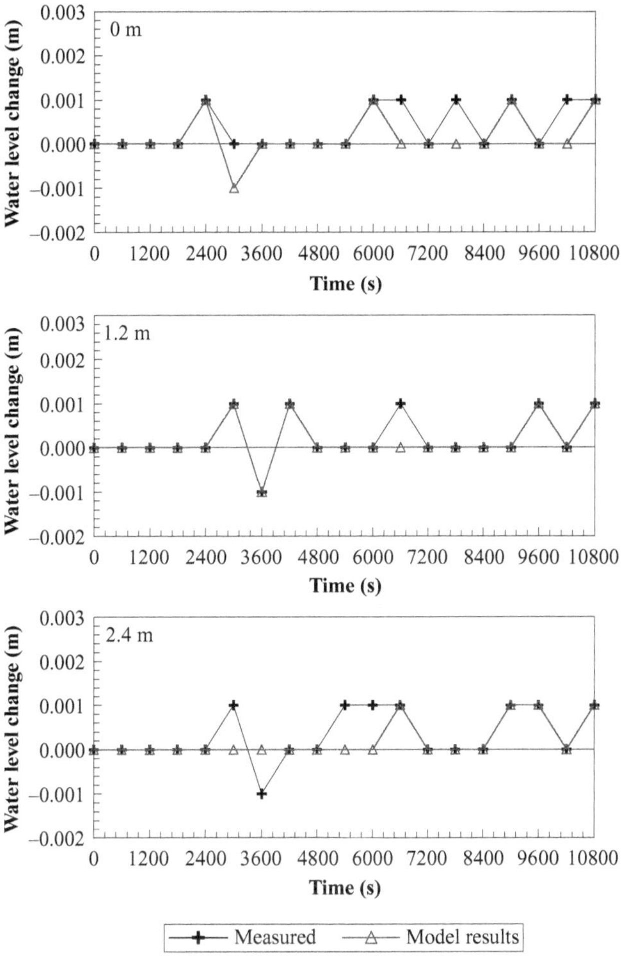

Figure 12.3. Field measurements and model results of water table changes for a rainfall event (4.7 mm occurring over 3 hours) at MW 1, 2 and 3 represented at 0, 1.2 and 2.4 m respectively from the left inlet boundary.

boundary condition is very difficult to deal with when part of the flow domain is unsaturated. Therefore, it was necessary to develop an algorithm for solving the problem using the COMSOL Multiphysics software. The numerical procedure is described in more detail in the Appendix.

The leakage from the bottom (clay) layer was assumed to be free drainage initially. Underlying this clay layer is a fractured rock formation according to Taylor and Howard (1998). During the research period (2002–present), it has been observed that bottom layer never dries out irrespective of season. With this observation, we concluded that there is no free drainage from the bottom of this layer as this would render the layer dry to an extent that does not occur. Hence the chosen value of hydraulic conductivity, which is less than the hydraulic conductivity of this layer that was imposed on the lower boundary of the model (10^{-9} m s^{-1}), is fairly arbitrary. The contamination plume is practically insensitive to values below $q_n = 10^{-9}$ m s^{-1}. Smaller values than that would make the bottom layer practically impervious.

With these settings, the model was then solved for steady state. Calibration of the flow model was with soil moisture measurements for samples taken at a depth of 0.3 m from the ground level (Kulabako *et al.*, 2010b) and resulted in the adjustment of the pore size distribution index, *n* for the top layer from 0.341 to 0.190 (Table 12.1). Prior to testing the different scenarios, *in-situ* logging measurements of the change in the water table elevation in each of the three monitoring wells (MW1–MW3, Fig. 12.1) represented at distances of 0, 1.2 and 2.4 m from the left inlet boundary during a rainfall event (using three PDCR1830 pressure transducers and a 52203 Model tipping bucket rain gauge (0.1 mm per tip) connected to a multichannel CR10X logger (Campbell Scientific Ltd., UK)) were compared to the model calculations (Fig. 12.3). Here, the measured and modeled water table changes generally match at the beginning and end of the rainfall event. However, mismatches in time observed in all the wells are expected given the nature of the fill material at the site (section 12.2) which appears to influence rain infiltration and hence the true water table elevation. Nevertheless, the current single-porosity model may be useful to test impacts of different scenarios, which provides valuable insight into the hydrodynamics of groundwater flow and solute (phosphorus) transport mechanisms and especially by comparison with the abundant data measured for the study area in Kulabako *et al.* (2007, 2008).

12.3.2 *Solute model*

12.3.2.1 *Soil phosphorus sorption*

The concentration of phosphorus in the soil solution is critical for losses of dissolved phosphorus. Therefore processes such as phosphorus sorption and release are important. Phosphorus sorption is a combination of adsorption and precipitation implying that some of the sorbed phosphorus is not readily desorbable (Börling, 2003).

Release of phosphorus from soil can occur through desorption of sorbed phosphorus to the soil surfaces, dissolution of minerals containing phosphates and mineralization of organic phosphorus. However, desorption of phosphorus from soil solid surfaces is a much slower process than sorption (Muhammad and Robinson, 2003).

In this study, phosphorus sorption was assumed to play an important role in the attenuation of this nutrient. This was based on the fact that high phosphorus loads generated in the area and subsequent concentrations in the shallow groundwater make microbial mineralization and immobilization within the subsurface a minor process in phosphorus regulation (Sobehrad, 1997).

At the model site, three soil profiles (about 1 m long) were extracted adjacent to the monitoring wells using fabricated steel cores (about 1200 mm long and diameter 100 mm). The extracted soil profiles were air dried at room temperature before being sliced into the different layers. The gross sample for each of the layers was divided using a riffle splitter to the required mass (500 g) to give a representative sample for that layer. This was then ground and homogenized to a fine powder with a mortar and pestle and passed through a 2 mm stainless steel sieve (Houba *et al.*, 1995).

Phosphate solutions used in the experiments (to cover the range observed in the monitoring wells) were prepared by dissolving anhydrous KH_2PO_4 in distilled water containing 0.01 M $CaCl_2$. Batch studies were conducted in the laboratory at a room temperature of 24 ± 0.05°C. The initial *pH* of the solution was not adjusted, as this was found to fall within the range measured *in-situ*. A series of 5 g air-dried soil samples were placed in 236 mL glass bottles. Each of these bottles was filled with 100 ml of different initial PO_4-P standard solutions (0.1, 1, 3, 5, 10 and 15 mg L^{-1}) in triplicates. The solutions were then equilibrated on an orbital shaker at 150 rpm for 24 h.

At the end of the equilibration time, samples were collected, filtered through a 0.45 μm sieve and phosphorus concentration determined using the ascorbic acid method (APHA/AWWA/WEF, 1998).

Calculation of the phosphorus adsorption amount and estimation of the most appropriate sorption isotherm were determined according to Gale *et al.* (1994).

12.3.2.2 *Solute transport model*

The transport of phosphorus from the ground surface with and without sorption to ascertain the breakthrough time given rainfall occurrence during the wet and dry season was modeled using equation (12.10) (COMSOL AB, 2007). This equation was solved simultaneously with the transient part of the flow model:

$$\left(\theta + \rho_b \frac{\partial c_p}{\partial c}\right)\frac{\partial c}{\partial t} + \nabla\cdot(-\theta D_L \nabla c + qc) = S_C \tag{12.10}$$

Equation (12.10) governs advection, dispersion, and sorption of solutes in a variably saturated porous medium. It describes the time rate change in two terms: c denotes the dissolved concentration [kg m^{-3}], and c_p the mass of adsorbed contaminant per dry unit weight of solid [mg kg^{-1}]. The parameter θ denotes the volume fluid fraction [m^3 m^{-3}] and ρ_b the bulk density [kg m^{-3}]. The first term of the time derivative gives the change in liquid concentration, and the second terms give the time rate change in the sorbed solute.

In the equation, D_L is the hydrodynamic dispersion tensor [m^2 s^{-1}]; q is the Darcy flux [m s^{-1}], which comes from the flow equation. The right hand side of equation (12.10) is the solute source S_c which equals solute added per unit volume of soil per unit time [kg m^{-3} s^{-1}]. The source takes on positive values for processes that increase concentrations (sources) and negative values for processes that decrease them (sinks).

Fitting of the measured sorption data at the model site (section 12.3.2.1), had the linear isotherm describe the sorption of the top layer better while the Langmuir isotherm equation fitted well the sorption data for the middle and bottom soil layers (Fig. 12.4). Hence the linear sorption with a partition coefficient was used for the top layer and non-linear sorption with the Langmuir isotherm for the middle and bottom layers. With the Langmuir sorption isotherm, solute mass per solid mass c_p is related to the dissolved concentration c for use in equation (12.11) as follows:

$$c_p = \frac{k_L \bar{s} c}{(1 + k_L c)}, \frac{\partial c_p}{\partial c} = \frac{k_L \bar{s}}{(1 + k_L c)^2} \tag{12.11}$$

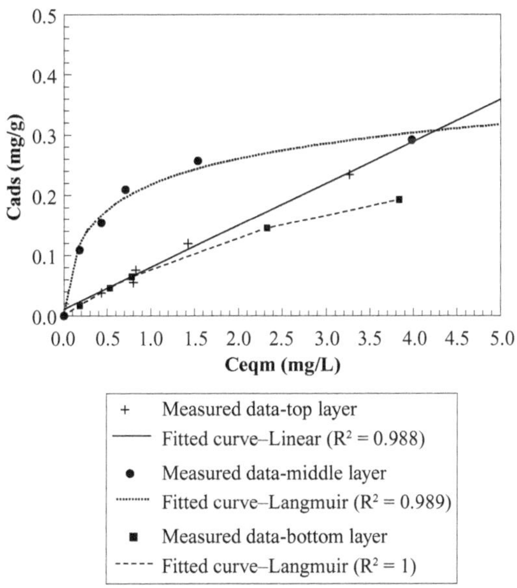

Figure 12.4. Measured phosphorus sorption data and fitted adsorption isotherms for the different layers of soil at the model site.

where k_L and $\bar{s}$ are the Langmuir constant [m³ kg⁻¹] and sorption maximum coefficient (dimensionless), respectively.

With linear sorption, the relationship between c_p and c in equation (12.12) is a proportionality constant equal to the partition coefficient k_p [$m^3\ kg^{-1}$] given as follows:

$$c_p = k_p c, \frac{\partial c_p}{\partial c} = k_p \tag{12.12}$$

A study by Wang *et al.* (2009) demonstrates that sorption coefficients obtained from batch experiments would be consistent with those from column experiments, which offer a better simulation of the real environment only when *S*/*L* ratio exceeds 0.25. To make a convincing safety assessment, batch experiments with *S*/*L* ratio > 0.25 are essential to obtain a reliable transport parameter. The same study observes that for *S*/*L* > 0.25, the sorption capacity of the solid is about an order of magnitude less than that in low *S*/*L* ratios (<0.25).

However, solid/liquid (*S*/*L*) ratios used in the batch experiments to ascertain the sorption coefficients (k_L, $\bar{s}$ and k_p) of the soils at the model site were less than 0.25 (section 12.3.2.1). Hence in this paper, the estimated sorption coefficients (k_L and k_p) following from the fitted sorption isotherms (Fig. 12.4) were adjusted to an order of magnitude lower (Table 12.2).

In the model, infiltrating water with a phosphorus concentration, c_0 (eq. 12.13) was prescribed at the upper boundary (Fig. 12.2):

Prescribed concentration

$$c = c_0 \tag{12.13}$$

Initially, the concentration was set to zero ($c = 0$ in eq. 12.13), but at the onset of the rains to ascertain how long it would take for the contamination to occur with sorption, a phosphorus concentration of 0.015 kg m⁻³, slightly more than the maximum measured in the shallow groundwater in the area (Kulabako *et al.*, 2007) was set in the influx (eq. 12.13). All the other water entering the cross section was without any contamination ($c = 0$ in eq. 12.13). To account for phosphorus transport out of the system, an advective flux boundary condition (eq. 12.14) was imposed on the right hand side of layer 2 (Fig. 12.2):

Advective flux

$$\mathbf{n} \cdot (-\theta_s D_L \nabla c) = 0 \tag{12.14}$$

where **n** is the normal to the boundary. The advective flux boundary condition equation (12.14) sets the dispersive flux to zero so that the advective flux across the boundary is determined as part of the solution.

12.4 SIMULATIONS

The simulations were designed to identify the time it would take to pollute the shallow groundwater (considering phosphorus sorption) with the discharge from the modeled cross section having a reactive phosphorus concentration exceeding the national effluent discharge standard of 0.005 kg m⁻³.

Transient flow processes were modeled with the initial condition for the transient simulations being the steady state flow solution. To assess remediation, the model solution following realization of contamination with the above criteria, were used as the initial condition.

In the transient models, the rainfall boundary had selected infiltration rates (corresponding to the different rainfall intensities during the wet and dry season and an imposed phosphorus concentration (0.015 kg m^{-3}) held constant with time to ascertain how long it would take to pollute and remediate the subsurface. A review of total daily rainfall records for the period 1943–2004 for the weather station located 1.5 km away from the model site, had on average 17 mm day^{-1} of rain every 5 days during the dry season (102 days year^{-1}) and 27 mm day^{-1} of rain every 2 days during the wet season (162 days year^{-1}).

However observations indicate that this amount of rain does not occur over the whole 24 h period but over times less than this. Typical heavy rain (>10 mm day^{-1}) occurs over 30 and 45 minutes (Watkins and Fiddes, 1984). Our basic assumption in this study was to adopt rains of 17 mm h^{-1} during the dry season and 27 mm h^{-1} during the wet season and imposing these for 30 minutes for each rainy day in each season. With this assumption, an annual rainfall amount of 1521 mm is realized which falls within the range for the average annual amount for the Makerere weather station (1498 ± 49 mm). Given the location of the model site (within the low lying part of Bwaise III), the infiltration rates are assumed to be much higher than the rainfall intensity (section 12.2) andthe prescribed seasonal infiltration rates were enhanced to about 5 times the rainfall intensity.

Model sensitivity to selected key parameters, the sorption coefficients (k_p and k_L), air entry value and pore size distribution index was carried out for how long it would take to pollute according to the described criteria (section 12.3.1). For the sorption coefficients, the values were reduced by two orders of magnitude while for air entry and pore size distribution given the nature of the soils in the three layers (section 12.3), ranges reported in literature for similar types of soils were used. Air entry values were changed to fall within 10.5–13.5 cm for sandy soils and 5.0–5.3 cm for clay soils. For pore size distribution, values were adjusted to fall within and outside 0.32–0.56 for sandy soils and 0.03–0.06 for clay soils (Lundmark, 2008).

12.5 RESULTS AND DISCUSSION

12.5.1 *Field measurements*

The results for the field measurements for monitoring wells MW1 and MW3 are presented in Figure 12.5. EC as a conservative tracer generally indicates that the higher EC levels in MW3 compared to MW1 suggests that the groundwater flow at the model site is from MW1 to MW3. That this is the case could be attributed to more mixing of MW1 water with groundwater from lateral inflow which is assumed to have zero contaminant concentration and or leaching of the contaminants given the nature of the subsurface material as described in section 12.3. However, the water levels appearing lower at MW1 than at MW3 implies otherwise. The latter is however unlikely due to the change in the ground surface level from which the water table was measured during the course of monitoring. It was common to find soil and or silt removed from an adjacent drain dumped at the site.

Considering groundwater flow from MW1 to MW3 therefore, the variation of the water quality between these wells is used here to assess the potential of natural remediation for total phosphorus (P, measured as ortho-phosphates). For both wells, P, *EC* and water levels varied with time (Fig. 12.5). *EC* is an indirect indicator of pollution levels (Rajib *et al.*, 2006). At MW1, *EC* exhibits a positive though insignificant correlation with P ($r = 0.181$, $p = 0.210$) while at MW3 it has a relatively strong negative correlation ($r = -0.405$, $p = 0.034$). A possible explanation for this variation could be the likely attenuation of P as the contaminated groundwater flows from MW1 to MW3 resulting in a decrease of this solute, in some cases as indicated in Figure 12.5. This attenuation occurs both during periods of low (<5 mm day^{-1}) or no rain and high rain amounts (>10 mm day^{-1}). The absence of rain or low rainfall amounts implies low flow rates of the shallow groundwater with longer contact times with the subsurface soils for retention of P to occur (Kulabako *et al.*, 2008). The reduction in P

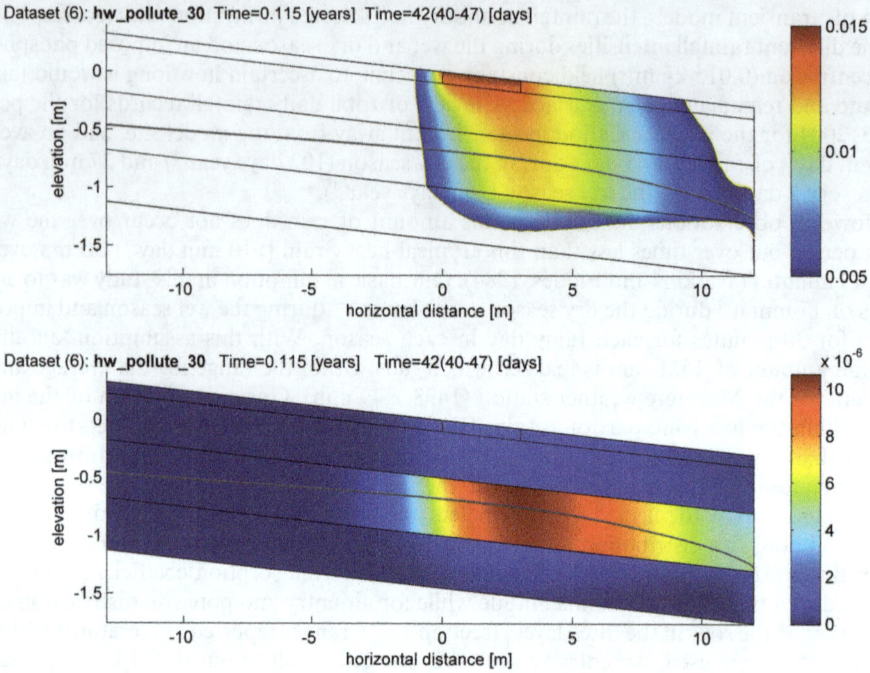

Figure 12.5. Simulation results for dissolved and corresponding sorbed phosphorus concentrations at 0.115 years when pollution is simulated by sorption coefficient values reduced by two orders of magnitude.

concentrations following heavy rains might be due to dilution. However, there are cases during which the P levels are higher at MW3 than at MW1 during high rain amounts e.g., 2003-08-07 (17.5 mm), 2004-04-29 (53.5 mm), and 2004-10-27 (36 mm).

This increase could be attributed to leaching of total-P from the soil media (particularly when the infiltrating rainwater contains lower dissolved P concentrations than that sorbed on the soils as a result of accumulation from waste disposal (section 12.2), McGechan, 2002) and or transport via macropore or preferential flow paths given the nature of the soils at the site (section 12.3) during sporadic waste disposal practices in the area (Kulabako *et al.*, 2004; Kulabako *et al.*, 2007). That leaching of P from the soil, transport via macropore or preferential flow paths is a possibility in the area, explains the rather limited P attenuation at the site with the P range at MW1 (0.474–1.995 mg L^{-1}) not being so different from that at MW3 (0.56–1.961 mg L^{-1}) over the monitoring period.

Worth noting is the effect of the water table rise on the P concentrations. A rise of the water table above the ground indicative of flooding as a result of heavy rains on 2004-11-25 (70.5 mm) at MW1 (–0.11 m bgl), had a peak P concentration (1.995 mg L^{-1}) unlike at MW3 where the water table did not rise above ground. This suggests that flooding conditions increase P concentrations in the area most likely due to changes in the system variables e.g., *pH*, redox conditions, concentrations of organic carbon, iron, geochemistry of the soil as a P source (Rao and Prasad, 1997; Carlye *et al.*, 2001; Kulabako *et al.*, 2008) reducing natural remediation of P in the area.

12.5.2 *Pollution and remediation simulation scenarios*

The simulation results for how long it takes to pollute and remediate the subsurface (with the pollutant source removed) are depicted in Figure 12.6. Figure 12.6a indicates that it takes >3 years for the groundwater discharging from the modeled section to have a concentration of ≥0.005 kg m^{-3} implying that the soils retention (sorption) capacity of P is relatively high. The attenuation by sorption is stronger in the middle layer than the top and bottom layers as indicated by the higher concentrations of the sorbed P (Fig. 12.6a). This could be attributed to the high sorption capacity there and lower velocity of flow compared to the more porous top layer (Kulabako *et al.*, 2008). Also, the low velocity of flow implies that during the simulation period, P is hardly transported to the bottom layer.

On the other hand, since linear sorption is occurring in the top layer (Table 12.2), the retardation factor decreases with increase in the moisture content of soils (moisture content, $\theta <$ saturated moisture content, θ_s) (Fetter, 2001). Hence during the wet period (occurs over many more days than the dry period, section 12.4) the P front travels faster in the shallow groundwater within this layer and also there is an increase in the dissolved P concentrations as a result of the reduced sorption rate (Fig. 12.6a).

Regarding natural remediation of P with the pollutant source removed, Figure 12.6b shows that it takes about 7.5 years for this to occur (P concentration < 0.005 kg m^{-3}). Here, it is observed that the top layer is without any sorbed P implying that it has a high P desorption potential. The relatively high potential for the soils within the top layer to desorb P has been confirmed in the previous study findings with respect to the amount of loosely bound P (Kulabako *et al.*, 2008) and dissolved ortho-P concentrations in a 1:1 soil (g) to water (mL) extract (Kulabako, 2010). The middle layer on the other hand still contains some sorbed P after the relatively long remediation period confirming the strong retention of P by the soils in this layer as noted earlier. Comparison with observations following field measurements (section 12.4.1), means that if these are taken into consideration in the solute transport modeling here, the time it takes to pollute would be shorter while the remediation times would be longer.

12.5.3 *Sensitivity analyses*

12.5.3.1 *Impact of change of sorption coefficients (K_L and K_{plin}) on pollution time*

The simulation results for how long it takes to pollute with the sorption coefficients reduced by two orders of magnitude are presented in Figure 12.6. Here, the time it takes to pollute with this change, occurs in 0.115 years, a reduction of more than 50% of the time it would take initially (Fig. 12.6a). This change is great and means that the model solution is sensitive to the sorption coefficient values given the data settings and boundary conditions.

12.5.3.2 *Impact of change of the pore size distribution values on pollution time*

The pore size distribution index (λ or n) influences the relative permeability of the soils. Hence given that the bottom layer is saturated all the time (Fig. 12.2), changing the pore size distribution index there would not have any impact on the pollution time (Tindall and Kunkel, 1999). Hence in this case, n for each of the top and bottom layers presented in Table 12.1 was reduced by 50% and increased by 20% to assess the effect on the pollution times. The time it takes to pollute given these changes is presented in Table 12.3.

The pollution times indicated in Table 12.3 on changing the n values in the top and middle layers do not vary much from the previous (3.43 years, Fig. 12.6). This means that

Table 12.3. Adjusted pore size distribution index (n) of top and middle soil layers and corresponding pollution times.

Layer	Initial n	Adjusted n	Pollution time (year)
Top	0.341	n × 1.2	3.4
(Layer 3)		n × 0.5	3.4
Middle	0.180	n × 1.2	3.6
(Layer 2)		n × 0.5	3.5

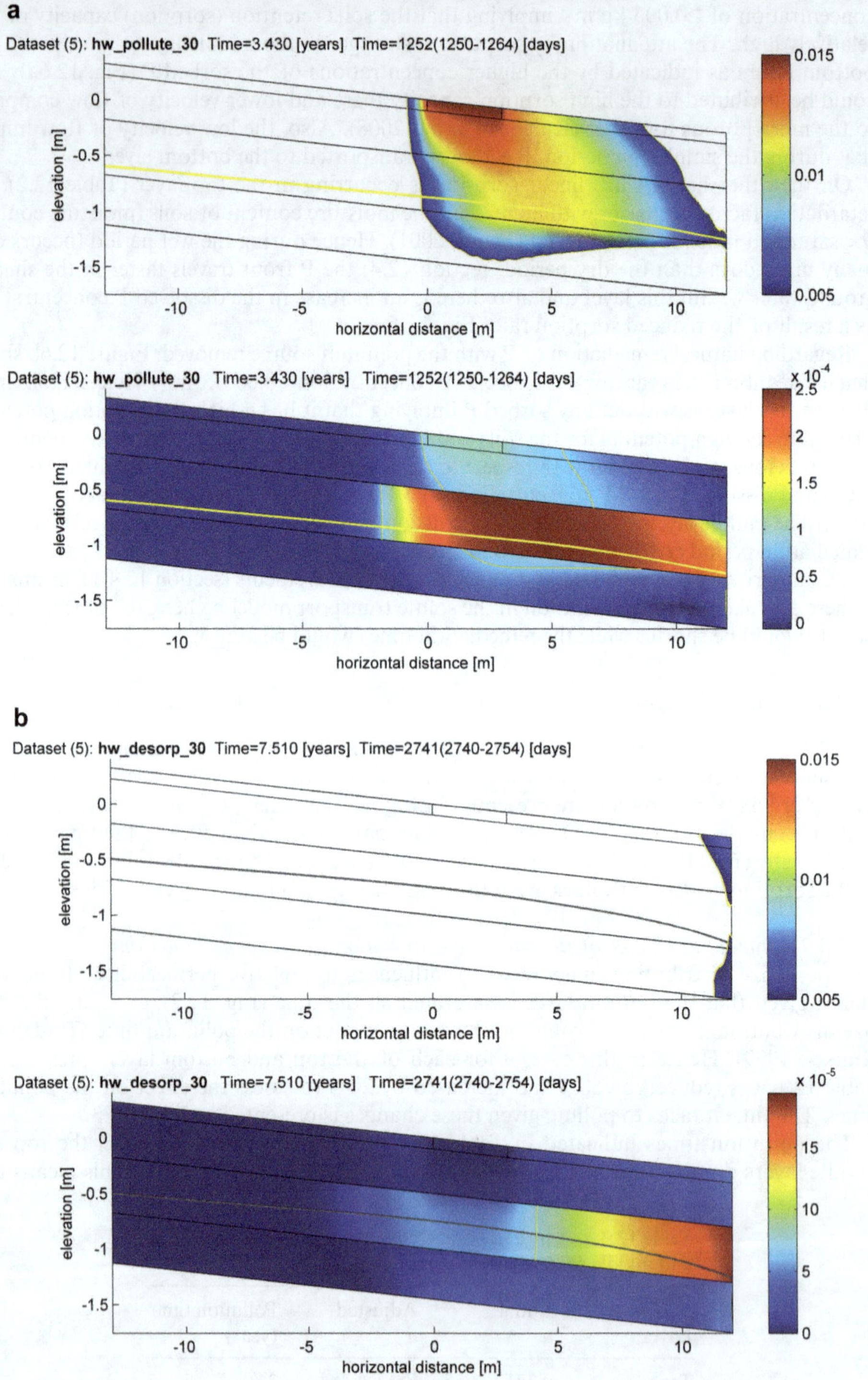

Figure 12.6. Simulation results for (a) dissolved and corresponding sorbed phosphorus concentrations at 3.43 years when pollution is realized and (b) dissolved and corresponding sorbed phosphorus concentrations at 7.51 years when remediation occurs following pollutant source removal.

the sensitivity of the model to this parameter for both the top and middle soil layers is relatively small.

12.5.3.3 *Impact of change of the air entry values on pollution time*
The air entry value was changed for all the soil layers. Here, the air entry values were reduced to 0.03, 0.05 and 0.12 m giving the inverse α [m^{-1}] as 33.33, 20 and 8.33 in the top, middle and bottom layers respectively. With this change, the pollution time was about 3.34 years which does not vary much from the previous (3.43 years, Fig. 12.6). Hence the sensitivity of the air-entry is relatively insignificant for the model setting and its boundary conditions.

12.6 CONCLUSIONS

The field measurements at the model site reveal that natural remediation occurs both during periods of low (<5 mm day^{-1}) or no rain and during periods of heavy rains (>10 mm day^{-1}). Here retention of phosphorus is attributed to sorption and dilution effects. However this natural remediation of phosphorus during transport in the subsurface is relatively low with the concentration ranges between the upgradient (0.474–1.995 mg P/L) and downgradient (0.56–1.961 mg P/L) wells sited 2.4 m away, not varying much. The complex hydrogeological setting of the area (high water table, nature of soils with fill material, flood occurrence) and sporadic poor waste disposal practices inhibit natural remediation processes.

The solute transport model shows that it takes 3.43 years to have the groundwater discharging from the modeled cross section having a phosphorus concentration of ≥ 0.005 kg m^{-3}, which is more than the national effluent discharge standard of this solute to receiving streams, rivers and lakes. The remediation time on removal of the pollutant source is in the region of 7.5 years. The field measurements used in this paper though not directly comparable with the model solutions, suggest that the contamination and remediation times can be shorter and longer respectively than modeled. That this is the case calls for further field measurements of higher frequency so as to capture information on the potential remediation processes and system variables. These data can be utilized to improve the conceptual model used here.

The sensitivity approach used here changes one factor at a time to see what effect this produces on the pollution times. It indicates that sorption coefficients contribute more to the output variability than the pore size distribution and air entry values given the observed data and the boundary conditions used in the model. However, there is need for further sensitivity analyses on the output especially with an improved conceptual model and validated numerical model so as to provide information on the most influential parameters. This information is useful in guiding data collection efforts.

REFERENCES

American Society for Testing and Materials (ASTM): Standard guide for remediation of groundwater by natural attenuation at petroleum release sites. ASTM E-1943–98. American Society for Testing and Materials, West Conshohocken, PA, 1998.

Anderson, M.P. & Woessner, W.W.: The role of postaudit in model validation. *Adv. Water Resour.* 15:3 (1992), pp. 167–173.

Börling, K.: *Phosphorus sorption, accumulation and leaching. Effects of long-term inorganic fertilization of cultivated soils.* PhD Thesis, Swedish University of Agricultural Sciences, Department of Soil Sciences, Uppsala, Sweden, 2003.

Carlye, C. & Hill, A.R.: Groundwater phosphate dynamics in a river riparian zone: effects of hydrological flow paths, lithology and redox chemistry. *J. Hydrol.* 247:3–4 (2001), pp. 151–168.

COMSOL AB: Earth Science Module Model Library, COMSOL Multiphysics Version 3.5. COMSOL AB Publication, Stockholm, Sweden, 2008.

Fetter, C.W.: *Applied hydrogeology*. 4th ed., Prentice-Hall, Inc, NJ, 2001.

Gale, P.M., Reddy, K.R. & Graetz, D.A.: Phosphorus retention by wetland soils used for treated wastewater disposal. *J. Environ. Qual.* 23:2 (1994), pp. 370–377.

Gard, C. & Torstensson, I.: *Hydrological characteristics of a settlement in peri-urban Kampala, Uganda*. MSc. Thesis, Lund University, Sweden, 2006.

Houba, V.J.G., Van der Lee, J.J. & Novozamsky, I.: Soil analysis procedures (soil and plant analysis, part 5B), a series of syllabi, 6th ed., Department of Science and Plant Nutrition, Wageningen Agricultural University, The Netherlands, 1995.

Herzog, A.: *Transient groundwater modelling in peri-urban Kampala, Uganda*. MSc. Thesis. Royal Institute of Technology, Sweden, 2007.

Kelbe, B.E., Bodenstein, B. & Mulder, G.J.: Investigating the hydrological response to informal settlements on small catchments in South Africa. *International Association of Hydrological Sciences Publication* 203 (1991), pp. 219–227.

Kulabako, R.: *Analysis of the impact of anthropogenic pollution of shallow groundwater in peri-urban Kampala*. Licentiate Thesis in Land and Water Resources Engineering Department, Royal Institute of Technology, Stockholm, Sweden, 2005.

Kulabako, R.: *Environmental sanitation situation and solute transport in variably saturated soil in peri-urban Kampala*. PhD Thesis, Royal Institute of Technology, Stockholm, Sweden/Makerere University, Kampala, Uganda, 2010.

Kulabako, N.R., Nalubega, M. & Thunvik, R.: Characterisation of peri-urban anthropogenic pollution in Kampala, Uganda. In: S. Godfrey (ed): *Proceedings of the 30th WEDC international conference, Vientianne, Lao PDR on people centred approaches to water and environmental sanitation*. WEDC, Loughborough, UK, 2004, pp. 474–482.

Kulabako, N.R., Nalubega, M. & Thunvik, R.: Study of the impact of land use and hydrogeological settings on shallow groundwater quality in a peri-urban area in Kampala, Uganda. *Sci. Total Environ.* 381:1–3 (2007), pp. 180–199.

Kulabako, N.R., Nalubega, M. &Thunvik, R.: Phosphorus transport in shallow groundwater in peri-urban Kampala—Results of field and laboratory measurements. *Environ. Geol.* 53:7 (2008), pp. 1535–1551.

Kulabako, N.R., Nalubega, M., Wozei, E. & Thunvik, R.: Environmental health practices, constraints and possible interventions in peri-urban settlements in developing countries—a review of Kampala, Uganda. *Int. J. Environ. Health Res.* 20:4 (2010a), pp. 231–257.

Kulabako, N.R., Thunvik, R., Soutter, L.A. & Nalubega, M.: Modelling of flow and phosphorus transport in variably saturated soil in a peri-urban settlement in Kampala, Uganda. Manuscript submitted to *J. Hydrol.* (2010b).

Looney, B.B. & Falta, R.W. (eds): *Vadose zone science and technology solutions*. Battelle Press, Columbus, OH, 2000.

Li, Q., Ito, K., Wu, Z., Lowry, C.S. & Loheide, S.P.: COMSOL Multiphysics: A novel approach to groundwater modelling. *Ground Water* 47:4 (2009), pp. 480–487.

Lundmark, A.: *Monitoring transport and fate of de-icing salts in the roadside environment—Modelling and field measurements*. PhD Thesis, Royal Institute of Technology, Stockholm, Sweden, 2008.

McGechan M.B.: Sorption of phosphorus by soil, part 2: measurement methods, results and model parameter values, a review paper. *Biosyst. Eng.* 82:2 (2002), pp. 115–130.

Muhammad, T.S. & Robinson, J.S.: Phosphorus sorption and availability in soils amended with animal manures and sewage sludge. *J. Environ. Qual.* 32:3 92003), pp. 1114–1121.

Nichols, E.M.: Monitored natural attenuation for groundwater remediation. Superfund and hazardous waste committee Newsletter Archive, 4: 2 (2003), pp. 7–12. Available athttp://www.abanet.org/environ/committees/superfundnatresdamages/newsletter/mar03/pdf/superfundnews0303.pdf (accessed March 2010).

Nyenje, P., Foppen, J.W., Uhlenbrook, S., Kulabako, R. & Muwanga, A. Eutrophication and nutrient release in urban areas of sub-Saharan Africa—A review. *Sci. Total Environ.* 408:3 (2010), pp. 447–455.

Rajib, D., Nihar, R.S., Pankaj, K.R. & Debojyoti, M.: Role of electrical conductivity as an indicator of pollution in shallow lakes. *Asian J. Water Environ. Pollut.* 3:1 (2006), pp. 143–146.

Rao, N.S. & Prasad, R.P.: Phosphate pollution in the groundwater of lower Vamsadhara River basin, India. *Environ Geol* 31:1–2 (1997), pp. 117–122.

Rügner, H., Finkel, M., Kaschl, A. & Bittens, M.: Application of monitored natural attenuation in contaminated land management—A review and recommended approach for Europe. *Environ. Sci. Policy* 9:6 (2006): pp. 568–576.

Sobehrad, S.: Phosphorus cycling in the estuarine environment. Research paper, 1997. Available at: http://www.co.bell.tx.us/bellnet/bellnetweb/web/phosphor.htm (accessed January 2009).

Taylor, G.R. & Howard, W.F.K.: The dynamics of groundwater flow in the regolith of Uganda. In: P. Dillon, P. and I. Simmers(eds): *International Contributions to Hydrogeology*, Series 18, *Shallow Groundwater Systems*. Balkema, The Netherlands, 1998, pp. 97–113.

Tecle, S.A.: *Modelling phosphorus and Arsenic contaminants using COMSOL Multiphysics*. MSc Thesis, Royal Institute of Technology, Stockholm, Sweden, 2008.

Tindall, J.A. & Kunkel, J.R.: *Unsaturated zone hydrology for scientists and engineers*. Prentice-Hall, Inc., NJ, 1999.

UBOS: 2002 Ugandan population and housing census-Kampala District Report. Uganda Bureau of Statistics (UBOS), Entebbe, Uganda, 2005.

U.S. NRC (National Research Council): Natural attenuation of groundwater remediation. Committee on Intrinsic Remediation, Water Science and Technology Board, Board on Radioactive Waste Management and, National Research Council. National Academies Press, Washington, D.C, USA, 2000.

Wang, T.H., Li, M.H. and Teng, S.P.: Bridging the gap between batch and column experiments: A case study of Cs adsorption on granite. *J. Hazard. Mater.* 161:1 (2009), pp. 409–415.

Watkins, L.H. and Fiddes, D.: *Hydrology in the Tropics*. Pentech Press Ltd, Plymouth, UK, 1984.

Xu, Y. & Usher, B. (eds): *Groundwater pollution in Africa*. United Nations Environment Programme (UNEP), Taylor and Francis, London, UK, 2006.

APPENDIX: NUMERICAL APPROACH TO THE SEEPAGE FACE BOUNDARY CONDITION

This appendix describes how to create a right hand boundary condition that varies from no flow condition above the water table to flow condition below the water table.

In the figure below, the equation for the normal flux represents the equation modeled by COMSOL Multiphysics (version 3.5a).

Here is how we use the procedure:

1. We modify the term R_b in the equation with logical operators so that we create a flow condition below the water table ($R_b > 0$) and a no flow condition above it ($R_b <= 0$). See **Physics>Boundary Settings** for definition of R_b

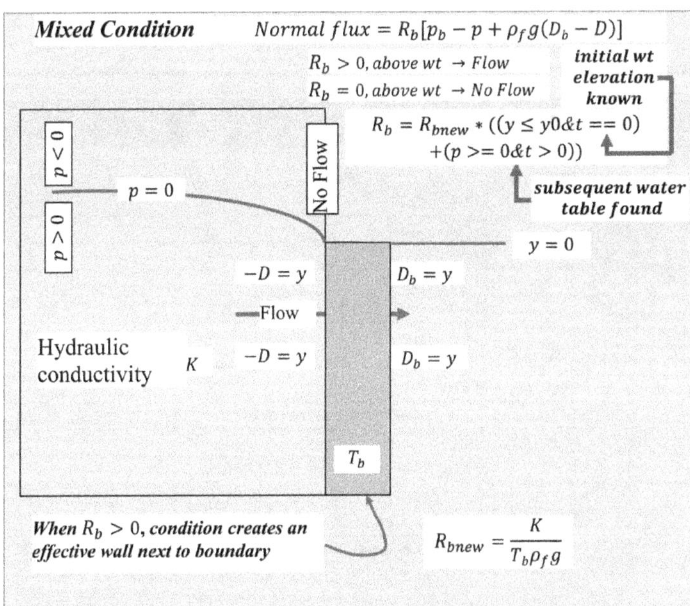

Figure 12A1. Numerical procedure for dealing with a seepage face in the present study *wt* = water table elevation *No* = inward flux—obviously zero in this case.

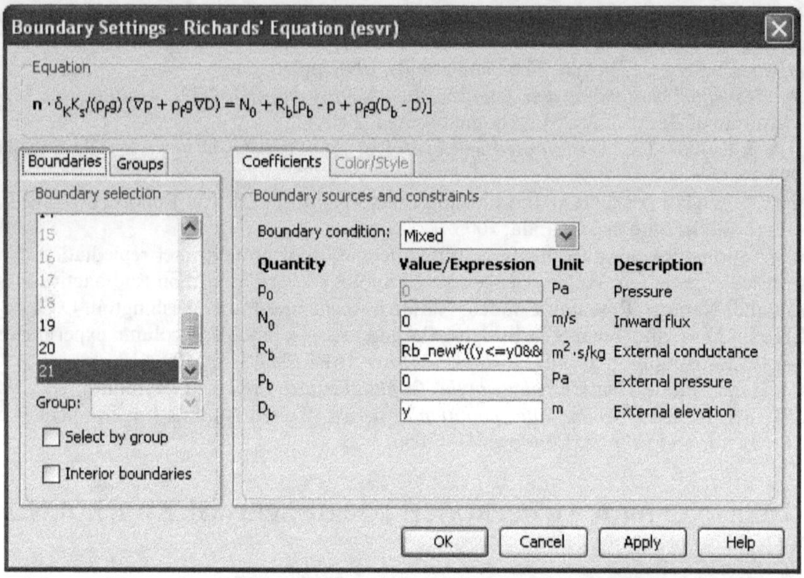

Boundary Settings - Richards' Equation (esvr)
Equation
n · δKKs/(ρfg) (∇p + ρfg∇D) = N0 + Rb[pb - p + ρfg(Db - D)]
Boundaries
Groups
Boundary selection
15
16
17
18
19
20
21
Group:
Select by group
Interior boundaries
Coefficients
Color/Style
Boundary sources and constraints
Boundary condition: Mixed
Quantity
Value/Expression
Unit
Description
p0
0
Pa
Pressure
N0
0
m/s
Inward flux
Rb
Rb_new*((y<=y0&&
m2·s/kg
External conductance
pb
0
Pa
External pressure
Db
y
m
External elevation
OK
Cancel
Apply
Help

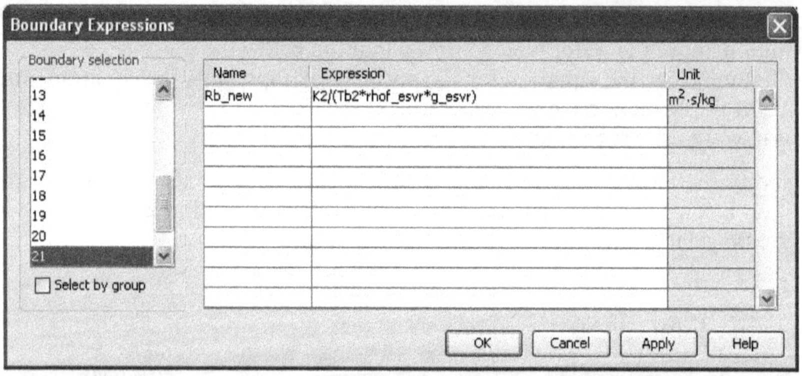

Boundary Expressions
Boundary selection
13
14
15
16
17
18
19
20
21
Select by group
Name
Expression
Unit
Rb_new
K2/(Tb2*rhof_esvr*g_esvr)
m2·s/kg
OK
Cancel
Apply
Help

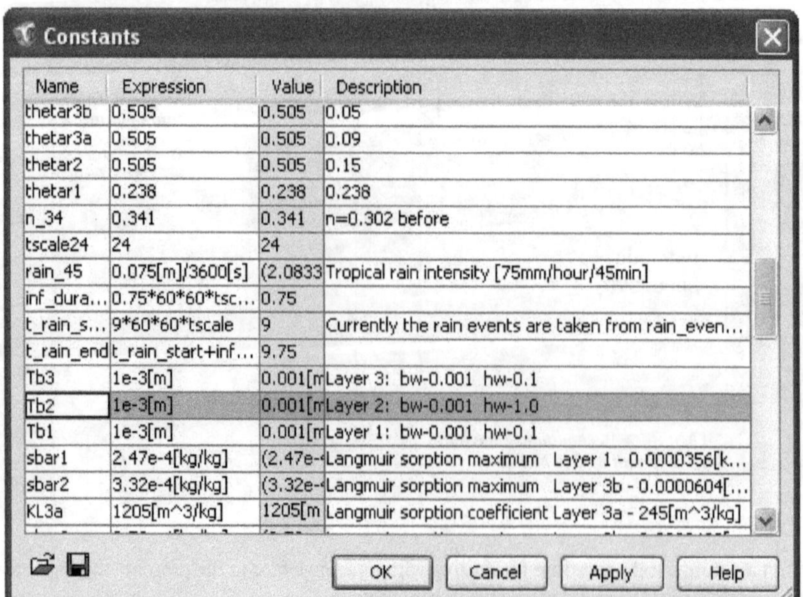

Constants
Name
Expression
Value
Description
thetar3b 0.505 0.505 0.05
thetar3a 0.505 0.505 0.09
thetar2 0.505 0.505 0.15
thetar1 0.238 0.238 0.238
n_34 0.341 0.341 n=0.302 before
tscale24 24 24
rain_45 0.075[m]/3600[s] (2.0833 Tropical rain intensity [75mm/hour/45min]
inf_dura... 0.75*60*60*tsc... 0.75
t_rain_s... 9*60*60*tscale 9 Currently the rain events are taken from rain_even...
t_rain_end t_rain_start+inf... 9.75
Tb3 1e-3[m] 0.001[m Layer 3: bw-0.001 hw-0.1
Tb2 1e-3[m] 0.001[m Layer 2: bw-0.001 hw-1.0
Tb1 1e-3[m] 0.001[m Layer 1: bw-0.001 hw-0.1
sbar1 2.47e-4[kg/kg] (2.47e- Langmuir sorption maximum Layer 1 - 0.0000356[k...
sbar2 3.32e-4[kg/kg] (3.32e- Langmuir sorption maximum Layer 3b - 0.0000604[...
KL3a 1205[m^3/kg] 1205[m Langmuir sorption coefficient Layer 3a - 245[m^3/kg]
OK
Cancel
Apply
Help

2. With the logical terms in (1) we establish that saturated conditions **inside the flow domain** exist below the elevation y0 at initial time ($t = 0$) and where $p >= 0$ subsequently ($t > 0$). See **Physics>Boundary Settings** for definition of R_b. In the Bwaise III study we have set the elevation y = elevation of top of bottom layer (in **Constants**).
3. Below the water table (where $R_b > 0$) the boundary condition mimics a wall with thickness *Tb* and hydraulic conductivity *K*. The elevation just inside the domain from the wall is *D*. The elevation just outside the wall is $D_b \cdot D = D_b = y$ since the wall is vertical. See **Physics > Boundary Settings** for definition of D_b.
4. The term Rb_new is the intrinsic value of R_b for the wall (i.e., without the logical operator modifications). Rb_new equals the hydraulic conductivity of the wall divided by the thickness and the specific weight of the fluid. Since the wall does not exist we use the *K* for the corresponding layer (*K*1, *K*2, or *K*3) and set T_b equal to a very small number (0.001 m). See options > constants for T_b and see **Options > Expressions > Boundary expressions** for Rb_new.
5. The water pressure outside the wall P_b is set to zero which implies that water is not ponding to the elevation of the water table outside of the wall and instead there is air outside the wall.See **Physics > Boundary Settings** for the definition of P_b.

Subject index

H

Multiphysics Modeling

Series Editors: Jochen Bundschuh & Mario César Suárez Arriaga

ISSN: 1877-0274

Publisher: CRC/Balkema, Taylor & Francis Group

1. Numerical Modeling of Coupled Phenomena in Science and Engineering
   Editors: M.C. Suárez Arriaga, J. Bundschuh & F.J. Domínguez-Mota
   2009
   ISBN: 978-0-415-47628-7

2. Introduction to the Numerical Modeling of Groundwater and Geothermal Systems: Fundamentals of Mass, Energy and Solute Transport in Poroelastic Rocks
   J. Bundschuh & M.C. Suárez Arriaga
   2010
   ISBN: 978-0-415-40167-8

3. Drilling and Completion in Petroleum Engineering
   Editors: Xinpu Shen, Mao Bai & William Standifird
   2011
   ISBN: 978-0-415-66527-8

4. Computational Modeling of Shallow Geothermal Systems
   Rafid Al-Khoury
   2011
   ISBN: 978-0-415-59627-5

5. Geochemical Modeling of Groundwater, Vadose and Geothermal Systems
   Editors: Jochen Bundschuh & Michael Zilberbrand
   2012
   ISBN: 978-0-415-66810-1

For Product Safety Concerns and Information please contact our EU representative GPSR@taylorandfrancis.com
Taylor & Francis Verlag GmbH, Kaufingerstraße 24, 80331 München, Germany

www.ingramcontent.com/pod-product-compliance
Lightning Source LLC
LaVergne TN
LVHW080641011026
849160LV00014B/1283

* 9 7 8 1 1 3 8 0 7 4 4 4 6 *